VOLUME ONE HUNDRED AND SEVENTEEN

ADVANCES IN CANCER RESEARCH

VOLUME ONE HUNDRED AND SEVENTEEN

ADVANCES IN CANCER RESEARCH

The Role of Sphingolipids in Cancer Development and Therapy

Series Editors

KENNETH D. TEW

Cell and Molecular Pharmacology,
John C. West Chair of Cancer Research,
Department of Cell and Molecular
Pharmacology & Experimental Therapeutics
at MUSCCharleston, South Carolina, USA

PAUL B. FISHER

Professor and Chair,
Department of Human & Molecular Genetics,
Director, VCU Institute of Molecular Medicine,
Thelma Newmeyer Corman Chair in Cancer Research,
VCU Massey Cancer Center,
Virginia Commonwealth University, School of Medicine,
Richmond, Virginia, USA

Edited by

JAMES S. NORRIS

Department of Microbiology and Immunology
Medical University of South Carolina
Charleston, South Carolina, USA

ELSEVIER

AMSTERDAM • BOSTON • HEIDELBERG • LONDON
NEW YORK • OXFORD • PARIS • SAN DIEGO
SAN FRANCISCO • SINGAPORE • SYDNEY • TOKYO
Academic Press is an imprint of Elsevier

Academic Press is an imprint of Elsevier
525 B Street, Suite 1900, San Diego, CA 92101-4495, USA
225 Wyman Street, Waltham, MA 02451, USA
32 Jamestown Road, London, NW1 7BY, UK
The Boulevard, Langford Lane, Kidlington, Oxford, OX51GB, UK
Radarweg 29, PO Box 211, 1000 AE Amsterdam, The Netherlands

First edition 2013

ISBN: 978-0-12-394274-6
ISSN: 0065-230X

For information on all Academic Press publications
visit our website at store.elsevier.com

Printed and bound by CPI Group (UK) Ltd, Croydon, CR0 4YY
Transferred to Digital Printing, 2013

CONTENTS

CONTRIBUTORS

José Luis Abad
Spanish National Research Council (Consejo Superior de Investigaciones Científicas, CSIC), Research Unit on Bioactive Molecules (RUBAM), and Departament de Química Biomèdica, Institut de Química Avançada de Catalunya (IQAC-CSIC), Jordi Girona 18-26, 08034-Barcelona, Spain

Isabelle Ader
CNRS, Institut de Pharmacologie et de Biologie Structurale, and Université de Toulouse, UPS, IPBS, Toulouse, France

Heba Alshaker
Department of Surgery and Cancer, Faculty of Medicine, Imperial College London, Hammersmith Hospital, London, United Kingdom

Thomas H. Beckham
Department of Microbiology and Immunology, Medical University of South Carolina, Charleston, South Carolina, USA

Torsten Böhler
Research and Development, Gambro, Hechingen, Germany

Pierre Bouquerel
CNRS, Institut de Pharmacologie et de Biologie Structurale, and Université de Toulouse, UPS, IPBS, Toulouse, France

Leyre Brizuela
CNRS, Institut de Pharmacologie et de Biologie Structurale, and Université de Toulouse, UPS, IPBS, Toulouse, France

Josefina Casas
Spanish National Research Council (Consejo Superior de Investigaciones Científicas, CSIC), Research Unit on Bioactive Molecules (RUBAM), and Departament de Química Biomèdica, Institut de Química Avançada de Catalunya (IQAC-CSIC), Jordi Girona 18-26, 08034-Barcelona, Spain

Joseph C. Cheng
Department of Microbiology and Immunology, Medical University of South Carolina, Charleston, South Carolina, USA

Olivier Cuvillier
CNRS, Institut de Pharmacologie et de Biologie Structurale, and Université de Toulouse, UPS, IPBS, Toulouse, France

Antonio Delgado
Spanish National Research Council (Consejo Superior de Investigaciones Científicas, CSIC), Research Unit on Bioactive Molecules (RUBAM); Departament de Química Biomèdica, Institut de Química Avançada de Catalunya (IQAC-CSIC), Jordi Girona 18-26, 08034-Barcelona, Spain, and Facultat de Farmàcia, Unitat de Química Farmacèutica

(Unitat Associada al CSIC), Universitat de Barcelona (UB), Avda. Joan XXIII, s/n, 08028-Barcelona, Spain

Gemma Fabriàs
Spanish National Research Council (Consejo Superior de Investigaciones Científicas, CSIC), Research Unit on Bioactive Molecules (RUBAM), and Departament de Química Biomèdica, Institut de Química Avançada de Catalunya (IQAC-CSIC), Jordi Girona 18-26, 08034-Barcelona, Spain

Cécile Gstalder
CNRS, Institut de Pharmacologie et de Biologie Structurale, and Université de Toulouse, UPS, IPBS, Toulouse, France

Linda A. Heffernan-Stroud
Molecular and Cellular Biology and Pathobiology Program, and College of Medicine, Medical University of South Carolina, Charleston, South Carolina, USA

Ronald A. Hill
Department of Basic Pharmaceutical Sciences, University of Louisiana at Monroe, Monroe, Louisiana, USA

Yu-Teh Li
Department of Biochemistry and Molecular Biology, Tulane University School of Medicine, New Orleans, Louisiana, USA

Xiang Liu
Department of Microbiology and Immunology, Medical University of South Carolina, Charleston, South Carolina, USA

Yong-Yu Liu
Department of Basic Pharmaceutical Sciences, University of Louisiana at Monroe, Monroe, Louisiana, USA

Bernard Malavaud
CNRS, Institut de Pharmacologie et de Biologie Structurale, and Université de Toulouse, UPS, IPBS, Toulouse, France

S. Tucker Marrison
Department of Microbiology and Immunology, Medical University of South Carolina, Charleston, South Carolina, USA

Danielle Monteil
Department of Surgery and Cancer, Faculty of Medicine, Imperial College London, Hammersmith Hospital, London, United Kingdom

James S. Norris
Department of Microbiology and Immunology, Medical University of South Carolina, Charleston, South Carolina, USA

Lina M. Obeid
Molecular and Cellular Biology and Pathobiology Program, and Department of Medicine, Medical University of South Carolina, Charleston, South Carolina, USA

Besim Ogretmen
Department of Biochemistry and Molecular Biology, and Hollings Cancer Center, Medical University of South Carolina, Charleston, South Carolina, USA

Silvia Ottaviani
Department of Surgery and Cancer, Faculty of Medicine, Imperial College London, Hammersmith Hospital, London, United Kingdom

Dmitri Pchejetski
Department of Surgery and Cancer, Faculty of Medicine, Imperial College London, Hammersmith Hospital, London, United Kingdom

Sahar A. Saddoughi
Department of Biochemistry and Molecular Biology, and Hollings Cancer Center, Medical University of South Carolina, Charleston, South Carolina, USA

Lysann Sauer
Department of Surgery and Cancer, Faculty of Medicine, Imperial College London, Hammersmith Hospital, London, United Kingdom

Radoslav Savić
Department of Genetics and Genomic Sciences, Mount Sinai School of Medicine, New York, USA

Edward H. Schuchman
Department of Genetics and Genomic Sciences, Mount Sinai School of Medicine, New York, USA

Shyam Srivats
Department of Surgery and Cancer, Faculty of Medicine, Imperial College London, Hammersmith Hospital, London, United Kingdom

PREFACE

Sphingolipid metabolism plays a number of key roles in the response of cancer to therapy; however, strategies to interdict sphingolipid metabolism are frequently overlooked as avenues to improve cancer response to therapy. This volume reviews the literature on the sphingolipid response to cancer therapy in order to provide a concise resource for translating the scientific principles of sphingolipid metabolism into realistic strategies for improvement of cancer treatment outcomes.

Chapter 1 introduces the subject by reviewing the enzymes that are involved in sphingolipid metabolism and relating them to therapeutic efficacy. It is now widely understood that ceramide is a mediator of apoptosis in both normal and cancer cells. Generally speaking, ceramide signaling winds up inducing apoptosis by activation of the mitochondrial intrinsic apoptosis pathway. Classically described resistance mechanisms to chemotherapy and radiation are frequently upstream of the mitochondria pathway. In many cases, overcoming these forms of resistance can be achieved experimentally by reestablishing ceramide-activated intrinsic apoptosis at the mitochondrial level. Converging with this knowledge are studies that reveal deregulation of sphingolipid metabolism is contributory to the inability of chemotherapy or radiation to effectively treat cancer. Enzymes that are associated with sphingolipid metabolism and linked with inhibition or stimulation of cancer cell growth include acid ceramidase (AC), glucosylceramide synthase (GCS), ceramide kinase, sphingosine kinase 1 and 2 (Spk1 and Spk2), ceramide synthase, S1P lyase, and S1P phosphatase.

Of particular importance is the pathway originally described by Cuvillier and Spiegel in 1996 that consists of ceramide–sphingosine–sphingosine 1-phosphate. Ceramide, the central molecule in the pathway, is considered to be a tumor-suppressive proapoptotic lipid with significant importance in controlling cell growth, apoptosis, and senescence. Ceramide can be deacylated by the ceramidases. The products of deacylation are sphingosine and a fatty acid. Sphingosine can be phosphorylated by Spk1 or Spk2 to yield the putative oncogenic lipid S1P. This pathway, known as the ceramide S1P rheostat, can be interdicted at several places. Specifically AC is perhaps the most important enzyme for producing sphingosine and is thus a prime target for inhibition. Spk1 and Spk2 are also prime targets for inhibition with ongoing clinical trials. Inhibition of S1P binding to their cellular receptors

(S1PR-1 through S1PR-5) is also under study and is in clinical trials. Although these targets are prime examples for regulation of sphingolipid metabolism in cancer therapy, other targets like GCS are under study and may have similar importance for regulating sphingolipids in response to therapy in a number of cancer models.

Chapter 2 covers another aspect of sphingolipid metabolism in cancer and specifically focuses on the family of enzymes known as the ceramide synthases. These enzymes are involved in both *de novo* synthesis of ceramide as well as the so-called salvage pathway of ceramide regeneration intracellularly. Specifically this chapter reviews the recent discoveries about the mechanistic details of *de novo* generated ceramides in cancer pathogenesis and potential translational opportunities these enzymes provide.

Chapter 3 covers the role of GCS and cancer drug resistance. This enzyme is overexpressed in a number of different tumors including breast, colon, and in some types of leukemia, and it has been clearly shown that overexpression leads to drug resistance through well-characterized mechanisms involving MDR1. In particular, it has also been shown that ceramide glycosylation is able to modulate expression of a variety of genes and to restore p53 expression by effecting RNA splicing mechanisms. This latter observation is particularly intriguing from both a scientific and a therapeutic point of view and has the potential to lead to novel therapeutic approaches for treating this group of cancers.

In Chapter 4, acid sphingomyelinase is reviewed in the context of Niemann Pick Disease (NPD), where it was first discovered, to its current role in cancer biology where it exhibits preclinical efficacy for treatment of solid tumors. Acid sphingomyelinase is considered to be a lysosomal protein based on its acidic pH optimum although secreted and other forms of the enzyme also exist. A recombinant form of the enzyme grown in CHO cells was granted orphan drug status for the treatment of type B NPD in 2000. It is the ability of acid sphingomyelinase to remodel cell membranes and induce cell signaling that has attracted the interest of scientists seeking to utilize it in cancer therapies. It would appear that, *in vivo*, the primary target of acid sphingomyelinase-generated ceramide is tumor vasculature. This has led to the use of exogenous acid sphingomyelinase as an adjunct in cancer therapy with the intent of improving ceramide production through sphingomyelin hydrolysis.

Chapter 5 introduces the importance of S1P in the cancer process. This chapter focuses on how inhibiting or neutralizing Spk1/S1P signaling can be used for enhancement of cancer therapy by regulating HIF-1 or HIF-2 activity in hypoxic regions of the tumor bed. Rapid growth of solid tumors causes cancer cells to expand away from the normal tissue vasculature that is required for provision of oxygen and nutrients to energy-hungry cancer cells. Neovascularization is a critical part of tumor progression; however, new tumor-supporting vessels are not well structured, leak, and suffer distorted network structure all of which culminate in local tissue hypoxia. Interestingly, hypoxic areas of the tumor have long been known to be resistant to both radio- and chemotherapy. Although hypoxia can be predicted to result in the death of both normal and cancer cells, it appears that cancer cells develop mechanisms as well as genetic changes that permit survival. The role of sphingolipids in this process has been studied extensively, and in particular, it has been shown that S1P is capable of regulating proliferation, survival, migration, inflammation, and angiogenesis in cancer cells. The subject in the role of S1P is extensively reviewed in this chapter. Finally, the role of S1P signaling in the context of an antihypoxic strategy target is revealed.

Chapter 6 is an extensive description of how therapeutic strategies directed at Spk1 in human tumors can have clinical applications. As described earlier, S1P is able to signal through a family of five receptors that were originally known as Edg receptors and have been renamed S1PR1-5. S1PR 1, 2, and 3 are widely expressed on most mammalian tissues and their signaling pathways via G-protein-coupled receptors have been extensively studied and shown to be of importance in many different cancers. The generation of S1P by Spk1 has been shown by Pitson et al. to be an oncogenic process making Spk1 a classical oncogene. Spk1 is believed to be predominantly cytosolic, whereas Spk2's location is more complex and is perhaps cell cycle dependent.

Spk1 has been linked with cancer cell growth and resistance to stress. This enzyme is upregulated in a variety of different types of cancer and its inhibition has been demonstrated to lead to improved cancer cell killing during therapy. This is not surprising as S1P can control migration, angiogenesis, and tumor inflammation. There are also ongoing studies examining the potential of S1P as a biomarker for human cancer and as an indicator of therapeutic responsiveness. Thus, inhibition of Spk1 becomes a prime target for drug development. Early studies on small molecules that inhibit Spk1 revealed that dihydrosphinganines and dimethylsphingosine were able to alter the ceramide/S1P biostat.

There are several categories of inhibitors of Spk1 and 2. Early studies demonstrated that dihydrosphingosine and dimethylsphingosine were able to shift the ceramide-S1P biostat in favor of ceramide. FTY720, another sphingosine analog, has recently been shown to be an Spk1 inhibitor and is approved for use for the treatment of multiple sclerosis and is now being studied in the context of cancer therapy. Finally, a phase 1 clinical trial using Safingol in combination with cisplatin in 43 cancer patients was undertaken. In that study, Safingol was noted to be associated with a dose-dependent reduction in S1P levels in plasma with at least one patient having a significant regression of liver and lung metastases derived from an adrenocortical cancer. Very recently, a novel Spk1 inhibitor has been described with low-nanomolar activity by Pfizer called PF-543. Studies with this new drug are ongoing. Other drugs have either been developed or are in the pipeline and are reviewed in this chapter as well including LCL351 (Sphingogene, Inc.) which has selectivity for Spk1 in the low nanomolar range. Using small molecule library screening approaches, French et al. were able to develop a group of Spk inhibitors which have selectivity for Spk2 and exhibit both anticancer and anti-inflammatory activity. One such drug, ABC294640 (Apogee, Inc.), is in clinical trials for a number of different diseases. The use and importance of Spk inhibitors is extensively reviewed in this chapter.

Continuing with the role of sphingosine kinase in cancer, Chapter 7 delves into the concept that Spk1 is an oncogene that is activated in multiple different cancers. Although Spk1 and 2 have some overlapping physiological functions, it is now rather clear that they serve multiple and distinct purposes associated with cell survival, angiogenesis, migration, and inflammation. Spk1 appears to function in the cytoplasm and plasma membrane, whereas Spk2 appears to have more of a nuclear localization. Studies described in this chapter demonstrate the important role of Spk1 in both colon and head and neck cancer ontogeny. It is also been linked to myelogenous leukemia and endometrial, prostate, and thyroid cancer. It is now reasonable to make the statement that Spk1 is an important complement in the deregulated states found in cancer and that Spk2 is contributory to the inflammatory state seen in many types of cancers and likely to other events associated with cancer cell growth. It is perhaps too early to tell whether it will be a significant biomarker in cancer biology, but evidence is mounting that this may be the case. With respect to therapeutic outcomes, elevated Spk1 correlates with resistance to chemotherapy which in combination with the elevation of AC suggested that the ceramide–S1P rheostat is an important pathway that regulates the clinical success of chemo- or radiation therapy making development of drugs that interdict sphingolipid metabolism important and necessary.

Chapter 8, the final chapter of the book, discusses the search for natural products that affect sphingolipid metabolism as potential anticancer agents. From the previous chapters in this book it should be clear by now that the ceramide–Spk–S1P rheostat plays a major role in the cancer cells' life and death. As discussed earlier in other sections of the book, the screening of natural libraries of small molecules or natural products constitutes a well-established mechanism for discovery of bioactive agents that interact with specific biomolecules of interest. This chapter discusses the search amongst natural products for such agents. The roots of this search lie with fungal metabolites long used for research purposes that inhibit various enzymatic pathways that result in ceramide generation. Examples that have arisen from such studies include Fingolimod (FTY720), discussed above and throughout the book. The utilization of myriocin for further development of drugs in this category is extensively reported in this chapter. Other fungal metabolites have also been used as the basis for development of additional drugs. Scyphostatin and related fungal metabolites are known to inhibit neutral sphingomyelinase and have been further developed for this purpose and are extensively discussed in this chapter.

Another category of fungal metabolites of which fumonisin and the related 1-deoxysphingolipids are prime examples is discussed. Modification of these molecules has led to some interesting molecular variance that show promise as inhibitors of ceramide synthase; there is also a group of these deoxysphingolipids of marine origin that are discussed. Although relatively less well studied than other families of natural products, it is clear that sphingolipids are important biomolecules that regulate a multitude of cellular processes. Alterations in sphingolipid metabolism are well known to alter the outcome and/or progression of disease and drug resistance. Thus, the search amongst natural products for agents that can regulate sphingolipid metabolism and restore normality to this process is important and is gaining ground.

In summary, this book covers most of the important enzymes of sphingolipid metabolism, with the focus on implicating these enzymes as the regulators of the ceramide S1P rheostat. The development of inhibitors of the enzymes in this pathway is an important step that is now being shown to influence sphingolipid metabolism in such a way as to promote the ability of the clinician to overcome resistance mechanisms elicited by alteration of the pathway. This is now resulting in the promise that the use of these small-molecule inhibitors may lead to better clinical outcomes for patients suffering from a multitude of different types of cancer. It is my hope this book will stir interest in this often-overlooked field.

James S. Norris

CHAPTER ONE

Interdiction of Sphingolipid Metabolism to Improve Standard Cancer Therapies

Thomas H. Beckham[1], Joseph C. Cheng, S. Tucker Marrison, James S. Norris, Xiang Liu

Department of Microbiology and Immunology, Medical University of South Carolina, Charleston, South Carolina, USA

[1]Corresponding author: e-mail address: beckham@musc.edu

Contents

Advances in Cancer Research, Volume 117
ISSN 0065-230X
http://dx.doi.org/10.1016/B978-0-12-394274-6.00001-7

Abstract

Non-surgical therapies for human malignancies must negotiate complex cell signaling pathways to impede cancer cell growth, ideally promoting death of cancer cells while sparing healthy tissue. For most of the past half century, medical approaches for treating cancer have relied primarily on cytotoxic chemotherapeutics that interfere with DNA replication and cell division, susceptibilities of rapidly dividing cancer cells. As a consequence, these therapies exert considerable cell stress, promoting the generation of ceramide through *de novo* synthesis and recycling of complex glycosphingolipids and sphingomyelin into apoptotic ceramide. Radiotherapy of cancer exerts similar geno- and cytotoxic cell stresses, and generation of ceramide following ionizing radiation therapy is a well-described feature of radiation-induced cell death. Emerging evidence now describes sphingolipids as mediators of death in response to newer targeted therapies, cementing ceramide generation as a common mechanism of cell death in response to cancer therapy. Many studies have now shown that dysregulation of ceramide accumulation—whether by reduced generation or accelerated metabolism—is a common mechanism of resistance to standard cancer therapies. The aims of this chapter will be to discuss described mechanisms of cancer resistance to therapy related to dysregulation of sphingolipid metabolism and to explore clinical and preclinical approaches to interdict sphingolipid metabolism to improve outcomes of standard cancer therapies.

1. INTRODUCTION

1.1. Ceramide as a mediator of apoptosis

Ceramide is the central molecule in sphingolipid metabolic pathways, and its generation and metabolism are key in understanding advantageous and dysregulated sphingolipid responses to cancer therapy. While many diverse functions have been ascribed to ceramide, for the purposes of this review, ceramide is best characterized to promote apoptosis and cell senescence. Among, the major effectors of ceramide signaling are protein phosphatases PP2A and PP1, which are activated by ceramide (Chalfant et al., 1999; Dobrowsky, Kamibayashi, Mumby, & Hannun, 1993; Galadari, Kishikawa, Kamibayashi, Mumby, & Hannun, 1998). Through activation of PP2A, ceramide promotes numerous signaling alterations including deactivation of Akt (Teruel, Hernandez, & Lorenzo, 2001), PKC (Lee, Hannun, & Obeid, 1996), and c-Jun (Shirakabe et al., 1997); destabilization of c-Myc (Mukhopadhyay et al., 2009); and disruption of the Bax/Bcl-2 interaction (Xin & Deng, 2006). PP1 activation causes dephosphorylation of SR proteins with subsequent alternative splicing of Caspase 9 and Bcl-X (Chalfant et al., 2002) and activation of retinoblastoma (Liu, Wang, &

Berndt, 2006). Apart from the myriad functions of ceramide downstream of PP2A and PP1 activation, ceramide generated in the lysosome by acid sphingomyelinase (ASMase) has also been shown to directly bind and induce autoproteolytic cleavage of Cathepsin D (Heinrich et al., 1999), promoting the cleavage-induced activation of proapoptotic Bid. These functions of ceramide, represented in Fig. 1.1, ultimately converge upon causing cell cycle arrest, senescence, and in many cases apoptosis and cell death. While the vast

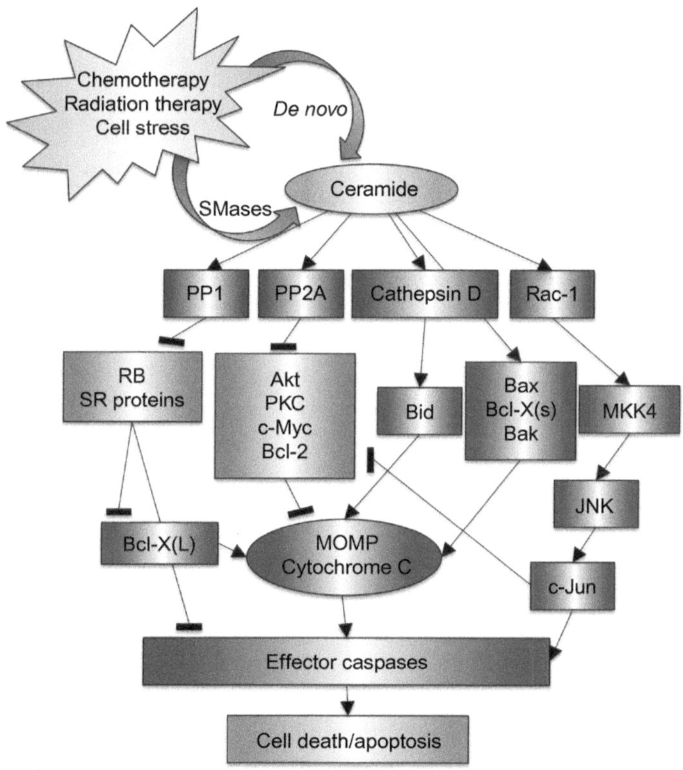

Figure 1.1 Ceramide in apoptosis. Several direct targets of ceramide have been identified, including Cathepsin D and the serine/threonine protein phosphatases PP1 and PP2A. These phosphatases act on several substrates such as the retinoblastoma gene product Rb, Bcl-2, PKC, Akt, and SR proteins. Increasingly, these pathways seem to be compartmentalized. In particular, Cathepsin D is activated by ceramide generated in lysosome membranes leading to activation of the proapoptotic protein Bid. Mitochondrial membrane potential can also be altered by these pathways, probably through disparate signal cascades affecting various Bcl-2 family members. Apoptosis induction through JNK activation is mediated by ceramide-dependent signaling by MKK4. These downstream effects can lead to changes in growth arrest, senescence, and apoptosis. (See Page 1 in Color Section at the back of the book.)

majority of the literature support these anticancer effects of ceramide, it is worthwhile to acknowledge that antiapoptotic roles have been described for some specific ceramide species (Hoye, Davoren, Wipf, Fink, & Kagan, 2008), highlighting the complexities of ceramide signaling that remain to be fully characterized. A more complete review by Ogretmen and Hannun (2004) details these functions of ceramide and more and explains why ceramide accumulation is a hallmark of diverse apoptotic stimuli, including chemo- and radiotherapy and lay the groundwork for why adjuvant therapeutics that promote accumulation of ceramide are promising approaches to improving response to standard therapies for cancer.

1.2. Ceramide generation and metabolism

A favorable sphingolipid response to cancer therapy is frequently a net accumulation of ceramide; therefore, it is convenient to view metabolic pathways of ceramide generation and consumption, outlined in Fig. 1.2, in terms of steps that generate ceramide and those that consume it. Ceramide is *N*-acylated sphingosine with naturally occurring acyl chain lengths from 14 to 26 carbons. It can be generated by *de novo* synthesis, a multistep process initiated by the condensation of serine and palmitoyl CoA (Xu et al., 1998); by the salvage pathway, which recycles cellular sphingosine (Kitatani, Idkowiak-Baldys, & Hannun, 2008); or by hydrolysis from complex sphingolipids such as sphingomyelin and cerebrosides (galactosyl- and glucosylceramide). *De novo* synthesis and hydrolysis of sphingomyelin frequently play key roles in generating ceramide in response to cancer therapy, and defects in generation of ceramide by these pathways are implicated in therapy resistance. The following sections will highlight known defects in ceramide generation and how they impact cancer therapy. Conversely, consumption of free ceramide by incorporation into sphingomyelin or cerebrosides or by deacylation of ceramide to form sphingosine is associated with defects in therapy-induced apoptosis. As will be discussed in this chapter, response of cancer cells and tumors to standard therapies like chemotherapy and radiation can be improved by potentiating accumulation of ceramide whether by promoting its generation or by inhibiting its catabolism. Importantly, deacylation of ceramide by ceramidases generates sphingosine, the substrate of sphingosine kinases. Sphingosine 1-phosphate (S1P) is a pleiotropic signaling lipid that frequently opposes apoptosis and promotes angiogenesis and cell migration, justifying prevention of S1P formation in addition to enhancement of ceramide generation as a worthy goal in interdiction of sphingolipid metabolism to improve response to cancer therapy. While this chapter will include discussion of sphingosine kinases and S1P, their roles in cancer promotion and therapy are dealt with more completely in Chapters 5, 6, and 7 of this volume.

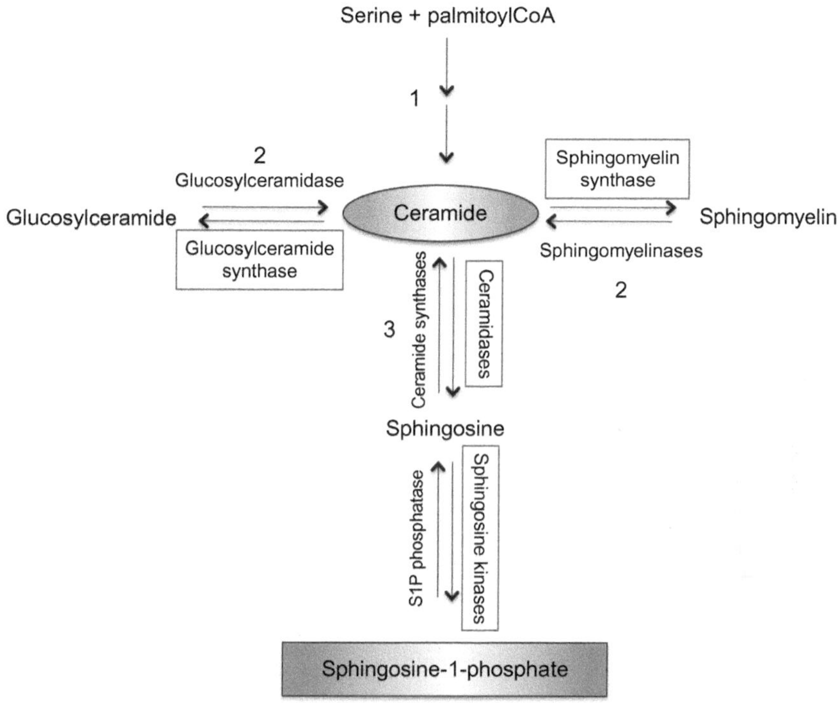

Figure 1.2 Basic sphingolipid metabolism and nodes of interdiction: Ceramide is generated by (1) *de novo* biosynthesis, a multistep process which is initiated by condensation of serine and palmitoyl CoA; (2) metabolism of existing complex sphingolipids such as glucosylceramide and sphingomyelin; and (3) salvage of free sphingosine by *N*-acylation by ceramide synthases. Enzymes that catalyze steps that reduce cellular ceramide and have been investigated as potential nodes of interdiction of ceramide metabolism to improve clinical response to standard therapies are boxed in red and will be further discussed. Note that other sphingolipids and sphingolipid metabolizing enzymes, such as ceramide-1-phosphate and ceramide kinase, are not further discussed in this chapter and are not included in this diagram. A complete review of sphingolipid metabolism is referenced (Gault, Obeid, & Hannun, 2010). (See Page 2 in Color Section at the back of the book.)

1.3. Dysregulation of sphingolipid metabolism in cancer

One of the defining characteristics of cancer cells is the ability to evade apoptosis by development of signaling mechanisms that prevent response to apoptotic stimuli (Hanahan & Weinberg, 2000). Numerous mechanisms are described for death resistance in cancer, including upregulation of pro-survival genes and downregulation of apoptotic genes. Cancer cells also frequently cultivate dysregulation of enzymes involved in sphingolipid metabolism to escape cell death. Dysregulation of sphingolipid metabolic pathways is a common theme in human cancers, perhaps because sphingolipid signaling operates upstream of several apoptotic mediators, suggesting that

Table 1.1 Summary of the literature demonstrating alterations in sphingolipid-metabolizing enzymes in primary human cancer tissues

Sphiningolipid enzymes	Expression in cancer	Cancer model	Reference
Glucosylceramide synthase (GCS)	Elevated	Bladder, breast	Liu et al. (2011) and Sun et al. (2010)
Acid ceramidase	Elevated	Head and neck, prostate, T-cell large granular lymphocyte (LGL) leukemia	Elojeimy et al. (2007), Norris et al. (2006), and Shah et al. (2008)
Sphingomyelinase	Decreased (Alkaline)	Colon, liver	Di Marzio et al. (2005) and Hertervig, Nilsson, Nyberg, and Duan (1997)
Sphingosine kinases	Elevated	Breast, astrocytomas, head and neck, prostate, thyroid, gastric	Erez-Roman, Pienik, and Futerman (2010), Guan et al. (2011), Li et al. (2008, 2009), Malavaud et al. (2010), and Shirai et al. (2011)
Ceramide synthases	Elevated (Cer2/4/6)	Breast	Erez-Roman et al. (2010) and Schiffmann et al. (2009)
Ceramide kinase	Elevated	Breast, non-small cell lung	Kossenkov et al. (2011) and Ruckhäberle et al. (2009)
S1P lyase	Decreased	Colon	Oskouian et al. (2006)
S1P phosphatase	Decreased	Colon	Oskouian et al. (2006)

alteration of sphingolipid metabolism may be an important step in tumorigenesis and, as such, that manipulation of sphingolipid metabolism may improve response to therapeutics. Multiple sphingolipid enzymes are known to be aberrantly expressed in human cancer (Table 1.1).

1.4. Susceptible nodes of sphingolipid metabolism for interdiction in cancer therapy

Study of sphingolipid metabolism in cancer has led to the identification of several nodes of sphingolipid metabolism that are susceptible to interference as we seek to improve outcomes of cancer therapy. These nodes can generally be divided into those that promote ceramide generation and those that prevent its consumption. Multiple therapeutic approaches cause generation of ceramide through *de novo* synthesis and sphingomyelin hydrolysis, and thus ceramide generation is an important part of standard cancer therapy. Increasingly, approaches that rely on inhibition of enzymes that consume ceramide such as glucosylceramide synthase (GCS), ceramidases, or sphingosine kinases are being combined with standard cancer therapies as known dysregulation of these enzymes occurs in cancer and contribute to therapy resistance. Each of the approaches outlined in Fig. 1.2 will be discussed in terms of both known defects in these pathways that contribute to resistance to standard cancer therapies as well as work that demonstrates the principles and feasibility of targeting these enzymes to improve patient response to standard of care therapies including cytotoxic chemotherapy, radiation therapy, and more recently introduced targeted therapies like tyrosine kinase inhibitors (TKIs), deacetylase inhibitors, and monoclonal antibodies (mAbs).

2. SPHINGOLIPIDS AND CYTOTOXIC CHEMOTHERAPIES

The past decade has brought on a new wave of therapeutic approaches to medical management of cancer including TKIs, mAbs, and cancer vaccines; however, mainstay therapy for most human cancer remains, as it has since the mid-century development of mustine (Mechlorethamine Hydrochloride) in the treatment of lymphoma, cytotoxic chemotherapy that targets DNA synthesis and cell division, thus, affecting rapidly proliferating cancer cells more than most normal tissues. Interestingly, several classes of cytotoxic chemotherapeutics cause accumulation of ceramide. These include vinca alkaloids (vincristine and vinblastine), anthracyclines (doxorubicin and daunorubicin), taxanes (paclitaxel), and topoisomerase inhibitors (irinotecan, etoposide), among others. Here, we will discuss how cytotoxic chemotherapeutics elicit ceramide generation, research that demonstrates dysregulation of sphingolipids as a mechanism of resistance to chemotherapy, and how augmentation of a favorable sphingolipid response to therapy could improve outcomes.

2.1. Chemotherapy-induced ceramide generation

Mammalian cells respond to diverse stressors including radiation, nutrient deprivation, and oxidative stress with ceramide generation, largely through hydrolysis of sphingomyelin and *de novo* synthesis (Hannun, 1996; Ogretmen & Hannun, 2001). Evidence of stress-induced ceramide generation in yeast points to ceramide generation as an evolutionarily ancient response to cell stress (Cowart & Obeid, 2007). While how exactly remains incompletely understood, it seems that cytotoxic chemotherapeutics co-opt this stress response, inducing ceramide generation and thereby the goal of such therapies: apoptosis. Ceramide accumulation in response to chemotherapy occurs both through *de novo* synthesis and hydrolysis of sphingomyelin. Danuorubicin, etoposide, camptothecin, and gemcitibine appear to cause *de novo* ceramide biosynthesis either by activation of dihydroceramide synthases or by increasing the activity of serine palmitoyl transferase, the first step in the *de novo* pathway (Bose et al., 1995; Chalfant et al., 2002; Chauvier, Morjani, & Manfait, 2002; Perry et al., 2000). Interestingly, inhibiting the *de novo* pathway of ceramide synthesis antagonizes the cytotoxic effects of these drugs, highlighting the role of *de novo* ceramide generation in chemotherapy-induced cell death. Chemotherapy also promotes ceramide generation through the sphingomyelinase (SMase) pathway, liberating death-inducing ceramide from the vast reservoir of membrane sphingomyelin. Treatment of HL-60 cells with DNA antimetabolite cytosine arabinoside increased ceramide levels following stimulation of neutral sphingomyelinase (NSMase) activity (Strum, Small, Pauig, & Daniel, 1994), demonstrating sphingomyelin hydrolysis in response to a chemotherapy drug. Jaffrezou et al. (1996) described biphasic generation of ceramide in leukemia cells in response to daunorubicin that was unaffected by the ceramide synthase inhibitor Fumonisin B1, again implicating SMase activation as a mechanism of chemotherapy-induced ceramide generation. In another study, etoposide was found to cause ceramide generation by activating NSMase in glioma cells (Sawada et al., 2000). Interestingly, overexpression of Bcl-2 suppressed NSMase activation, preventing ceramide accumulation and cell death. This study is an excellent example of how cancers can suppress ceramide generation following chemotherapy by inhibiting SMase, thereby escaping treatment-induced cell death.

2.2. Glucosylceramide synthase dysregulation

While mechanisms employed by cancer cells to prevent generation of ceramide to avoid therapy-induced death are interesting proof of concept that ceramide is a potent mediator of successful chemotherapy, inhibition of enzymes that catabolize ceramide is ultimately more likely to result in treatment protocols with sphingolipid metabolizing enzyme inhibitors as chemosensitizing agents. Perhaps the best-studied mechanism of chemotherapy resistance due to sphingolipid dysregulation is the increased activity of GCS and accumulation of glucosylceramide in chemotherapy resistant cells.

Early in the study of sphingolipid dysregulation as a mechanism of chemotherapy resistance, it was discovered that glucosylceramide was markedly elevated in multidrug resistant cells compared to their drug-sensitive counterparts (Lavie et al., 1997), as well as the presence of glucosylceramide in the tissues of patients who did not respond to chemotherapy, but not in the tissues of patients who had chemotherapy response (Lucci et al., 1998). These studies led to the identification of GCS as a mediator of chemotherapy resistance. Inhibition of GCS restored paclitaxel, etoposide, and doxorubicin sensitivity to drug-resistant MCF-7 cells (Liu et al., 2004; Liu, Han, Giuliano, & Cabot, 2001). GCS has since been shown to be overexpressed in metastatic breast cancer and bladder cancer (Liu et al., 2011; Sun et al., 2010) and has been found by multiple groups to regulate multidrug resistance through regulation of p-glycoprotein and MDR1 expression (Gouaze et al., 2005). Interference with glucosylceramide generation by targeting GCS with small molecule inhibition and RNA interference techniques has consistently shown that inhibition of this enzyme can restore chemotherapy sensitivity and promote cell death and apoptosis by reversing upregulation of MDR1 (Sun et al., 2010) and p-glycoprotein (Gouaze et al., 2005) and causing ceramide accumulation (Baran, Bielawski, Gunduz, & Ogretmen, 2011). Thus, at least in preclinical models, GCS is a prototypical ceramide-metabolizing enzyme that is a prime target for overcoming cancer resistance to standard chemotherapy.

2.3. GCS as a target for chemosensitization

Given the strong body of the literature implicating GCS in multidrug resistance in cancer cell lines, it is surprising and perhaps disappointing how few studies have examined the feasibility and efficacy of targeting GCS *in vivo*. The data that are reported, however, are encouraging. Treatment of ganglioside-rich murine melanoma cells with the GCS inhibitor 1-phenyl-2-hexadecanoylamino-3-pyrrolidino-1-propanol drastically reduced their tumor-forming ability and incidence of metastasis when injected into mice (Deng, Li, & Ladisch, 2000). The same group reported similar results using GCS antisense (Deng, Li, Guerrera, Liu, & Ladisch, 2002), confirming that targeting GCS inhibition is efficacious in inhibiting cancer *in vivo*. Miglustat is an FDA-approved inhibitor of GCS used in some patients with Gaucher's disease, who lack glucosylceramidase, as substrate reduction therapy (Machaczka et al., 2012). Though this drug has not been evaluated in cancer and is often poorly tolerated, its approval lends proof of concept to treating patients with GCS inhibitors as chemosensitizing agents.

2.4. Ceramidase dysregulation

Another means of preventing the accumulation of ceramide that is necessary for chemotherapy-induced apoptosis is by action of ceramidases. These enzymes, of which there are five in humans, deacylate ceramide to form sphingosine and a fatty acid. This positions them at a critical juncture in the balance between proapoptotic ceramide and antiapoptotic S1P. As such, ceramidases have been intensively studied as mediators of sphingolipid metabolism and cancer. Much of the focus of the roles of ceramidases in cancer has been on lysosomal acid ceramidase. Our group and others have demonstrated that acid ceramidase is frequently dysregulated in human cancer including >60% of human prostate tumors (Saad et al., 2007) and >70% of head and neck squamous cell carcinoma (Elojeimy et al., 2007). Acid ceramidase has been shown to confer aggressive characteristics to cancer cells *in vitro* and *in vivo*, promoting cell proliferation and xenograft growth (Saad et al., 2007), and cancer cell invasion (Beckham et al., 2012). Additionally, acid ceramidase has been shown to mediate resistance to a number of treatment modalities including cytotoxic chemotherapy. Our group has shown that acid ceramidase overexpression promoted prostate cancer cell resistance to chemotherapy with etoposide, cisplatin, doxorubicin, and gemcitabine by preventing accumulation of proapoptotic C_{16}-ceramide following administration of these drugs. Importantly, targeting acid ceramidase with siRNA restored sensitivity to chemotherapy, further highlighting the potential of targeting ceramidases to overcome chemotherapy resistance. Interestingly, evidence suggests that acid ceramidase is upregulated in cancer cells as a response to chemotherapy. Morales et al. (2007) showed that treatment of hepatoma cells, but not normal hepatocytes, with daunorubicin posttranscriptionally upregulated acid ceramidase. When acid ceramidase was pharmacologically inhibited or targeted with siRNA in these cells, ceramide accumulated more than when cells were treated with daunorubicin alone, and the hepatoma cells were sensitized to cell death. We have shown that acid ceramidase overexpression induced by radiation therapy (discussed in Section 3) protected prostate cancer cells from subsequent taxol chemotherapy, an effect that was reversed using a small molecule inhibitor of acid ceramidase, LCL385 (Mahdy et al., 2009). Small molecule inhibitors of acid ceramidase based on the structure of the prototypical ceramidase inhibitors (1*R*, 2*R*)-*N*-myristoylamino-4′-nitro-phenylpropandiol-1,3 (B13) and (1*S*, 2*R*)-*N*-myristoylamino-phenylpropanol-1 (D-e-MAPP) have been developed (Bielawska et al., 2008; Szulc et al., 2008) and tested *in vitro* and *in vivo* against cancer cell lines and xenografts with promising results (Holman et al., 2008).

Ceramidases generate sphingosine which is phosphorylated by sphingosine kinases to form S1P. Thus, sphingosine kinase inhibition has been evaluated as a mechanism to prevent generation of frequently oncogenic S1P, but also as a means to prevent consumption of cellular ceramide and sphingosine. Indeed, sphingosine kinases have been well studied as mediators of cancer. Sphingosine kinase 1 (SK1) is overexpressed at the message or protein level in lung, breast, stomach, thyroid, brain, colon, and kidney cancer (Pyne & Pyne, 2010). Myriad studies demonstrate a clear oncogenic role for sphingosine kinases and S1P, which are discussed more completely in Chapters 5, 6, and 7 of this volume and in a number of excellent reviews (Pyne & Pyne, 2010; Takabe, Paugh, Milstien, & Spiegel, 2008).

With the well-described oncogenic roles of sphingosine kinases, it is perhaps unsurprising that sphingosine kinase dysregulation has been observed as a mechanism for resistance to conventional chemotherapy. Resistance of prostate cancer cells to the topoisomerase inhibitor camptothecin was found to be due to upregulation of both SK1 and S1P receptors 1 and 3 in PC3 cells in response to camptothecin treatment. Camptothecin-sensitive LNCaP cells did not exhibit upregulation of these proteins. Knockdown of SK1 with siRNA and inhibition of S1P G-protein-coupled receptor signaling with pertussis toxin predictably resensitized PC3 cells to camptothecin (Akao et al., 2006). These results were confirmed by another group, which also showed that LNCaP resistance to docetaxel was similarly due to upregulation of SK1. Again, inhibition of sphingosine kinases, this time with small molecule inhibition, sensitized PC3 and LNCaP cells to chemotherapy with camptothecin and docetaxel, respectively (Pchejetski et al., 2008). Further work by this group demonstrated that docetaxel chemotherapy in hormone-resistant prostate cancer cells could be potentiated by pharmacological and RNAi targeting of SK1, with inhibition resulting in a fourfold improvement in docetaxel IC50 (Sauer et al., 2009). In notoriously difficult to treat pancreatic cancer, manipulation of the ceramide/S1P ratio using ceramide analogs and SK1 inhibition improved sensitivity to gemicitibine therapy, whereas upregulating SK1 caused gemcitibine resistance. In another interesting study, SK1 overexpression was frequent in human non-small cell lung carcinoma tissues, and positively correlated with nodal involvement and metastasis whereas it negatively correlated with patient survival (Song et al., 2011). The same study determined that SK1 expression mediated resistance to docetaxel and doxorubicin, and that targeting SK1 expression with RNAi sensitized cancer cells to these therapies. While much focus of sphingosine kinase in cancer has been on SK1, sphingosine kinase 2 (SK2) is also considered a target as its inhibition has been effective in *in vitro* models of cancer (Gao & Smith, 2011). In fact, an isozyme-specific SK2 inhibitor, ABC294640, is in a phase I clinical trial for patients with solid tumors making it the most clinically advanced sphingosine kinase inhibitor to date.

2.5. Sphingomyelin synthase and ceramide kinase

While somewhat less studied than chemotherapy resistance by metabolism of ceramide by GCS, ceramidases, and sphingosine kinases, activation of sphingomyelin synthase activity is another means by which cancers have been shown to escape death by chemotherapy. Sphingomyelin synthase and GCS activities were found to be elevated in doxorubicin-resistant leukemia cells compared to doxorubicin-sensitive cells, and resistant cells had lower levels of ceramide (Itoh et al., 2003). This study highlights the possibility that sphingomyelin synthesis could be a mechanism of escape from treatment sensitivity in leukemia and perhaps other cancers, and that sphingomyelin synthase inhibition may restore ceramide accumulation and apoptosis in response to chemotherapy. Conversely, a recent study has shown that the efficacy of the novel anticancer drug 2-hydroxyoleic acid is effective in inducing cell cycle arrest and apoptosis glioma cells by virtue of its ability to activate sphingomyelin synthase (Barcelo-Coblijn et al., 2011). The authors point to the critical role of sphingomyelin in organizing lipid raft signaling events that are frequently necessary for apoptosis. These conflicting reports on the impact of sphingomyelin synthase in cancer chemotherapy remind us of the complexity of sphingolipid signaling and the diverse roles that ceramide plays both as a signaling lipid and as a component of more complex structural components of cell membranes. Another ceramide-modifying enzyme, ceramide kinase, and its product ceramide 1-phosphate have been shown to promote cancer in *in vitro* models (Mitra et al., 2007). While no studies implicating ceramide kinase in resistance to cancer therapy have been published to date, ceramide kinase has been proposed as a potential target for cancer therapy, and could, theoretically, improve outcomes to standard chemotherapies in certain contexts (Lamour & Chalfant, 2008).

The rationale for targeting ceramide-metabolizing enzymes to enhance response to cytotoxic chemotherapies is clear: generation of ceramide is a consequence of cancer chemotherapy, and mechanisms that prevent ceramide accumulation in response to chemotherapy cause resistance. Several enzymes have been identified as worth targets of interdiction of ceramide metabolism in chemotherapy including GCS, acid ceramidase, and sphingosine kinases. With a wealth of data justifying targeting these enzymes to improve standard cancer chemotherapy, it is our hope that clinically viable agents will be developed in the near future.

3. SPHINGOLIPIDS AND RADIATION THERAPY

3.1. Radiation therapy in the standard of care

Radiation therapy aims at deteriorating the malignant cells and tumor bed via delivery of physical lesions resultant of high energy liberation of electrons from vital cellular structures, which in turn activate pathways of programmed cell death. Therefore, unlike the systemic administration of cytotoxic chemotherapy, the efficacy of most radiation modalities is based on targeted delivery to discernible sites of tumor growth, and is limited by toxic exposure to neighboring, noncancerous organs and tissues. A classical understanding of radiation-induced cellular injury centers on DNA damage as a primary target of programmed cell death pathway activation manifesting cellular lethality from double-strand breaks or reproductive/mitosis-associated cell death from unrepaired and misrepaired breaks (Radford, 1986; Yu, Long, Fu, Zhang, & Ni, 2003). Yet, while the p53-mediated pathway is the most widely recognized mechanism of radiation-induced apoptosis (Ali, Schriml, & Dean, 1999), the past two decades have yielded a burgeoning crop of the literature, much of which grows from the work of Richard Kolesnick and colleagues at the Memorial-Sloan Kettering Cancer Center, that continues to elucidate the p53-independent contributions of sphingolipid metabolism and signaling to cellular fate upon irradiation.

3.2. Induction of p53-independent apoptosis

The first foray into defining the DNA damage-independent context of sphingolipids in γ-radiation-induced apoptosis described the rapid, SMase-mediated hydrolysis of sphingomyelin into ceramide (Haimovitz-Friedman et al., 1994). This process was sensitive to phorbol ester and coincided with the induction of apoptosis. Moreover, preparations of nuclei-free cell membrane isolate generated ceramide upon stimulation with ionizing radiation, thereby demonstrating a functional apoptotic response without the involvement of DNA damage. While this phenomenon occurs generally, the literature suggests that the specific SMase isoenzymes driving sphingomyelin metabolism may be cell type specific. For instance, rapid, radiation-induced NSMase activation is preferred in bovine aortic endothelial (BAEC; Haimovitz-Friedman et al., 1994), WEHI-231 murine lymphoid (Chmura, Nodzenski, et al., 1997), and TF-1 human leukemia cells (Bruno et al., 1998), whereas ASMase activation plays a predominant ceramide-generating role in irradiated Rat-1 MycER fibroblasts (Zundel & Giaccia, 1998), human B lymphoblastoid cells, and lung epithelium (Santana et al., 1996). SQ-20B human head and neck squamous carcinoma cells demonstrated involvement of both NSMase and ASMase in the response to radiation (Chmura, Nodzenski, et al., 1997).

Experimental support of the SMase-mediated mechanisms came from the genetic ablation of ASMase in human tissues, immortalized from type I Niemann–Pick Disease patient B lymphoblastoid cells (Santana et al., 1996), or murine models (Lozano et al., 2001; Santana et al., 1996). Here, defects in radiation-stimulated ceramide generation and apoptosis were shown in each model, which was rescued by ectopic restoration of ASMase expression and function. Importantly, radiosensitization of cells was also restored by the exogenous administration of natural ceramides in combination with irradiation, indicating the defect of apoptosis in these cells was caused not by the dysfunction of a particular SMase isoenzyme, but results from the general failure of radiation-induced ceramide generation, thereby fingering ceramide as the obligate molecule necessary for radiation-induced apoptosis.

3.3. DNA damage incites ceramide generation by ceramide synthases

Apart from the ceramide generated via SMase activation, ceramide synthases (CerS) were implicated in radiation-induced apoptosis via cellular responses to DNA damage. Studies by the Kolesnick group demonstrated irradiation of BAEC and HeLa cells induced fumonisin-sensitive ceramide generation and apoptosis (Liao et al., 1999). In the same report, it was shown that the ataxia telangiectasia-mutated (ATM) *gene* product regulates the apoptotic pathway activated by radiation-induced DNA damage (Liao et al., 1999). Epstein Barr Virus (EBV)-immortalized B-cell lines from six ataxia telangiectasia (AT) patients with different mutations of the ATM gene exhibited radiation-induced CerS activation, ceramide generation, and apoptosis, whereas irradiation of three cell lines from normal patients did not produce these same responses. Ectopic expression of wild-type ATM reversed the response of the AT patient-derived cells, while antisense targeting of

ATM in normal B cells produced the AT phenotype. These data provided evidence that signals from DNA damage and ATM can reciprocally regulate the CerS activity and the proclivity to apoptosis upon radiation exposure.

Vit and Rosselli (2003) integrated the previous evidence into a model of biphasic ceramide accumulation within the radiation-exposed human lymphoblast cell lines, AHH-1 and HSC-93, consisting of transitory, DNA damage-independent SMase activity followed by a lagging, CerS-dependent generation of ceramide. The late phase ceramide accumulation was found to be dependent upon the early phase ceramide-generating process, but independent of caspase activity. Ultimately, the late phase generation of ceramide depends on the DNA damage machinery, supporting earlier results involving the ATM gene, and CerS-mediated ceramide represents the rate-limiting apoptotic step of radiation-induced cell death. Similar biphasic patterns of ceramide generation were demonstrated in Jurkat leukemia cells (Ardail et al., 2009). The dependence of late phase ceramide generation on the earlier phase was supported by this study which found sphingomyelin-sourced ceramide in plasma membrane lipid rafts only 15 min postirradiation; a second peak in ceramide 4 h later was evidenced at the mitochondria via ASMase targeting of this organelle.

3.4. Downstream effectors of radiation-induced ceramide generation

The downstream signaling of ceramide generation upon radiation exposure harkens to well-characterized responses to other forms of cell stress. The stress-activated protein kinase/c-jun kinase (JNK) cascade and apoptosis are activated in irradiated BAEC and human monocytic U937 cells (Verheij et al., 1996). Dominant negative kinases of the JNK cascade abrogated the apoptotic response to either ionizing radiation or exogenous ceramide administration. Basic fibroblast growth factor (bFGF) inhibited radiation-induced apoptosis through JNK activity (Verheij et al., 1996) by activating the diacylglycerol (DAG)-PKC pathway (Cairns et al., 1997; Haimovitz-Friedman et al., 1994). Zundel and Giaccia (1998) demonstrated in Rat-1 Myc-ER cells ASMase-dependent PI3K inhibition via stress- or ceramide-induced coupling to caveolin which resulted in PI3K inactivation, Akt inhibition, and diminution of death effector phospho-Bad (Zundel & Giaccia, 1998). In addition, cells from Niemann–Pick disease patients were deficient in radiation-induced ceramide generation, PI3K inhibition, and apoptosis.

A definitive study demonstrating the necessity of ceramide biogenesis in radiation-induced apoptosis was reported in the *Caenorhabditis elegans* germline, which required mitochondrial ceramide generation for ultimate execution (Turner et al., 2011). Investigations of the mitochondrial effectors of cell survival and death on an evolutionary higher order were first plumbed by a study that, on the one hand, generated CD95 (FAS/APO-1) resistance in Jurkat T cells, which resulted in cross-resistance to either ionizing radiation or etoposide, and was resensitized by administration of ceramide analogs (Tepper, de Vries, van Blitterswijk, & Borst, 1999). On the other hand, the report showed that forced Bcl-2 overexpression in wild-type Jurkat cells manifested lower levels of ceramide accumulation upon exposure to etoposide or ionizing radiation. Although other reports demonstrate that Bcl-2 does not block ceramide generation upon stress (Martin et al., 1995; Pesche et al., 1998; Susin et al., 1997), this study established ceramide downstream of radiation and upstream of mitochondrial control of apoptosis execution. In HeLa and BAEC cells, the engagement of Bax downstream of ceramide was first suggested (Bold, Chandra, & McConkey, 1999) and later reported (Lee et al., 2011) by the Kolesnick group. This study showed that radiation-induced fumonisin-sensitive ceramide generation at the mitochondrial outer membrane forms a platform—termed a mitochondrial ceramide-rich macrodomain—into which Bax inserts, oligomerizes, and functionalizes mitochondrial outer membrane permeabilization leading to cytochrome c release.

3.5. Ceramide in γ-irradiation damage *in vivo*

Ionizing radiation was shown to activate ceramide-mediated endothelial cell apoptosis in several organs *in vivo*, including respiratory alveolar septi (Fuks, Alfieri, Haimovitz-Friedman, Seddon, & Cordon-Cardo, 1995; Santana et al., 1996), intestinal mucosa (Paris et al., 2001), and central nervous system (Pena, Fuks, & Kolesnick, 2000). Histopathological analyses seemed to indicate the endothelial damage was focused in the microvasculature, while larger vessels remained intact. Endothelial apoptosis developed as an early, dose-dependent event after exposure, peaking at 4–10 h after irradiation. Maximal apoptotic effects were observed in the intestinal mucosa at 15 Gy, in the lung at 25 Gy, and in the CNS at 40 Gy (Fuks, Haimovitz-Friedman, & Kolesnick, 1995; Paris et al., 2001; Pena et al., 2000; Santana et al., 1996). Ceramide kinetics in tissue extracts from the lung demonstrated that the apoptosis induction was preceded by ceramide elevation which peaked at twofold of control by 15 min after 10 Gy (Santana et al., 1996). ASMase was indicated as the obligatory mediator of this response based on the reports of the asmase −/− mice demonstrating abrogated ceramide generation and apoptosis in the endothelium of the lung and intestines and brain after whole body radiation (Paris et al., 2001; Pena et al., 2000; Santana et al., 1996). Interestingly, these animals showed normal, p53-mediated apoptosis in the thymus, whereas p53 −/− mice manifested not only normal ceramide generation and apoptosis in the lung and intestines, but also abolition of thymic apoptosis, cementing the independence of the p53 and sphingomyelin pathways of radiation-induced apoptosis. Consistent with the *in vitro* studies above, intravenous administration of bFGF inhibited both a ceramide generation and an endothelial apoptotic response in the intestines, brain, and lung of wild-type mice (Haimovitz-Friedman et al., 1997; Paris et al., 2001; Pena et al., 2000; Santana et al., 1996).

The microvasculature became the focus of radiation-related organ injury because of evidence from the genetic and pharmacologic studies in intestinal and lung models. These studies posited the microvascular endothelium, rather than tissue stem cells, as the primary target for radiation. Whole-body irradiated C57BL/6 mice receiving 10–13 Gy died between 10 and 13 days with an intact gastrointestinal tract and depleted bone marrow, while mice irradiated with 15 Gy died at 61 days with denuded GI mucosa and moderately damaged marrow (Paris et al., 2001). TUNEL staining demonstrated apoptosis in the crypt/villus microvasculature as early as 4-h postirradiation and the risk of death from the GI syndrome correlated with the severity of endothelial apoptosis. Intravenous bFGF administration abrogated endothelial apoptosis and prevented radiation-induced crypt damage, organ failure, and GI death. Inhibition of endothelial apoptosis and protection from GI death were also found in asmase −/− mice treated with 15 Gy radiation. *In situ* hybridization showed that the endothelium, and not epithelial crypt cells, expressed high affinity bFGF receptors, thereby stratifying the endothelial lesion upstream of crypt stem cell damage in the manifestation of the radiation-induced GI syndrome. Consistent with this notion, bFGF administration also protected C3H/HeJ mice from lethal radiation pneumonitis (Fuks, Alfieri, et al., 1995).

Exquisite sensitivity to ASMase-mediated apoptosis upon irradiation was demonstrated in the ovary of fertile female mice. Tilly et al. demonstrate a defect in asmase −/− mice of normal apoptotic deletion of fetal oocytes during embryogenesis, leading to neonatal ovarian hyperplasia (Morita et al., 2000). The cell autonomy of this defect was confirmed *ex vivo* in oocytes from asmase−/− mice, or wild-type oocytes treated with S1P-resisted daunorubicin-induced apoptosis. Of note, Bax was found to function downstream of ceramide in this pathway. Irradiation at only 0.1 Gy induced 90% loss of oocytes and infertility in wild-type female mice, which was abrogated *in vivo* by administration of S1P. Furthermore, oocytes superovulated from S1P-pretreated and irradiated animals generated normal blastocysts after *in vitro* fertilization, while oocytes from vehicle-treated animals yielded defective blastocysts, indicating preservation of organ function by S1P.

3.6. Restoration and perturbation of sphingolipid balance in radiotherapeutic enhancement

Armed with an understanding of the ceramide response to ionizing radiation, we have witnessed a number of pharmacologic interventions in sphingolipid metabolism that produce predictable functional outcomes to alter the balance between ceramide generation and its removal, thereby influencing radiation-induced apoptotic signaling and its pathologic consequences. Chmura, Nodzenski, et al. (1997) selected a population of WEHI-231 cells after long-term selection with the acid ceramidase inhibitor *N*-oleoylethanolamine, which purportedly selects for ceramide-intolerant cells. The resultant WEHI-231 subline exhibited resistance to SMase-mediated apoptosis in response to radiation.

Much attention has been paid to prostate cancer radiotherapy, because it is a leading indication for local solid organ radiation modalities (Nguyen et al., 2011). Kimura, Bowen, Spiegel, and Gelmann (1999) found that androgen-sensitive human LNCaP prostate cancer cells produced neither a ceramide nor an apoptotic response to ionizing radiation, but were apoptotic under the influence of TNF-α (Kimura et al., 1999). However, the combination of radiation and TNF-α affected synergistically ceramide elevation and apoptosis. Moreover, substitution of TNF-α with ceramide analogs, at doses that did not induce apoptosis under singular treatment, yielded synergistic cell killing in combination with radiation. An alternative approach to target radiation resistance in LNCaP cells was reported by Garzotto et al. (Facher & Law, 1998; Garzotto et al., 1999). Although most cell types treated with phorbol ester signal antiapoptotic signals via PKC activation (Cairns et al., 1997; Haimovitz-Friedman et al., 1994), in LNCaP cells, TPA treatment demonstrated a CerS-mediated apoptotic response in combination with radiation (Garzotto et al., 1999). Pretreatment of LNCaP cells with TPA significantly enhanced ceramide generation by CerS activation, and resulted in a synergistic, fumonisin-sensitive apoptotic response. Orthotopic transplantation into the prostates of nude mice, recapitulated the responses of LNCaP cell-derived tumors to combination treatment within 24 h of irradiation. An additional benefit was noted, in that neighboring rectal tissues were protected by TPA administration, thereby improving the therapeutic window of radiation therapy in this small animal model.

Norris and colleagues have targeted prostate cancer cells undergoing radiation therapy with rationally designed ceramide analog inhibitors of acid ceramidase to effect radiation sensitization and enhanced tumor killing (Mahdy et al., 2009). The strategy is based on the observation that ionizing radiation stimulates overexpression of the acid ceramidase and alleviates this choke point of ceramide accumulation by conversion to sphingosine and rapid phosphorylation to S1P. Rational design of ceramide analogs based on the structures of d-e-MAPP and B13 has led to development of several lead compounds (Bai et al., 2009; Bielawska et al., 2008; Szulc et al., 2008) that demonstrate efficacious inhibition of acid ceramidase activity and enhancement of cancer cell killing either with monotherapeutic administration (Holman et al., 2008; Samsel et al., 2004; Selzner et al., 1999) of the inhibitor or in combination with other modalities (Elojeimy et al., 2007; Liu, Elojeimy, et al., 2006; Norris et al., 2006), including radiation therapy (Mahdy et al., 2009).

Targeting the sphingosine kinases in order to elevate ceramide and sphingosine levels, as well as shut down S1P production and signaling, is another mechanistic node for investigation. Cuvillier and colleagues utilized the sphingosine analog, FTY720 (fingolimod), in combination with ionizing radiation on prostate cancer cell lines *in vitro* and *in vivo* (Pchejetski et al., 2010). In prostate cancer cell lines, FTY720 inhibited SK1 and induced prostate cancer cell apoptosis in a manner independent of S1P receptors. Ionizing radiation did not affect SK1 activity in prostate cancer cells; yet, synergized with FTY720 to inhibit SK1. Combination therapy in orthotopic and subcutaneous mouse models demonstrated marked radiosensitization of the tumors. *N,N*-dimethyl-D-erythro-sphingosine also has demonstrable applications as a sphingosine kinase inhibitor that, when combined with radiation, produced increased sphingosine accumulation and apoptosis in LNCaP prostate adenocarcinoma cells (Nava et al., 2000) and Lewis lung carcinoma cells (Park et al., 2004).

Several studies have shown that PKC modulation affects the apoptotic response to radiation (Cairns et al., 1997; Haimovitz-Friedman et al.,

1994), likely through the blockade of radiation-induced DAG generation (Bruno et al., 1998; Chmura, Mauceri, et al., 1997; Okahara, Ikawa, Kanaho, & Maehama, 2004). As mentioned above, PKC activation with phorbol ester blocked ceramide generation and apoptosis in most cell types tested with radiation exposure (Bruno et al., 1998; Cairns et al., 1997; Haimovitz-Friedman et al., 1994; Nakamura et al., 2000; Okahara et al., 2004; Vazquez, Ramaswamy, Nakamura, & Sellers, 2000). Conversely, pharmacologic inhibition of PKC augmented radiation-induced cell killing (Hallahan et al., 1992). The ceramide-related mechanism by which phorbol ester and DAG affect apoptosis apparently likely occurs through SMase activation (Nakamura et al., 2000). As mentioned above, bFGF similarly protected endothelial cells against radiation-induced apoptosis, in part through activation of PKC and MAPK cascades (Cairns et al., 1997; Verheij et al., 1996).

Studies by Rodriguez-Lafrasse and colleagues have explored mechanisms of achieving high levels of ceramide accumulation in order to sensitize a range of cancer cell lines. In a study of the radioresistant head and neck squamous cell carcinoma line SQ20B versus a radiosensitive cell line of the same histological type, SCC61, the group found that the resistant cell line lacked the typical ceramide response to ionizing radiation and was demonstrably insensitive to SMase-generated ceramide (Alphonse et al., 2002, 2004). The SQ20B cell line was found to lack ASMase activity and translocation to produce ceramide-associated membrane rafts (Bionda et al., 2007). Poly-drug pretreatment of SQ20B cells to bolster the endogenous ceramide levels consisted of: (1) an inhibitor of GCS, DL-PDMP; (2) an inhibitor of ceramidase, D-MAPP; and (3) an amphiphillic amine that disturbs lipid turnover in biological membranes, imipramine (Alphonse et al., 2004; Rodriguez-Lafrasse et al., 2002). This potent cocktail radiosensitized SQ20B to produce clonogenic and apoptotic cell death upon exposure to 10 Gy ionizing radiation, and demonstrated a threshold of intracellular ceramide that was necessary for cell death. Polydrug treatment was similarly effective in Jurkat leukemia cells (Rodriguez-Lafrasse et al., 2002). In addition, attacking the antioxidant defenses of SQ20B cells with either peroxide or GSH depletion also triggered ASMase activation and translocation, lipid raft coalescence, and apoptosis induction (Bionda et al., 2007).

Additionally, untested hypotheses have been posited to drive cancer toward radiosensitization through interdiction of ceramide-related pathways. Pharmacological and biochemical studies suggest that investigating compounds capable of interfering with SMase activity is of potential benefit

(Claus, Dorer, Bunck, & Deigner, 2009). Exploitation of a positive feedback loop between PLA2-mediated arachidonic acid generation and ceramide and/or ceramide 1-phosphate has also been hypothesized to provide a viable strategy to mediate ceramide-induced radiosensitivity (Eng, 2003; Huwiler, Johansen, Skarstad, & Pfeilschifter, 2001; Nakamura, Hirabayashi, Shimizu, & Murayama, 2006; Shimizu et al., 2009).

3.7. Swords to ploughshares

As research continues into increasing tumor radiosensitization and improving radiotherapeutic response, radioprotective strategies have also been developed from our understanding of sphingolipid metabolism. Interest in protecting normal tissues from collateral injury during radiation therapy, or, from general radiation contamination, highlighted by recent interest ignited by radiation concerns after the 2011 Japanese earthquake (Bhattacharjee, 2011), has led to investigation of sphingolipid interdiction as well. Richard Kolesnick is leading development of a humanized monoclonal anticeramide antibody, 2A2, for efficacious protection against radiation injury, particularly GI syndrome (Bhattacharjee, 2011). Interference of the acute response to IR via sphingomyelin was investigated with the administration of sphingomyelin synthase inhibitor tricyclodecan-9-yl-xanthogenate (D609), which caused diminution of ionizing radiation (IR)-induced (1) production of reactive oxygen species, (2) decrease in intracellular reduced glutathione, (3) oxidative damage to proteins and lipids, and (4) activation of nuclear factor-kappaB (Davies, Jans, & Wagstaff, 2010). S1P analogs, such as FTY720, (S)-FTY720-phosphonate (fTyS), and SEW-2871, have recently shown efficacy in attenuating radiation-induced lung injury in a murine models (Mathew et al., 2011). At the step of S1P metabolism, oral administration of an S1P Lyase (SPL) inhibitor to mice prolonged their survival after exposure to a lethal dose of total body IR (Kumar et al., 2011). Successful induction of sphingolipid-modulated radioprotection combined with precise delivery to noncancerous tissues present a feasible, alternative strategy to improving the therapeutic window for radiation therapy of solid tumors.

4. SPHINGOLIPIDS AND TARGETED ANTICANCER AGENTS

The discovery of the first human oncogene, Ras, in 1982 by Weinberg and colleagues brought on boundless energy and enthusiasm for the concept of cancer therapies that target cancer-driving signaling pathways specifically rather than relying on cytotoxic chemotherapies, whose tumor specificity was centered around rapidity of cell division (Hanahan & Weinberg, 2000). Since then, myriad oncogenes have been discovered and, in turn, inhibited with high hopes for transforming cancer therapy. Major classes of targeted therapies now approved for cancer include TKIs, histone deacetylase inhibitors (HDACIs), and mAbs. These therapies certainly have improved therapeutic outcome for many cancer patients. Disappointingly, however, targeted therapies infrequently cure patients, as development of resistance is nearly universal. Interestingly, as with cytotoxic chemotherapy and radiotherapy, study of the mechanisms of cell death and resistance to targeted therapies has again implicated sphingolipid metabolism as integral to both favorable outcomes and development of resistance. In this section, we will review the literature on sphingolipid metabolism in response to newer targeted therapies, both in terms of eliciting favorable treatment responses and the development of resistance.

4.1. Tyrosine kinase inhibitors

TKIs are a class of anticancer drugs that function by inhibiting many signal transducing protein kinases. The classic TKI is imatinib, which targets the bcr–abl fusion kinase that drives chronic myelogenous leukemia (CML). Constitutive bcr–abl signaling activates well-characterized oncogenic pathways including Ras/ERK, Jak Stat, PI3K, and c-MYC. The development of imatinib to target the bcr–abl tyrosine kinase by competitive interference at the ATP-binding pocket has been one of the great advancements of cancer therapy. Interestingly, sphingolipid metabolism has been found to be a key element in imatinib effectiveness and resistance against CML. Treatment of CML cells with imatinib caused an accumulation of ceramide in imatinib-sensitive cells, but not in cells that had been cultured to become imatinib resistant (Baran et al., 2007). Ceramide generation was found to occur through CerS1, increasing primarily C18 ceramide and inducing apoptosis. Analysis of gene expression in imatinib-resistant cells led to the discovery that resistant cells had elevated SK1 and S1P. Targeting SK1 restored imatinib sensitivity, whereas SK1 overexpression conferred resistance, demonstrating a clear role for SK1 in imatinib resistance in CML. The same group later showed that imatinib-resistant cells also have elevated GCS activity, and that ceramide accumulation and therapeutic response could be augmented by treatment with the GCS inhibitor PDMP (Baran et al., 2011). Multiple groups have reported similarly that imatinib resistance is mediated by SK1 (Bonhoure et al., 2008; Marfe et al., 2011) and GCS (Huang et al., 2011). Second generation TKIs used in imatinib-resistant CML have also been investigated with regards to sphingolipid signaling. Gencer, Ural, Avcu, and Baran (2011) showed that dasatinib cell killing was due to increased ceramide generation through CerS1 and downregulation of SK1. Similar results were observed by the same group with nilotinib (Camgoz, Gencer, Ural, Avcu, & Baran, 2011), suggesting that ceramide is indeed a critical mediator of TKI-induced treatment of CML.

Sphingolipid manipulation may be further important in combination with TKIs during CML blast crisis. It has been found that during blast crisis, PP2A, which performs many of the apoptotic signaling functions induced by ceramide accumulation, is functionally inactive due to CML-driven expression of the PP2A inhibitor SET (Neviani et al., 2005). FTY720, an S1P analog that induces S1P receptor downregulation and is an approved therapy for the treatment of relapsing multiple sclerosis and with preclinical data supporting its potential use as an antineoplastic agent (Wallington-Beddoe, Hewson, Bradstock, & Bendall, 2011), can lead to the reactivation of PP2A (Perrotti & Neviani, 2008). Additional work has shown that the mechanism of SK1 and S1P in conferring TKI resistance in CML is through inhibition of PP2A (Salas et al., 2011). These studies implicate that targeting PP2A stability and activation, perhaps with the clinical agent FTY720 or with a novel clinical approach to elevating cancer cell ceramide, are worthy of further study in the treatment of CML.

In addition to the large body of work examining sphingolipids as key in CML resistance to TKIs, ceramide generation has been shown to occur downstream of multiple other TKIs. Gefitinib inhibits EGFR by competing at the ATP-binding site of the EGFR tyrosine kinase domain (Albitar et al., 2010). Gefitinib combination therapy with tamoxifen and etoposide (Mimeault, Venkatraman, et al., 2007) or gefitinib with cyclopamine and doxetaxel (Mimeault, Johansson, et al., 2007) in metastatic prostate cancer cells enhanced apoptotic cell death in part through ceramide accumulation. Sorafenib, a multikinase inhibitor with antiVEGF, -PDGF, and -Raf effects was shown to be effective in combination with the SK2 inhibitor ABC294640 in hepatocellular carcinoma by antagonizing ERK signaling (Beljanski, Knaak, Zhuang, & Smith, 2011). Sunitinib, a VEGFR/PDGFR kinase inhibitor approved for the treatment of renal cell carcinoma and gastrointestinal tumors, in combination with FTY720, was shown to potentiate the sunitinib response in breast cancer models, with the combination therapy decreasing tumor growth compared to single agent controls

(Mousseau et al., 2011). This effect is perhaps due to further suppression of PDGFR by FTY720, as S1P receptor 3 crosstalks with PDGFR (Brunati et al., 2008). The FTY720-mediated enhancement in TKI cell killing may be able to be more broadly applied to the other TKIs, and further evaluation of mechanisms of resistance to TKIs may uncover conservation of ceramide metabolism as an approach to potentiating therapy.

4.2. Histone deacetylase inhibitors

HDACIs are a class of newly developed antineoplastic agents that inhibit histone deacetylases and thereby modulate gene expression by affecting the relative accessibility of targeted regions of DNA to transcription machinery (Imanishi et al., 2002). HDACIs have been heavily investigated in the treatment of cancer. Vorinostat is clinically approved for the treatment of T-cell lymphomas, and many other agents are in clinical trials. Interestingly, links have been found between chemotherapy with HDACIs and sphingolipids. *In vitro* HDACI treatment with MS-275 enhanced leukemia cell killing with fludarabine, with a large increase in ceramide over single agent treatment (Maggio et al., 2004). Delivery of exogenous cell permeable C6 ceramide and the HDACI trichostatin A synergistically enhanced cell death and decreased tumor proliferation *in vivo* and *in vitro* in ovarian and pancreatic models (Zhu et al., 2011). Interestingly, trichostatin A itself elicited ceramide generation that failed to result in apoptosis due to rapid glycosylation. Treatment with PDMP, the GCS inhibitor, markedly increased trichostatin A-induced ceramide accumulation and potentiated cell death. In another study which combined vorinistat with the Akt inhibitor Perifosine, it was shown that efficacy of the combination in human leukemia cells was dependent on ASMase-generated ceramide, as inhibition of this enzyme prevented ceramide accumulation and downstream reactive oxygen species generation and cell death (Rahmani et al., 2005).

Combination regimens with TKIs and HDACIs have also been evaluated. Low dose sorafenib and vorinostat treatment enhanced cell killing and inhibited tumor growth *in vivo* in renal cell carcinoma, hepatocellular carcinoma, and pancreatic adenocarcinomas (Park et al., 2010, 2008; Walker et al., 2009). Sorafenib and vorinostat cotreatment was initially found to activate CD95, an effect that was found to be the result of ceramide generation through ASMase and *de novo* ceramide synthesis. Interestingly, it appears that LASS6, the gene for CerS6, was hyperacetylated with vorinostat treatment, suggesting a mechanism by which ceramide and dihydroceramide were generated in the combination therapy. Inhibition of ASMase and *de novo* ceramide generation reduced the reduced ROS production, intracellular calcium accumulation, and synergistic cell death. This combination is currently being evaluated in a Phase I clinical trial for renal cell carcinoma and non-small cell carcinoma.

4.3. Monoclonal antibodies

mAbs are a mechanism to target and neutralize specific cellular antigens. These therapies are advantageous in that they allow very specific targeting of proteins or molecules, and thus can be tumor specific. An excellent example is trastuzumab which inhibits the her2/neu translocation kinase that drives many breast cancers. While relatively little is known about sphingolipid metabolism in response to treatment with approved mAbs, it is worth noting that rituximab, an antiCD20 mAb used to treat B-cell dyscrasias, results in a rapid and transient increase in ceramide associated with ASMase activation and colocalization with the CD20 receptor (Bezombes et al., 2002). Additional work conducted using milatuzumab, an antiCD74 mAb in Phase I/II clinical trials, demonstrated synergistic killing with FTY720 in mantle cell lymphoma (Alinari, Mahoney, et al., 2011; Alinari, Yu, et al., 2011). These studies uncovered an interesting mechanism through which FTY720 disrupted autophagy resulting in cell death through leakage of lysosomal hydrolases into the cytoplasm. Disruption of autophagy led to an accumulation of CD74, which is normally recycled in the lysosome, thus mechanistically explaining potentiation of milatuzumab effectiveness by FTY720.

Targeted therapies are likely to play a large role in the future of cancer therapy as techniques to identify specific molecular alterations that drive patient-specific malignancies improve. The oncogene era has defined hundreds of potential targets for cancer therapy, so the task remains to develop methods to attack those targets as well as to predict and respond to mechanism of resistance. The potential for interdiction of sphingolipid metabolism to be useful in the treatment of cancer extends beyond cytotoxic chemotherapy and radiation into the new age of targeted therapies. Preclinical evidence shows that these therapies frequently upregulate ceramide generation and that resistance often occurs through ceramide consumption. Development of clinical agents that inhibit sphingosine kinases, ceramidases, GCS, and perhaps other ceramide consumers holds promise in the potentiation of targeted cancer therapy.

5. CONCLUDING REMARKS

Decades of study have shown that ceramide is a critical mediator of apoptosis in response to standard cancer therapies including radiation, chemotherapy, and emerging targeted inhibitors. As we have delineated in this review, it is critical that the mechanisms that result in its accumulation of ceramide remain intact for fully efficacious therapy, as altered metabolism of sphingolipids can cause treatment resistance and result in poor therapeutic outcomes. Interdiction of sphingolipid metabolism represents a unique opportunity to sensitize cancers to standard therapy for several reasons:

1. Consumption of ceramide and generation of S1P are mechanisms of resistance to diverse therapies. We have outlined generation of ceramide as a critical part of therapy-induced cell death in response to cytotoxic chemotherapies, radiation therapy, and newer targeted therapies. This confirms a conserved role for sphingolipids in apoptosis.
2. Ceramide induces apoptosis through redundant mechanism. Ceramides mediates apoptosis through effectors upstream of multiple Bcl-2 family members, Akt, Myc, SR proteins, Rb, PKC, and JNK, among others. Therefore, in contrast to therapies that signal specifically at distal points in cell death cascades, ceramides have the potential to be effective even in cells with multiple apoptosis signaling defects by affecting intact pathways.
3. Multiple sphingolipid metabolizing enzymes have been identified as mechanisms of resistance to cancer therapy. Inhibition of acid ceramidase, sphingosine kinases, and GCS is well supported as means of sensitizing cancer cells to standard treatment.

The challenge is to develop feasible ways to inhibit these enzymes in patients. Clinically tested agents that target sphingolipid metabolism are few. As discussed above, there is an SK2 inhibitor in clinical trials, and multiple groups are actively searching for clinically relevant inhibitors of ceramidase (Draper et al., 2011), sphingosine kinases (French et al., 2003), and GCS (Nietupski et al., 2012). Successful development of these therapies is an exciting prospect for the future of cancer therapy. With decades of study justifying cotargeting sphingolipid metabolism with standard therapies, the time has come to test these ideas in human disease.

REFERENCES

Akao, Y., Banno, Y., Nakagawa, Y., Hasegawa, N., Kim, T.-J., Murate, T., et al. (2006). High expression of sphingosine kinase 1 and S1P receptors in chemotherapy-resistant prostate cancer PC3 cells and their camptothecin-induced up-regulation. *Biochemical and Biophysical Research Communications, 342*, 1284–1290.

Albitar, L., Pickett, G., Morgan, M., Wilken, J. A., Maihle, N. J., & Leslie, K. K. (2010). EGFR isoforms and gene regulation in human endometrial cancer cells. *Molecular Cancer, 9*, 166.

Ali, I. U., Schriml, L. M., & Dean, M. (1999). Mutational spectra of PTEN/MMAC1 gene: A tumor suppressor with lipid phosphatase activity. *Journal of the National Cancer Institute, 91*, 1922–1932.

Alinari, L., Mahoney, E., Patton, J., Zhang, X., Huynh, L., Earl, C. T., et al. (2011). FTY720 increases CD74 expression and sensitizes mantle cell lymphoma cells to milatuzumab-mediated cell death. *Blood, 118*, 6893–6903.

Alinari, L., Yu, B., Christian, B. A., Yan, F., Shin, J., Lapalombella, R., et al. (2011). Combination anti-CD74 (milatuzumab) and anti-CD20 (rituximab) monoclonal antibody therapy has in vitro and in vivo activity in mantle cell lymphoma. *Blood, 117*, 4530–4541.

Alphonse, G., Aloy, M. T., Broquet, P., Gerard, J. P., Louisot, P., Rousson, R., et al. (2002). Ceramide induces activation of the mitochondrial/caspases pathway in Jurkat and SCC61 cells sensitive to gamma-radiation but activation of this sequence is defective in radioresistant SQ20B cells. *International Journal of Radiation Biology, 78*, 821–835.

Alphonse, G., Bionda, C., Aloy, M. T., Ardail, D., Rousson, R., & Rodriguez-Lafrasse, C. (2004). Overcoming resistance to gamma-rays in squamous carcinoma cells by poly-drug elevation of ceramide levels. *Oncogene, 23*, 2703–2715.

Ardail, D., Maalouf, M., Boivin, A., Chapet, O., Bodennec, J., Rousson, R., et al. (2009). Diversity and complexity of ceramide generation after exposure of jurkat leukemia cells to irradiation. *International Journal of Radiation Oncology, Biology, Physics, 73*, 1211–1218.

Bai, A., Szulc, Z. M., Bielawski, J., Mayroo, N., Liu, X., Norris, J., et al. (2009). Synthesis and bioevaluation of [omega]-N-amino analogs of B13. *Bioorganic & Medicinal Chemistry, 17*, 1840–1848.

Baran, Y., Bielawski, J., Gunduz, U., & Ogretmen, B. (2011). Targeting glucosylceramide synthase sensitizes imatinib-resistant chronic myeloid leukemia cells via endogenous ceramide accumulation. *Journal of Cancer Research and Clinical Oncology, 137*, 1535–1544.

Baran, Y., Salas, A., Senkal, C. E., Gunduz, U., Bielawski, J., Obeid, L. M., et al. (2007). Alterations of ceramide/sphingosine 1-phosphate rheostat involved in the regulation of resistance to imatinib-induced apoptosis in K562 human chronic myeloid leukemia cells. *The Journal of Biological Chemistry, 282*, 10922–10934.

Barcelo-Coblijn, G., Martin, M. L., de Almeida, R. F., Noguera-Salva, M. A., Marcilla-Etxenike, A., Guardiola-Serrano, F., et al. (2011). Sphingomyelin and sphingomyelin synthase (SMS) in the malignant transformation of glioma cells and in 2-hydroxyoleic acid therapy. *Proceedings of the National Academy of Sciences of the United States of America, 108*, 19569–19574.

Beckham, T. H., Lu, P., Cheng, J. C., Zhao, D., Turner, L. S., Zhang, X., et al. (2012). Acid ceramidase-mediated production of sphingosine 1-phosphate promotes prostate cancer invasion through upregulation of cathepsin B. *International Journal of Cancer, 131*, 2034–2043.

Beljanski, V., Knaak, C., Zhuang, Y., & Smith, C. D. (2011). Combined anticancer effects of sphingosine kinase inhibitors and sorafenib. *Investigational New Drugs, 29*, 1132–1142.

Bezombes, C., de Thonel, A., Apostolou, A., Louat, T., Jaffrezou, J.-P., Laurent, G., et al. (2002). Overexpression of protein kinase czeta confers protection against antileukemic drugs by inhibiting the redox-dependent sphingomyelinase activation. *Molecular Pharmacology, 62*, 1446–1455.

Bhattacharjee, Y. (2011). Devastation in Japan. Candidate radiation drugs inch forward. *Science, 331*, 1505.

Bielawska, A., Bielawski, J., Szulc, Z. M., Mayroo, N., Liu, X., Bai, A., et al. (2008). Novel analogs of D-e-MAPP and B13. Part 2: Signature effects on bioactive sphingolipids. *Bioorganic & Medicinal Chemistry, 16*, 1032–1045.

Bionda, C., Hadchity, E., Alphonse, G., Chapet, O., Rousson, R., Rodriguez-Lafrasse, C., et al. (2007). Radioresistance of human carcinoma cells is correlated to a defect in raft membrane clustering. *Free Radical Biology & Medicine, 43*, 681–694.

Bold, R., Chandra, J., & McConkey, D. (1999). Gemcitabine-Induced programmed cell death (apoptosis) of human pancreatic carcinoma is determined by Bcl-2 content. *Annals of Surgical Oncology, 6*, 279–285.

Bonhoure, E., Lauret, A., Barnes, D. J., Martin, C., Malavaud, B., Kohama, T., et al. (2008). Sphingosine kinase-1 is a downstream regulator of imatinib-induced apoptosis in chronic myeloid leukemia cells. *Leukemia, 22*, 971–979.

Bose, R., Verheij, M., Haimovitz-Friedman, A., Scotto, K., Fuks, Z., & Kolesnick, R. (1995). Ceramide synthase mediates daunorubicin-induced apoptosis: An alternative mechanism for generating death signals. *Cell, 82*, 405–414.

Brunati, A. M., Tibaldi, E., Carraro, A., Gringeri, E., D'Amico, F., Jr., Toninello, A., et al. (2008). Cross-talk between PDGF and S1P signalling elucidates the inhibitory effect and potential antifibrotic action of the immunomodulator FTY720 in activated HSC-cultures. *Biochimica et Biophysica Acta, 1783*, 347–359.

Bruno, A. P., Laurent, G., Averbeck, D., Demur, C., Bonnet, J., Bettaieb, A., et al. (1998). Lack of ceramide generation in TF-1 human myeloid leukemic cells resistant to ionizing radiation. *Cell Death and Differentiation, 5*, 172–182.

Cairns, P., Okami, K., Halachmi, S., Halachmi, N., Esteller, M., Herman, J. G., et al. (1997). Frequent Inactivation of PTEN/MMAC1 in Primary Prostate Cancer. *Cancer Research, 57*, 4997–5000.

Camgoz, A., Gencer, E. B., Ural, A. U., Avcu, F., & Baran, Y. (2011). Roles of ceramide synthase and ceramide clearence genes in nilotinib-induced cell death in chronic myeloid leukemia cells. *Leukemia & Lymphoma, 52*, 1574–1584.

Chalfant, C. E., Kishikawa, K., Mumby, M. C., Kamibayashi, C., Bielawska, A., & Hannun, Y. A. (1999). Long chain ceramides activate protein phosphatase-1 and protein phosphatase-2A. Activation is stereospecific and regulated by phosphatidic acid. *The Journal of Biological Chemistry, 274*, 20313–20317.

Chalfant, C. E., Rathman, K., Pinkerman, R. L., Wood, R. E., Obeid, L. M., Ogretmen, B., et al. (2002). De novo ceramide regulates the alternative splicing of caspase 9 and Bcl-x in A549 lung adenocarcinoma cells. Dependence on protein phosphatase-1. *The Journal of Biological Chemistry, 277*, 12587–12595.

Chauvier, D., Morjani, H., & Manfait, M. (2002). Ceramide involvement in homocamptothecin- and camptothecin-induced cytotoxicity and apoptosis in colon HT29 cells. *International Journal of Oncology, 20*, 855–863.

Chmura, S. J., Mauceri, H. J., Advani, S., Heimann, R., Beckett, M. A., Nodzenski, E., et al. (1997). Decreasing the apoptotic threshold of tumor cells through protein kinase C inhibition and sphingomyelinase activation increases tumor killing by ionizing radiation. *Cancer Research, 57*, 4340–4347.

Chmura, S. J., Nodzenski, E., Beckett, M. A., Kufe, D. W., Quintans, J., & Weichselbaum, R. R. (1997). Loss of ceramide production confers resistance to radiation-induced apoptosis. *Cancer Research, 57*, 1270–1275.

Claus, R. A., Dorer, M. J., Bunck, A. C., & Deigner, H. P. (2009). Inhibition of sphingomyelin hydrolysis: targeting the lipid mediator ceramide as a key regulator of cellular fate. *Current Medicinal Chemistry, 16*, 1978–2000.

Cowart, L. A., & Obeid, L. M. (2007). Yeast sphingolipids: Recent developments in understanding biosynthesis, regulation, and function. *Biochimica et Biophysica Acta (BBA) - Molecular and Cell Biology of Lipids, 1771*, 421–431.

Davies, R. G., Jans, D. A., & Wagstaff, K. M. (2010). Use of fluorescence photobleaching techniques to measure the kinetics of intracellular transport. *Microscopy: Science, technology, applications and education*. Spain: Formatex Research Center 756–763.

Deng, W., Li, R., Guerrera, M., Liu, Y., & Ladisch, S. (2002). Transfection of glucosylceramide synthase antisense inhibits mouse melanoma formation. *Glycobiology, 12*, 145–152.

Deng, W., Li, R., & Ladisch, S. (2000). Influence of cellular ganglioside depletion on tumor formation. *Journal of the National Cancer Institute, 92*, 912–917.

Di Marzio, L., Di Leo, A., Cinque, B., Fanini, D., Agnifili, A., Berloco, P., et al. (2005). Detection of alkaline sphingomyelinase activity in human stool: Proposed role as a new diagnostic and prognostic marker of colorectal cancer. *Cancer Epidemiology, Biomarkers & Prevention, 14*, 856–862.

Dobrowsky, R. T., Kamibayashi, C., Mumby, M. C., & Hannun, Y. A. (1993). Ceramide activates heterotrimeric protein phosphatase 2A. *The Journal of Biological Chemistry, 268*, 15523–15530.

Draper, J. M., Xia, Z., Smith, R. A., Zhuang, Y., Wang, W., & Smith, C. D. (2011). Discovery and evaluation of inhibitors of human ceramidase. *Molecular Cancer Therapeutics, 10*, 2052–2061.

Elojeimy, S., Liu, X., McKillop, J. C., El-Zawahry, A. M., Holman, D. H., Cheng, J. Y., et al. (2007). Role of acid ceramidase in resistance to FasL: therapeutic approaches based on acid ceramidase inhibitors and FasL gene therapy. *Molecular Therapy, 15*, 1259–1263.

Eng, C. (2003). PTEN: one gene, many syndromes. *Human Mutation, 22*, 183–198.

Erez-Roman, R., Pienik, R., & Futerman, A. H. (2010). Increased ceramide synthase 2 and 6 mRNA levels in breast cancer tissues and correlation with sphingosine kinase expression. *Biochemical and Biophysical Research Communications, 391*, 219–223.

Facher, E. A., & Law, J. C. (1998). PTEN and prostate cancer. *Journal of Medical Genetics, 35*, 790.

French, K. J., Schrecengost, R. S., Lee, B. D., Zhuang, Y., Smith, S. N., Eberly, J. L., et al. (2003). Discovery and evaluation of inhibitors of human sphingosine kinase. *Cancer Research, 63*, 5962–5969.

Fuks, Z., Alfieri, A., Haimovitz-Friedman, A., Seddon, A., & Cordon-Cardo, C. (1995). Intravenous basic fibroblast growth factor protects the lung but not mediastinal organs against radiation-induced apoptosis in vivo. *The Cancer Journal from Scientific American, 1*, 62–72.

Fuks, Z., Haimovitz-Friedman, A., & Kolesnick, R. N. (1995). The role of the sphingomyelin pathway and protein kinase C in radiation-induced cell kill. *Important Advances in Oncology*, 19–31.

Galadari, S., Kishikawa, K., Kamibayashi, C., Mumby, M. C., & Hannun, Y. A. (1998). Purification and characterization of ceramide-activated protein phosphatases. *Biochemistry, 37*, 11232–11238.

Gao, P., & Smith, C. D. (2011). Ablation of sphingosine kinase-2 inhibits tumor cell proliferation and migration. *Molecular Cancer Research, 9*, 1509–1519.

Garzotto, M., Haimovitz-Friedman, A., Liao, W. C., White-Jones, M., Huryk, R., Heston, W. D., et al. (1999). Reversal of radiation resistance in LNCaP cells by targeting apoptosis through ceramide synthase. *Cancer Research, 59*, 5194–5201.

Gault, C. R., Obeid, L. M., & Hannun, Y. A. (2010). An overview of sphingolipid metabolism: from synthesis to breakdown. *Advances in Experimental Medicine and Biology, 688*, 1–23.

Gencer, E. B., Ural, A. U., Avcu, F., & Baran, Y. (2011). A novel mechanism of dasatinib-induced apoptosis in chronic myeloid leukemia; ceramide synthase and ceramide clearance genes. *Annals of Hematology, 90*, 1265–1275.

Gouaze, V., Liu, Y. Y., Prickett, C. S., Yu, J. Y., Giuliano, A. E., & Cabot, M. C. (2005). Glucosylceramide synthase blockade down-regulates P-glycoprotein and resensitizes multidrug-resistant breast cancer cells to anticancer drugs. *Cancer Research, 65*, 3861–3867.

Guan, H., Liu, L., Cai, J., Liu, J., Ye, C., Li, M., et al. (2011). Sphingosine kinase 1 is over-expressed and promotes proliferation in human thyroid cancer. *Molecular Endocrinology, 25*, 1858–1866.

Haimovitz-Friedman, A., Cordon-Cardo, C., Bayoumy, S., Garzotto, M., McLoughlin, M., Gallily, R., et al. (1997). Lipopolysaccharide induces disseminated endothelial apoptosis requiring ceramide generation. *The Journal of Experimental Medicine, 186*, 1831–1841.

Haimovitz-Friedman, A., Kan, C. C., Ehleiter, D., Persaud, R. S., McLoughlin, M., Fuks, Z., et al. (1994). Ionizing radiation acts on cellular membranes to generate ceramide and initiate apoptosis. *The Journal of Experimental Medicine, 180*, 525–535.

Hallahan, D. E., Virudachalam, S., Schwartz, J. L., Panje, N., Mustafi, R., & Weichselbaum, R. R. (1992). Inhibition of protein kinases sensitizes human tumor cells to ionizing radiation. *Radiation Research, 129*, 345–350.

Hanahan, D., & Weinberg, R. A. (2000). The hallmarks of cancer. *Cell, 100*, 57–70.

Hannun, Y. A. (1996). Functions of ceramide in coordinating cellular responses to stress. *Science, 274*, 1855–1859.

Heinrich, M., Wickel, M., Schneider-Brachert, W., Sandberg, C., Gahr, J., Schwandner, R., et al. (1999). Cathepsin D targeted by acid sphingomyelinase-derived ceramide.[erratum appears in EMBO J 2000 Jan 17;19(2):315]. *The EMBO Journal, 18*, 5252–5263.

Hertervig, E., Nilsson, A., Nyberg, L., & Duan, R. D. (1997). Alkaline sphingomyelinase activity is decreased in human colorectal carcinoma. *Cancer, 79*, 448–453.

Holman, D. H., Turner, L., El-Zawahry, A., Elojeimy, S., Liu, X., Bielawski, J., et al. (2008). Lysosomotropic acid ceramidase inhibitor induces apoptosis in prostate cancer cells. *Cancer Chemotherapy and Pharmacology, 61*, 231–242.

Hoye, A. T., Davoren, J. E., Wipf, P., Fink, M. P., & Kagan, V. E. (2008). Targeting Mitochondria. *Accounts of Chemical Research, 41*, 87–97.

Huang, W. C., Tsai, C. C., Chen, C. L., Chen, T. Y., Chen, Y. P., Lin, Y. S., et al. (2011). Glucosylceramide synthase inhibitor PDMP sensitizes chronic myeloid leukemia T315I mutant to Bcr-Abl inhibitor and cooperatively induces glycogen synthase kinase-3-regulated apoptosis. *The FASEB Journal, 25*, 3661–3673.

Huwiler, A., Johansen, B., Skarstad, A., & Pfeilschifter, J. (2001). Ceramide binds to the CaLB domain of cytosolic phospholipase A2 and facilitates its membrane docking and arachidonic acid release. *The FASEB Journal, 15*, 7–9.

Imanishi, R., Ohtsuru, A., Iwamatsu, M., Iioka, T., Namba, H., Seto, S., et al. (2002). A histone deacetylase inhibitor enhances killing of undifferentiated thyroid carcinoma cells by p53 gene therapy. *The Journal of Clinical Endocrinology and Metabolism, 87*, 4821–4824.

Itoh, M., Kitano, T., Watanabe, M., Kondo, T., Yabu, T., Taguchi, Y., et al. (2003). Possible role of ceramide as an indicator of chemoresistance: decrease of the ceramide content via activation of glucosylceramide synthase and sphingomyelin synthase in chemoresistant leukemia. *Clinical Cancer Research, 9*, 415–423.

Jaffrezou, J. P., Levade, T., Bettaieb, A., Andrieu, N., Bezombes, C., Maestre, N., et al. (1996). Daunorubicin-induced apoptosis: triggering of ceramide generation through sphingomyelin hydrolysis. *The EMBO Journal, 15*, 2417–2424.

Kimura, K., Bowen, C., Spiegel, S., & Gelmann, E. P. (1999). Tumor necrosis factor-alpha sensitizes prostate cancer cells to gamma-irradiation-induced apoptosis. *Cancer Research, 59*, 1606–1614.

Kitatani, K., Idkowiak-Baldys, J., & Hannun, Y. A. (2008). The sphingolipid salvage pathway in ceramide metabolism and signaling. *Cellular Signalling, 20*, 1010–1018.

Kossenkov, A. V., Vachani, A., Chang, C., Nichols, C., Billouin, S., Horng, W., et al. (2011). Resection of non–small cell lung cancers reverses tumor-induced gene expression changes in the peripheral immune system. *Clinical Cancer Research, 17*, 5867–5877.

Kumar, A., Oskouian, B., Fyrst, H., Zhang, M., Paris, F., & Saba, J. D. (2011). S1P lyase regulates DNA damage responses through a novel sphingolipid feedback mechanism. *Cell Death & Disease, 2*, e119.

Lamour, N. F., & Chalfant, C. E. (2008). Ceramide kinase and the ceramide-1-phosphate/cPLA2alpha interaction as a therapeutic target. *Current Drug Targets, 9*, 674–682.

Lavie, Y., Cao, H. T., Volner, A., Lucci, A., Han, T. Y., Geffen, V., et al. (1997). Agents that reverse multidrug resistance, tamoxifen, verapamil, and cyclosporin A, block glycosphingolipid metabolism by inhibiting ceramide glycosylation in human cancer cells. *The Journal of Biological Chemistry, 272*, 1682–1687.

Lee, J. Y., Hannun, Y. A., & Obeid, L. M. (1996). Ceramide inactivates cellular protein kinase Calpha. *The Journal of Biological Chemistry, 271*, 13169–13174.

Lee, H., Rotolo, J. A., Mesicek, J., Penate-Medina, T., Rimner, A., Liao, W. C., et al. (2011). Mitochondrial ceramide-rich macrodomains functionalize bax upon irradiation. *PLoS One, 6*, e19783.

Li, J., Guan, H.-Y., Gong, L.-Y., Song, L.-B., Zhang, N., Wu, J., et al. (2008). Clinical significance of sphingosine kinase-1 expression in human astrocytomas progression and overall patient survival. *Clinical Cancer Research, 14*, 6996–7003.

Li, W., Yu, C. P., Xia, J. T., Zhang, L., Weng, G. X., Zheng, H. Q., et al. (2009). Sphingosine kinase 1 is associated with gastric cancer progression and poor survival of patients. *Clinical Cancer Research, 15*, 1393–1399.

Liao, W. C., Haimovitz-Friedman, A., Persaud, R. S., McLoughlin, M., Ehleiter, D., Zhang, N., et al. (1999). Ataxia telangiectasia-mutated gene product inhibits DNA damage-induced apoptosis via ceramide synthase. *The Journal of Biological Chemistry, 274*, 17908–17917.

Liu, X., Elojeimy, S., El-Zawahry, A. M., Holman, D. H., Bielawska, A., Bielawski, J., et al. (2006). Modulation of ceramide metabolism enhances viral protein apoptin's cytotoxicity in prostate cancer. *Molecular Therapy, 14*, 637–646.

Liu, Y. Y., Han, T. Y., Giuliano, A. E., & Cabot, M. C. (2001). Ceramide glycosylation potentiates cellular multidrug resistance. *The FASEB Journal, 15*, 719–730.

Liu, Y.-Y., Han, T. Y., Yu, J. Y., Bitterman, A., Le, A., Giuliano, A. E., et al. (2004). Oligonucleotides blocking glucosylceramide synthase expression selectively reverse drug resistance in cancer cells. *Journal of Lipid Research, 45*, 933–940.

Liu, Y. Y., Patwardhan, G. A., Xie, P., Gu, X., Giuliano, A. E., & Cabot, M. C. (2011). Glucosylceramide synthase, a factor in modulating drug resistance, is overexpressed in metastatic breast carcinoma. *International Journal of Oncology, 39*, 425–431.

Liu, C. W. Y., Wang, R.-H., & Berndt, N. (2006). Protein phosphatase 1alpha activity prevents oncogenic transformation. *Molecular Carcinogenesis, 45*, 648–656.

Lozano, J., Menendez, S., Morales, A., Ehleiter, D., Liao, W.-C., Wagman, R., et al. (2001). Cell autonomous apoptosis defects in acid sphingomyelinase knockout fibroblasts. *The Journal of Biological Chemistry, 276*, 442–448.

Lucci, A., Cho, W. I., Han, T. Y., Giuliano, A. E., Morton, D. L., & Cabot, M. C. (1998). Glucosylceramide: a marker for multiple-drug resistant cancers. *Anticancer Research, 18*, 475–480.

Machaczka, M., Hast, R., Dahlman, I., Lerner, R., Klimkowska, M., Engvall, M., et al. (2012). Substrate reduction therapy with miglustat for type 1 Gaucher disease: a retrospective analysis from a single institution. *Upsala Journal of Medical Sciences*, *117*, 28–34.

Maggio, S. C., Rosato, R. R., Kramer, L. B., Dai, Y., Rahmani, M., Paik, D. S., et al. (2004). The histone deacetylase inhibitor MS-275 interacts synergistically with fludarabine to induce apoptosis in human leukemia cells. *Cancer Research*, *64*, 2590–2600.

Mahdy, A. E., Cheng, J. C., Li, J., Elojeimy, S., Meacham, W. D., Turner, L. S., et al. (2009). Acid ceramidase upregulation in prostate cancer cells confers resistance to radiation: AC inhibition, a potential radiosensitizer. *Molecular Therapy*, *17*, 430–438.

Malavaud, B., Pchejetski, D., Mazerolles, C., de Paiva, G. R., Calvet, C., Doumerc, N., et al. (2010). Sphingosine kinase-1 activity and expression in human prostate cancer resection specimens. *European Journal of Cancer*, *46*, 3417–3424.

Marfe, G., Di Stefano, C., Gambacurta, A., Ottone, T., Martini, V., Abruzzese, E., et al. (2011). Sphingosine kinase 1 overexpression is regulated by signaling through PI3K, AKT2, and mTOR in imatinib-resistant chronic myeloid leukemia cells. *Experimental Hematology*, *39*, 653–665 e6.

Martin, S. J., Newmeyer, D. D., Mathias, S., Farschon, D. M., Wang, H. G., Reed, J. C., et al. (1995). Cell-free reconstitution of Fas-, UV radiation- and ceramide-induced apoptosis. *The EMBO Journal*, *14*, 5191–5200.

Mathew, B., Jacobson, J. R., Berdyshev, E., Huang, Y., Sun, X., Zhao, Y., et al. (2011). Role of sphingolipids in murine radiation-induced lung injury: protection by sphingosine 1-phosphate analogs. *The FASEB Journal*, *25*, 3388–3400.

Mimeault, M., Johansson, S. L., Vankatraman, G., Moore, E., Henichart, J.-P., Depreux, P., et al. (2007). Combined targeting of epidermal growth factor receptor and hedgehog signaling by gefitinib and cyclopamine cooperatively improves the cytotoxic effects of docetaxel on metastatic prostate cancer cells. *Molecular Cancer Therapeutics*, *6*, 967–978.

Mimeault, M., Venkatraman, G., Johansson, S. L., Moore, E., Henichart, J. P., Depreux, P., et al. (2007). Novel combination therapy against metastatic and androgen-independent prostate cancer by using gefitinib, tamoxifen and etoposide. *International Journal of Cancer*, *120*, 160–169.

Mitra, P., Maceyka, M., Payne, S. G., Lamour, N., Milstien, S., Chalfant, C. E., et al. (2007). Ceramide kinase regulates growth and survival of A549 human lung adenocarcinoma cells. *FEBS Letters*, *581*, 735–740.

Morales, A., Paris, R., Villanueva, A., Llacuna, L., Garcia-Ruiz, C., & Fernandez-Checa, J. C. (2007). Pharmacological inhibition or small interfering RNA targeting acid ceramidase sensitizes hepatoma cells to chemotherapy and reduces tumor growth in vivo. *Oncogene*, *26*, 905–916.

Morita, Y., Perez, G. I., Paris, F., Miranda, S. R., Ehleiter, D., Haimovitz-Friedman, A., et al. (2000). Oocyte apoptosis is suppressed by disruption of the acid sphingomyelinase gene or by sphingosine-1-phosphate therapy. *Nature Medicine*, *6*, 1109–1114.

Mousseau, Y., Mollard, S., Faucher-Durand, K., Richard, L., Nizou, A., Cook-Moreau, J., et al. (2011). Fingolimod potentiates the effects of sunitinib malate in a rat breast cancer model. *Breast Cancer Research and Treatment*, *134*, 31–40.

Mukhopadhyay, A., Saddoughi, S. A., Song, P., Sultan, I., Ponnusamy, S., Senkal, C. E., et al. (2009). Direct interaction between the inhibitor 2 and ceramide via sphingolipid-protein binding is involved in the regulation of protein phosphatase 2A activity and signaling. *The FASEB Journal*, *23*, 751–763.

Nakamura, H., Hirabayashi, T., Shimizu, M., & Murayama, T. (2006). Ceramide-1-phosphate activates cytosolic phospholipase A2alpha directly and by PKC pathway. *Biochemical Pharmacology*, *71*, 850–857.

Nakamura, N., Ramaswamy, S., Vazquez, F., Signoretti, S., Loda, M., & Sellers, W. R. (2000). Forkhead transcription factors are critical effectors of cell death and cell cycle arrest downstream of PTEN. *Molecular and Cellular Biology*, *20*, 8969–8982.

Nava, V. E., Cuvillier, O., Edsall, L. C., Kimura, K., Milstien, S., Gelmann, E. P., et al. (2000). Sphingosine Enhances Apoptosis of Radiation-resistant Prostate Cancer Cells. *Cancer Research*, *60*, 4468–4474.

Neviani, P., Santhanam, R., Trotta, R., Notari, M., Blaser, B. W., Liu, S., et al. (2005). The tumor suppressor PP2A is functionally inactivated in blast crisis CML through the inhibitory activity of the BCR/ABL-regulated SET protein. *Cancer Cell*, *8*, 355–368.

Nguyen, P. L., Gu, X., Lipsitz, S. R., Choueiri, T. K., Choi, W. W., Lei, Y., et al. (2011). Cost implications of the rapid adoption of newer technologies for treating prostate cancer. *Journal of Clinical Oncology*, *29*, 1517–1524.

Nietupski, J. B., Pacheco, J. J., Chuang, W. L., Maratea, K., Li, L., Foley, J., et al. (2012). Iminosugar-based inhibitors of glucosylceramide synthase prolong survival but paradoxically increase brain glucosylceramide levels in Niemann-Pick C mice. *Molecular Genetics and Metabolism*, *105*, 621–628.

Norris, J. S., Bielawska, A., Day, T., El-Zawahri, A., Elojeimy, S., Hannun, Y., et al. (2006). Combined therapeutic use of AdGFPFasL and small molecule inhibitors of ceramide metabolism in prostate and head and neck cancers: a status report. *Cancer Gene Therapy*, *13*, 1045–1051.

Ogretmen, B., & Hannun, Y. A. (2001). Updates on functions of ceramide in chemotherapy-induced cell death and in multidrug resistance. *Drug Resistance Updates*, *4*, 368–377.

Ogretmen, B., & Hannun, Y. A. (2004). Biologically active sphingolipids in cancer pathogenesis and treatment. *Nature Reviews. Cancer*, *4*, 604–616.

Okahara, F., Ikawa, H., Kanaho, Y., & Maehama, T. (2004). Regulation of PTEN phosphorylation and stability by a tumor suppressor candidate protein. *The Journal of Biological Chemistry*, *279*, 45300–45303.

Oskouian, B., Sooriyakumaran, P., Borowsky, A. D., Crans, A., Dillard-Telm, L., Tam, Y. Y., et al. (2006). Sphingosine-1-phosphate lyase potentiates apoptosis via p53- and p38-dependent pathways and is down-regulated in colon cancer. *Proceedings of the National Academy of Sciences*, *103*, 17384–17389.

Paris, F., Grassme, H., Cremesti, A., Zager, J., Fong, Y., Haimovitz-Friedman, A., et al. (2001). Natural ceramide reverses Fas resistance of acid sphingomyelinase(-/-) hepatocytes. *The Journal of Biological Chemistry*, *276*, 8297–8305.

Park, M. A., Mitchell, C., Zhang, G., Yacoub, A., Allegood, J., Haussinger, D., et al. (2010). Vorinostat and sorafenib increase CD95 activation in gastrointestinal tumor cells through a Ca(2+)-de novo ceramide-PP2A-reactive oxygen species-dependent signaling pathway. *Cancer Research*, *70*, 6313–6324.

Park, H. W., Song, J. Y., Kim, K. S., Han, Y., Kim, C. W., Yi, S. Y., et al. (2004). Enhancement of radiosensitivity by combined ceramide and dimethylsphingosine treatment in lung cancer cells. *Experimental & Molecular Medicine*, *36*, 411–419.

Park, M. A., Zhang, G., Martin, A. P., Hamed, H., Mitchell, C., Hylemon, P. B., et al. (2008). Vorinostat and sorafenib increase ER stress, autophagy and apoptosis via ceramide-dependent CD95 and PERK activation. *Cancer Biology & Therapy*, *7*, 1648–1662.

Pchejetski, D., Bohler, T., Brizuela, L., Sauer, L., Doumerc, N., Golzio, M., et al. (2010). FTY720 (fingolimod) sensitizes prostate cancer cells to radiotherapy by inhibition of sphingosine kinase-1. *Cancer Research*, *70*, 8651–8661.

Pchejetski, D., Doumerc, N., Golzio, M., Naymark, M., Teissie, J., Kohama, T., et al. (2008). Chemosensitizing effects of sphingosine kinase-1 inhibition in prostate cancer cell and animal models. *Molecular Cancer Therapeutics*, 7, 1836–1845.

Pena, L. A., Fuks, Z., & Kolesnick, R. N. (2000). Radiation-induced apoptosis of endothelial cells in the murine central nervous system: Protection by fibroblast growth factor and sphingomyelinase deficiency. *Cancer Research, 60*, 321–327.

Perrotti, D., & Neviani, P. (2008). Protein phosphatase 2A (PP2A), a drugable tumor suppressor in Ph1(+) leukemias. *Cancer Metastasis Reviews, 27*, 159–168.

Perry, D. K., Carton, J., Shah, A. K., Meredith, F., Uhlinger, D. J., & Hannun, Y. A. (2000). Serine palmitoyltransferase regulates de novo ceramide generation during etoposide-induced apoptosis. *The Journal of Biological Chemistry, 275*, 9078–9084.

Pesche, S., Latil, A., Muzeau, F., Cussenot, O., Fournier, G., Longy, M., et al. (1998). PTEN/MMAC1/TEP1 involvement in primary prostate cancers. *Oncogene, 16*, 2879–2883.

Pyne, N. J., & Pyne, S. (2010). Sphingosine 1-phosphate and cancer. *Nature Reviews. Cancer, 10*, 489–503.

Radford, I. R. (1986). Evidence for a general relationship between the induced level of DNA double-strand breakage and cell-killing after X-irradiation of mammalian cells. *International Journal of Radiation Biology and Related Studies in Physics, Chemistry, and Medicine, 49*, 611–620.

Rahmani, M., Reese, E., Dai, Y., Bauer, C., Payne, S. G., Dent, P., et al. (2005). Coadministration of histone deacetylase inhibitors and perifosine synergistically induces apoptosis in human leukemia cells through Akt and ERK1/2 inactivation and the generation of ceramide and reactive oxygen species. *Cancer Research, 65*, 2422–2432.

Rodriguez-Lafrasse, C., Alphonse, G., Aloy, M. T., Ardail, D., Gerard, J. P., Louisot, P., et al. (2002). Increasing endogenous ceramide using inhibitors of sphingolipid metabolism maximizes ionizing radiation-induced mitochondrial injury and apoptotic cell killing. *International Journal of Cancer, 101*, 589–598.

Ruckhäberle, E., Karn, T., Rody, A., Hanker, L., Gätje, R., Metzler, D., et al. (2009). Gene expression of ceramide kinase, galactosyl ceramide synthase and ganglioside GD3 synthase is associated with prognosis in breast cancer. *Journal of Cancer Research and Clinical Oncology, 135*, 1005–1013.

Saad, A. F., Meacham, W. D., Bai, A., Anelli, V., Elojeimy, S., Mahdy, A. E., et al. (2007). The functional effects of acid ceramidase overexpression in prostate cancer progression and resistance to chemotherapy. *Cancer Biology & Therapy, 6*, 1455–1460.

Salas, A., Ponnusamy, S., Senkal, C. E., Meyers-Needham, M., Selvam, S. P., Saddoughi, S. A., et al. (2011). Sphingosine kinase-1 and sphingosine 1-phosphate receptor 2 mediate Bcr-Abl1 stability and drug resistance by modulation of protein phosphatase 2A. *Blood, 117*, 5941–5952.

Samsel, L., Zaidel, G., Drumgoole, H. M., Jelovac, D., Drachenberg, C., Rhee, J. G., et al. (2004). The ceramide analog, B13, induces apoptosis in prostate cancer cell lines and inhibits tumor growth in prostate cancer xenografts. *The Prostate, 58*, 382–393.

Santana, P., Pena, L. A., Haimovitz-Friedman, A., Martin, S., Green, D., McLoughlin, M., et al. (1996). Acid sphingomyelinase-deficient human lymphoblasts and mice are defective in radiation-induced apoptosis. *Cell, 86*, 189–199.

Sauer, L., Nunes, J., Salunkhe, V., Skalska, L., Kohama, T., Cuvillier, O., et al. (2009). Sphingosine kinase 1 inhibition sensitizes hormone-resistant prostate cancer to docetaxel. *International Journal of Cancer, 125*, 2728–2736.

Sawada, M., Nakashima, S., Banno, Y., Yamakawa, H., Takenaka, K., Shinoda, J., et al. (2000). Influence of Bax or Bcl-2 overexpression on the ceramide-dependent apoptotic pathway in glioma cells. *Oncogene, 19*, 3508–3520.

Schiffmann, S., Sandner, J., Birod, K., Wobst, I., Angioni, C., Ruckhaberle, E., et al. (2009). Ceramide synthases and ceramide levels are increased in breast cancer tissue. *Carcinogenesis, 30*, 745–752.

Selzner, M., Morse, M. A., Sindram, D., Bielawska, A., Hannun, Y. A., & Clavien, P.-A. (1999). Ceramide analogue B13 induces apoptosis and completely inhibits growth of liver metastases from colon cancer in vivo a new strategy for the treatment of metastatic liver disease. *50th Annual American Association for the Study of Liver Disease (AAASLD)* (p. 4) Abstract #1.

Shah, M. V., Zhang, R., Irby, R., Kothapalli, R., Liu, X., Arrington, T., et al. (2008). Molecular profiling of LGL leukemia reveals role of sphingolipid signaling in survival of cytotoxic lymphocytes. *Blood, 112*, 770–781.

Shimizu, M., Tada, E., Makiyama, T., Yasufuku, K., Moriyama, Y., Fujino, H., et al. (2009). Effects of ceramide, ceramidase inhibition and expression of ceramide kinase on cytosolic phospholipase A2alpha; additional role of ceramide-1-phosphate in phosphorylation and Ca2+ signaling. *Cellular Signalling, 21*, 440–447.

Shirai, K., Kaneshiro, T., Wada, M., Furuya, H., Bielawski, J., Hannun, Y. A., et al. (2011). A Role of Sphingosine Kinase 1 in Head and Neck Carcinogenesis. *Cancer Prevention Research, 4*, 454–462.

Shirakabe, K., Yamaguchi, K., Shibuya, H., Irie, K., Matsuda, S., Moriguchi, T., et al. (1997). TAK1 mediates the ceramide signaling to stress-activated protein kinase/c-Jun N-terminal kinase. *The Journal of Biological Chemistry, 272*, 8141–8144.

Song, L., Xiong, H., Li, J., Liao, W., Wang, L., Wu, J., et al. (2011). Sphingosine kinase-1 enhances resistance to apoptosis through activation of PI3K/Akt/NF-kappaB pathway in human non-small cell lung cancer. *Clinical Cancer Research, 17*, 1839–1849.

Strum, J. C., Small, G. W., Pauig, S. B., & Daniel, L. W. (1994). 1-beta-D-Arabinofuranosylcytosine stimulates ceramide and diglyceride formation in HL-60 cells. *The Journal of Biological Chemistry, 269*, 15493–15497.

Sun, C. C., Zhang, Z., Zhang, S. Y., Li, J., Li, Z. L., & Kong, C. Z. (2010). Up-regulation of glucosylceramide synthase in urinary bladder neoplasms. *Urologic Oncology, 30*, 444–449.

Susin, S. A., Zamzami, N., Castedo, M., Daugas, E., Wang, H. G., Geley, S., et al. (1997). The central executioner of apoptosis - multiple connections between protease activation and mitochondria in fas/apo-1/cd95- and ceramide-induced apoptosis. *The Journal of Experimental Medicine, 186*, 25–37.

Szulc, Z. M., Mayroo, N., Bai, A., Bielawski, J., Liu, X., Norris, J. S., et al. (2008). Novel analogs of D-e-MAPP and B13. Part 1: Synthesis and evaluation as potential anticancer agents. *Bioorganic & Medicinal Chemistry, 16*, 1015–1031.

Takabe, K., Paugh, S. W., Milstien, S., & Spiegel, S. (2008). "Inside-out" signaling of sphingosine-1-phosphate: therapeutic targets. *Pharmacological Reviews, 60*, 181–195.

Tepper, A. D., de Vries, E., van Blitterswijk, W. J., & Borst, J. (1999). Ordering of ceramide formation, caspase activation, and mitochondrial changes during CD95- and DNA damage-induced apoptosis.[erratum appears in J Clin Invest 1999 May;103(9):1363]. *The Journal of Clinical Investigation, 103*, 971–978.

Teruel, T., Hernandez, R., & Lorenzo, M. (2001). Ceramide mediates insulin resistance by tumor necrosis factor-alpha in brown adipocytes by maintaining Akt in an inactive dephosphorylated state. *Diabetes, 50*, 2563–2571.

Turner, L. S., Cheng, J. C., Beckham, T. H., Keane, T. E., Norris, J. S., & Liu, X. (2011). Autophagy is increased in prostate cancer cells overexpressing acid ceramidase and enhances resistance to C6 ceramide. *Prostate Cancer and Prostatic Diseases, 14*, 30–37.

Vazquez, F., Ramaswamy, S., Nakamura, N., & Sellers, W. R. (2000). Phosphorylation of the PTEN tail regulates protein stability and function. *Molecular and Cellular Biology, 20*, 5010–5018.

Verheij, M., Bose, R., Lin, X. H., Yao, B., Jarvis, W. D., Grant, S., et al. (1996). Requirement for ceramide-initiated SAPK/JNK signalling in stress-induced apoptosis. *Nature, 380*, 75–79.

Vit, J. P., & Rosselli, F. (2003). Role of the ceramide-signaling pathways in ionizing radiation-induced apoptosis. *Oncogene, 22*, 8645–8652.

Walker, T., Mitchell, C., Park, M. A., Yacoub, A., Graf, M., Rahmani, M., et al. (2009). Sorafenib and vorinostat kill colon cancer cells by CD95-dependent and -independent mechanisms. *Molecular Pharmacology, 76*, 342–355.

Wallington-Beddoe, C. T., Hewson, J., Bradstock, K. F., & Bendall, L. J. (2011). FTY720 produces caspase-independent cell death of acute lymphoblastic leukemia cells. *Autophagy*, 7, 707–715.

Xin, M., & Deng, X. (2006). Protein phosphatase 2A enhances the proapoptotic function of Bax through dephosphorylation. *The Journal of Biological Chemistry, 281*, 18859–18867.

Xu, J., Yeh, C. H., Chen, S., He, L., Sensi, S. L., Canzoniero, L. M., et al. (1998). Involvement of de novo ceramide biosynthesis in tumor necrosis factor-alpha/cycloheximide-induced cerebral endothelial cell death. *The Journal of Biological Chemistry, 273*, 16521–16526.

Yu, X. J., Long, J., Fu, D. L., Zhang, Q. H., & Ni, Q. X. (2003). Analysis of gene expression profiles in pancreatic carcinoma by using cDNA microarray. *Hepatobiliary & Pancreatic Diseases International, 2*, 467–470.

Zhu, Q. Y., Wang, Z., Ji, C., Cheng, L., Yang, Y. L., Ren, J., et al. (2011). C6-ceramide synergistically potentiates the anti-tumor effects of histone deacetylase inhibitors via AKT dephosphorylation and alpha-tubulin hyperacetylation both in vitro and in vivo. *Cell Death & Disease, 2*, e117.

Zundel, W., & Giaccia, A. (1998). Inhibition of the anti-apoptotic PI(3)K/Akt/Bad pathway by stress. *Genes & Development, 12*, 1941–1946.

CHAPTER TWO

Diverse Functions of Ceramide in Cancer Cell Death and Proliferation

Sahar A. Saddoughi[*,†], **Besim Ogretmen**[*,†,1]
[*]Department of Biochemistry and Molecular Biology, Medical University of South Carolina, Charleston, South Carolina, USA
[†]Hollings Cancer Center, Medical University of South Carolina, Charleston, South Carolina, USA
[1]Corresponding author: e-mail address: ogretmen@musc.edu

Contents

Abstract

Ceramide, a bioactive sphingolipid, is now at the forefront of cancer research. Classically, ceramide is thought to induce death, growth inhibition, and senescence in cancer cells. However, it is now clear that this simple picture of ceramide no longer holds true. Recent studies suggest that there are diverse functions of endogenously generated ceramides, which seem to be context dependent, regulated by subcellular/membrane localization and presence/absence of direct targets of these lipid molecules. For example, different fatty-acid chain lengths of ceramide, such as C_{16}-ceramide that can be generated by ceramide synthase 6 (CerS6), have been implicated in cancer cell proliferation, whereas CerS1-generated C_{18}-ceramide mediates cell death. The dichotomy of ceramides' function in cancer cells makes some of the metabolic enzymes of ceramide synthesis potential drug targets (such as Cers6) to prevent cancer growth in breast and head and neck cancers. Conversely, activation of CerS1 could be a new therapeutic option for the development of novel strategies against lung and head and neck cancers. This chapter focuses on recent discoveries about the mechanistic details of mainly *de novo*-generated ceramides and their

Advances in Cancer Research, Volume 117
ISSN 0065-230X
http://dx.doi.org/10.1016/B978-0-12-394274-6.00002-9

signaling functions in cancer pathogenesis, and about how these mechanistic information can be translated into clinically relevant therapeutic options for the treatment of cancer.

1. STRUCTURE AND METABOLISM OF CERAMIDE

Bioactive sphingolipid ceramide can be synthesized through multiple different pathways within the cell. Ceramide is composed of a sphingosine backbone, which is amide-linked to a fatty acyl chain, varying in length from C_{14} to C_{26} (Fig. 2.1) (Ogretmen & Hannun, 2004; Ponnusamy et al., 2010). Ceramide serves as a structural precursor for complex sphingolipids, such as sphingomyelin (SM), ceramide-1-phosphate (C1P), or glucosylceramide (GlcCer) and galactosylceramide (GalCer), and precursors for the generation

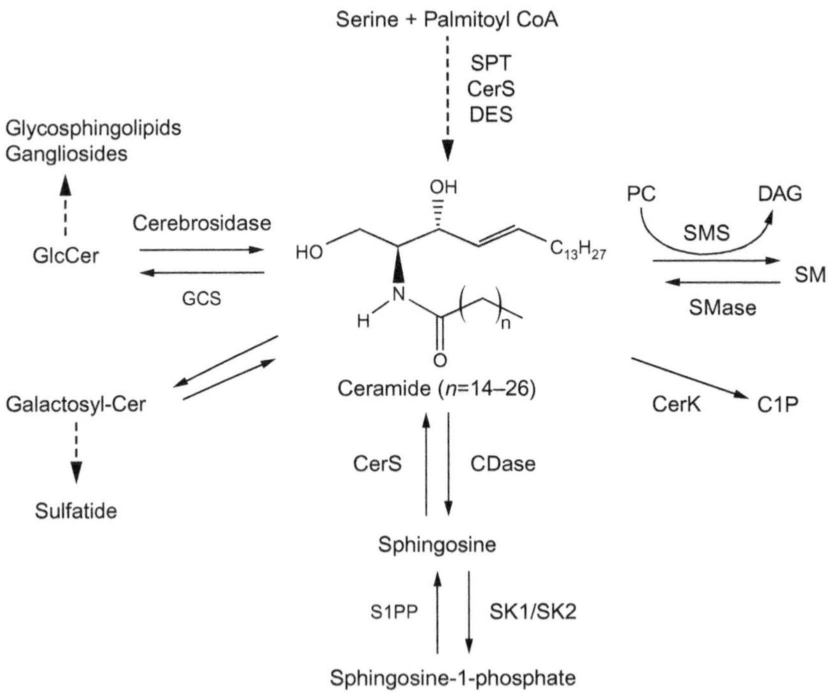

Figure 2.1 Pathways of ceramide metabolism. Ceramide can be generated endogenously through various different pathways, involving multiple enzymes and metabolic reactions. Complex sphingolipids can be hydrolyzed to generate ceramide, or ceramide can be generated *de novo*, which begins with serine and palmitoyl CoA. In addition, ceramide acts as a precursor for the generation of S1P, and complex sphingolipids.

of complex glycosphingolipids and gangliosides (Fig. 2.1). Ceramide can also be generated by breakdown of these complex sphingolipids, which involves a number of specialized enzymes (Fig. 2.1). For example, activation of sphingomyelinases (SMases) that hydrolyze SM, and cerebrosidases, which breakdown GlcCer or GalCer, generates ceramide. Ceramide can then be utilized as a substrate by ceramidases (CDases) to liberate sphingosine, which is phosphorylated to generate S1P, a pro-survival lipid. Ceramide is also metabolized by the functions of ceramide kinase (CK), or SM synthase (Fig. 2.1), which requires the transport of ceramide from the endoplasmic reticulum (ER) to the Golgi apparatus by ceramide transporter protein, CERT, via nonvesicular transport (Hanada, Kumagai, Tomishige, & Yamaji, 2009; Hanada et al., 2003). Ceramide can also be converted into GlcCer in the Golgi; however, this process is CERT independent. Importantly, nonvesicular transport of GlcCer from its site of synthesis (early Golgi) to distal Golgi compartments is carried out by FAPP2, four-phosphate adaptor protein, controlling the synthesis of glycosphingolipids, which might essentially play crucial roles in determining the lipid composition of the plasma membrane (D'Angelo et al., 2007).

Endogenous ceramides are synthesized via the *de novo* pathway, which begins with serine and palmitoyl CoA condensation to form 3-ketosphinganine by serine palmitoyl transferase, leading to the synthesis of dihydroceramide by dihydroceramide synthases (dhCerS1–6) (Pewzner-Jung et al., 2010) (Fig. 2.1). Then, dihydroceramide is desaturated to yield ceramide by dihydroceramide desaturase (DES), which is responsible for inserting a double bond between carbons 4 and 5 in the sphingosine backbone (Kraveka et al., 2007).

In addition, the salvage pathway generates endogenous ceramides. In the salvage pathway, ceramides can be hydrolyzed to sphingosine, which is reacylated by CerS to regenerate ceramides (Kitatani, Idkowiak-Baldys, & Hannun, 2008). To this end, exogenous ceramides, such as C_6-ceramide, are recycled to generate endogenous long-chain ceramides after its sphingosine backbone is reacylated (Ogretmen et al., 2002; Sultan et al., 2006).

2. DISTINCT FUNCTIONS OF CERS1–6

Recent studies indicate that different fatty-acid chain lengths of ceramide may have different functions in cancer pathogenesis, highlighting the importance of ceramide synthases (CerS) in sphingolipid metabolism. CerS, identified as the yeast *longevity assurance gene 1* (*LAG1*), is known to regulate life span/longevity in *Saccharomyces cerevisiae*, and its deletion prolongs the replicative lifespan of yeast (Guillas et al., 2001; Jazwinski & Conzelmann,

2002). Within *Caenorhabditis elegans*, CerS homologues are called *hyl-1* and *lagr-1*, and they are essential for hypoxia- or radiation-induced apoptosis (Deng et al., 2010; Menuz et al., 2009). The CerS homologue in Drosophila is referred to as schlank, and it was shown to be important in larval growth (Bauer et al., 2009). The discovery of the mouse homologue of *LAG1*, also known as *LASS1*, or the upstream of growth and differentiation factor 1 demonstrated that it specifically regulates the synthesis of C_{18}-ceramide with a high degree of fatty-acid chain length specificity (Venkataraman et al., 2002). Further studies showed that there are six LASS proteins (LASS1–6), which were later renamed as CerS1–6 (Pewzner-Jung, Ben-Dor, & Futerman, 2006). CerS1–6 are associated with the ER membrane and contain a crucial TRAM–Lag1p–CLN8 (TLC) domain (Tidhar et al., 2012). The TLC domain constitutes the catalytic activity of CerS and is required for the generation of ceramide (Mesika, Ben-Dor, Laviad, & Futerman, 2007). All of the CerS proteins, except CerS1, contain a homeobox transcription factor (HOX) domain found at the N-terminus, which might be important for the enzymatic activity of CerS2–6 (Mesika et al., 2007). However, the physiological roles of the HOX domain of CerS2–5 are still unclear.

Importantly, each CerS enzyme exerts some specificity for the generation of endogenous ceramides with distinct fatty-acid chain lengths (Fig. 2.2). For example, CerS1 and CerS4 mainly generate ceramide with an 18-carbon-containing fatty-acid chain (C_{18}-ceramide and to a lesser extend C_{20}-

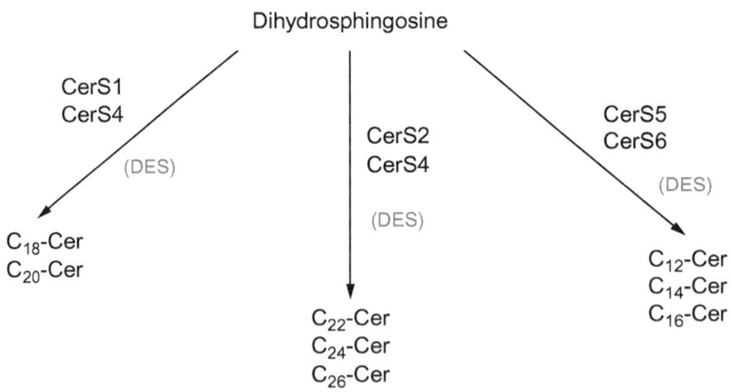

Figure 2.2 Roles of CerS1–6 in the generation of ceramides. CerS1–6 are responsible for incorporating different fatty-acid chains to a sphingosine backbone to form endogenous C_{12}–C_{26}-ceramides. CerS1 and 4 mainly generate C_{18}- and C_{20}-ceramides, while CerS2 and 4 are responsible to generate C_{22}-, C_{24}-, and C_{26}-ceramides. CerS5 and 6 are required to generate C_{12}-, C_{14}-, and C_{16}-ceramides. (For color version of this figure, the reader is referred to the online version of this chapter.)

ceramide), whereas CerS5 and CerS6 mainly generate C_{16}-ceramide and, to a lesser extent, C_{12}- and C_{14}-ceramides (Gault, Obeid, & Hannun, 2010; Pewzner-Jung et al., 2006). CerS2 generates very long-chain ceramides, particularly C_{24}-ceramide (Imgrund et al., 2009; Pewzner-Jung et al., 2010), and CerS3 is responsible for generating ultra-long-chain ceramides (C_{26}-ceramide and longer) (Jennemann et al., 2012).

There are important studies which indicate diverse biological functions for *de novo*-generated ceramides *in vivo*. For example, genetic loss of CerS2, which reduced C_{24}-ceramide generation in CerS2 −/− mice, was reported to have significant negative effects on the brain, such as loss of myelin and degeneration of cerebellum (Ben-David et al., 2011; Imgrund et al., 2009; Pewzner-Jung et al., 2010). Interestingly, after 7 months of age, these CerS2 −/− mice developed hepatocarcinomas (Pewzner-Jung et al., 2010). Moreover, global knockout of CerS3 reduced ultra-long-chain ceramides ($>C_{26}$), which phenotypically resulted in disruption to the epidermis (Jennemann et al., 2012), leading to death shortly after birth due to transepidermal water loss (Jennemann et al., 2012). There are also two mouse strains that contain mutations in CerS1, making the enzyme catalytically inactive, referred to as flincher (fln) and toppler (to) (Zhao et al., 2011). These mice develop cerebellar ataxia, Purkinje cell degeneration, and lipofusion accumulation. Collectively, these mice models demonstrate that the ceramides with different fatty-acid chain lengths generated by CerS have distinct and important biological functions *in vivo*, which might not be compensated by other CerS or metabolic enzymes.

Similarly, C_{18}- and C_{16}-ceramides generated by CerS1 or CerS6, respectively, play distinct roles in the regulation of cancer cell death or proliferation, respectively. In head and neck cancer cells, CerS1 overexpression or C_{18}-ceramide treatment suppresses tumor growth (Karahatay et al., 2007; Koybasi et al., 2004; Senkal et al., 2007). Conversely, CerS6/C_{16}-ceramide induces head and neck squamous cell carcinoma (HNSCC) tumor proliferation in SCID mice (Senkal, Ponnusamy, Bielawski, Hannun, & Ogretmen, 2010; Senkal et al., 2011). Mechanistically, it was shown that CerS6-generated C_{16}-ceramide plays key roles in the regulation of ER stress, via controlling ER–Golgi membrane network (Senkal et al., 2011). Downregulation of CerS6 resulted in ATF6 activation, regulated by ER Ca^{2+} release, and alterations of Golgi/ER membrane integrity in head and neck cancer cells. Consistent with these, recent data suggest that elevated C_{16}-ceramide associates with a positive lymph node status in breast cancer patients, indicating the metastatic potential of C_{16}-ceramide in the clinic (Schiffmann et al., 2009). In another published report, elevated *CerS2* and *CerS6* mRNA were observed in breast cancer tumors (Erez-Roman, Pienik, & Futerman, 2010). Thus, these data indicate that certain chain lengths of ceramide may have an unusual pro-survival role in cancer pathogenesis.

On the other hand, consistent with its antiproliferative effects, down-regulation of C_{18}-ceramide was associated with poor survival and metastatic disease in patients with head and neck cancers (Karahatay et al., 2007). Reconstitution of CerS1/C_{18}-ceramide induced cell death (Koybasi et al., 2004; Senkal et al., 2007) possibly via induction of mitophagy (Sentelle et al., 2012). Mechanistically, data showed that *CerS1* expression is repressed in head and neck cancer cells by concerted functions of HDAC1 and microRNA-574-5p (Meyers-Needham, Lewis, et al., 2012; Meyers-Needham, Ponnusamy, et al., 2012). Thus, these distinct roles of different chain lengths of ceramide demonstrate the complexity of its metabolism and roles in cancer pathogenesis and therapy.

3. NOVEL CERAMIDE–PROTEIN INTERACTIONS THAT REGULATE CERAMIDE METABOLISM AND SIGNALING

Diverse and distinct functions of ceramides can be partially explained by the discovery of novel ceramide–protein interactions (Table 2.1). One of the recently identified proteins known to bind to ceramide is CERT.

Table 2.1 Ceramide-binding proteins

Sphingolipid	Protein	Biological function
Ceramide	KSR	Transactivation of c-Raf
Ceramide	c-Raf	Activation of the MAPK cascade
Ceramide	Cathepsin D	Autocatalytic proteolysis of cathepsin D, leading to apoptosis
Ceramide	PKCz	Interaction with SAPK leading to growth suppression
Ceramide	I2PP2A	Increase in PP2A activity, leading to the regulation of c-Myc stability
Ceramide	CERT	Transports ceramide from the ER to the trans-Golgi, and it is involved in drug resistance
Ceramide-1-phosphate	$cPLA_2$	Arachidonic acid release

CERT specifically transports ceramide from the ER to the trans-Golgi for SM synthesis (Hanada et al., 2009, 2003). The crystal structure of CERT suggests that ceramide binds within the hydrophobic core of the protein for transport (Hanada et al., 2009, 2003). Interestingly, downregulation of CERT sensitizes cancer cells to chemotherapy agents, such as paclitaxel, and this may be due to an increased accumulation of ceramide leading to ER stress and cell death (Swanton et al., 2007).

One of the downstream targets of ceramide is a protein phosphatase, such as PP2A and PP1 (Chalfant et al., 1999). PP2A is a well-known tumor suppressor in cancer. PP2A diverse functions stem from the multitude of downstream targets that it regulates such as Akt, c-Myc, and Bcr-Abl (Salas et al., 2011; Yeh et al., 2004). The mechanism by which ceramide directly activates PP2A is elusive, and whether there is an intermediate protein involved in ceramide-mediated activation of PP2A remains unknown. Most recently, ceramide was shown to directly bind to the SET/I2PP2A oncoprotein, a biological inhibitor of PP2A, leading to PP2A activation in A549 lung cancer cells (Mukhopadhyay et al., 2009). Our studies determined that SET/I2PP2A, which also plays important roles in drug resistance of CML cells (Salas et al., 2011), preferentially associates with exogenous C_6-ceramide and endogenous C_{18}-ceramide generated in A549 cells (Mukhopadhyay et al., 2009). This degree of fatty-acid chain length preference of I2PP2A for ceramide binding was also observed in CERT–ceramide interaction, which preferentially binds C_{16}-ceramide, as well as C_6- and C_{18}-ceramides but not very long-chain C_{24}-ceramide (Muto et al. 2007). These data suggest that, similar to CERT, I2PP2A also preferentially interacts with C_6- and C_{18}-ceramides, but, unlike CERT, it does not interact with C_{16}-ceramide.

SET/I2PP2A is a nuclear protein and a known inhibitor of PP2A activity (Katayose, Li, Al-Murrani, Shenolikar, & Damuni, 2000). Thus, the binding relationship between ceramide and I2PP2A provided more mechanistic detail by which ceramide regulates PP2A activity. For example, ceramide–I2PP2A interaction reactivates PP2A and further induces the degradation of c-Myc (Mukhopadhyay et al., 2009). However, transient overexpression of I2PP2A, as it is observed in various cancer tissues and cells (Switzer et al., 2011), inhibits PP2A activity and binds ceramide as a biological sponge, leading to the protection of ceramide-mediated c-Myc degradation and resistance to ceramide-mediated antiproliferation (Mukhopadhyay et al., 2009). Thus, new strategies to decrease SET/I2PP2A expression and promoting ceramide generation might be advantageous for tumor suppression (Mukhopadhyay et al., 2009). Ceramide is known to play other important functional roles in the nucleus, such as the regulation of SP3–HDAC1 suppressor function for the repression of human telomerase reverse transcriptase (Wooten & Ogretmen, 2005; Wooten-Blanks, Song, Senkal, & Ogretmen, 2007). Therefore, it is important to determine subcellular regulation of how exogenous and endogenous ceramides reach nuclear proteins, such as SET/I2PP2A for binding in cells, and whether this interaction regulates only the nuclear functions of ceramide and PP2A to suppress growth and/or proliferation.

In addition to PP2A, PP1 is also an important protein phosphatase in cancer. There is a strong connection between PP1 and ceramide in regulating the retinoblastoma (RB) protein. It is well known that RB plays a critical role in cell cycle regulation, and ceramide treatment has been shown to lead to dephosphorylation of RB, leading to growth arrest in cancer cells (Alesse et al., 1998). Whether direct interaction between PP1 and ceramide is involved in the regulation of PP1 activity is unclear. However, it was shown that the PP1-dependent mechanism of ceramide was required for the generation of proapoptotic splice variants in lung cancer cells (Chalfant et al., 2001).

The protein kinases that interact with ceramide are kinase suppressor of RAS (KSR) and c-Raf. KSR was actually the first kinase identified to bind to ceramide (Zhang et al., 1997). Ceramide stimulates KSR autophosphorylation, through binding to the CA3 domain of KSR resulting in the transactivation of c-Raf (Yin et al., 2009). Additionally, c-Raf itself was identified as a ceramide-binding protein. IL-1b stimulation results in ceramide production that binds to c-Raf, and increases the activity of this mitogen-activated protein kinase (Huwiler et al., 1996). These data signify that ceramide can modulate the Ras–Raf–MAPK pathway via specific interactions with KSR and c-Raf. The last ceramide-activated protein kinase implicated in cancer pathogenesis is PKCz, which is involved in cell survival. Studies have shown that ceramide directly activates PKCz, via binding to the cysteine-rich domain, a putative ceramide-binding region (Bourbon, Yun, & Kester, 2000). This novel ceramide–protein association induces PKCz to interact with stress-activated protein kinase (SAPK), thereby leading to growth suppression.

Ceramide is known to activate cathepsin D. Interestingly, the interaction between ceramide and cathepsin D induced autocatalytic proteolysis, leading to cathepsin D activation, resulting in Bid cleavage and apoptosis (Heinrich et al., 2004, 1999).

Ceramide metabolite, C1P has been shown to induce arachidonic acid release in lung cancer cells, which is regulated by the direct binding between C1P and cytosolic phospholipase A_2 ($cPLA_2$) (Pettus et al., 2004). It was later identified that the cationic b-groove of the C2 domain of cPLA2a was the interaction site for C1P (Stahelin, Subramanian, Vora, Cho, & Chalfant, 2007). Thus, it may be advantageous to produce inhibitors of CK to block unwanted arachidonic acid formation, and inflammation in cancer cells. Additionally, a CK inhibitor may result in an increase in ceramide, thereby promoting pro-cell death pathways. Similarly, recent studies uncovered some intracellular targets of sphingosine kinase 1 (SK1)- and SK2-generated S1P for TRAF2 E3 ubiquitin ligase and HDAC regulation, respectively (Alvarez et al., 2010; Hait et al., 2009).

Studies, which revealed that ceramide and other sphingolipids have intracellular protein targets in cancer, also provide mechanistic insight into the pro-cell death versus pro-survival roles of these sphingolipid molecules in various cancers. These mechanistic data will be used to develop sphingolipid-based therapeutics to regulate tumor growth and help identify new potential drug targets.

4. CLINICAL RELEVANCE OF CERAMIDE METABOLISM AND SIGNALING IN CANCER PATHOGENESIS, THERAPY, OR DRUG RESISTANCE

Recent studies suggest that ceramide has important clinical significance in cancer pathogenesis and prognosis (Fig. 2.3). In 2004, data from our laboratory showed that total ceramide levels were elevated in the majority of HNSCC tissues compared with noncancerous adjacent head and neck tissues (Karahatay et al., 2007). Interestingly, only C_{18}-ceramide,

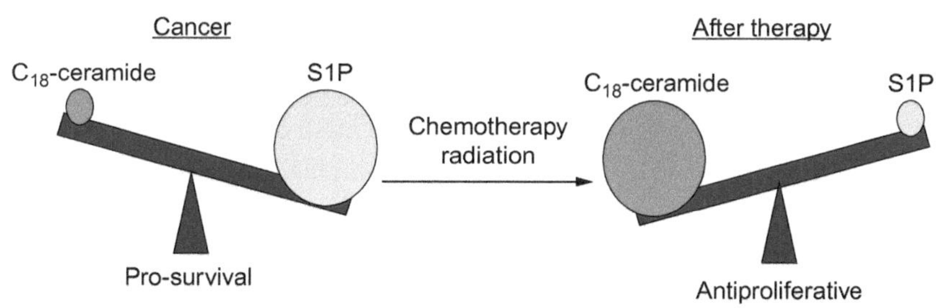

Figure 2.3 Metabolic balance between antiproliferation ceramide and pro-survival S1P determines cell fate for tumor suppression or proliferation. Generally, C_{18}-ceramide, a pro-cell death ceramide, tends to be low in cancer cells or tissues, while pro-survival sphingolipid, sphingosine-1-phosphate, appears to be elevated. With certain treatments, this balance can shift to be pro-cell death where C_{18}-ceramide is elevated and S1P is decreased, leading to tumor suppression. (See Page 2 in Color Section at the back of the book.)

and not other ceramide species, was significantly lower in ~70% of the tumor tissues of HNSCC patients sampled, when compared to controls, and decreased C_{18}-ceramide significantly correlated with lymphovascular invasion, and nodal metastasis in HNSCC patients (Karahatay et al., 2007). Thus, these studies indicated the clinical significance of CerS-generated ceramide with a specific fatty-acid chain length, especially CerS1/C_{18}-ceramide modulation, in the regulation of HNSCC pathogenesis and overall survival. Conversely, overexpression of CerS1 confers a decrease in HNSCC cell growth, concomitant with inhibition of telomerase activity, one of the downstream targets of ceramide involved in the suppression of cancer cell growth and/or proliferation. Upon chemotherapeutic treatment with gemcitabine (GEM) and doxorubicin (DOX) in HNSCC cells, CerS1 mRNA and activity were increased (Senkal et al., 2007, 2006). This was accompanied by an increase in caspase-3 and 9 activation, which was prevented by the silencing of CerS1 using small interfering RNAs (siRNAs) when treated with these two drugs. Importantly, GEM/DOX treatment *in vivo* decreased HNSCC xenograft tumor growth and progression in SCID mice, which was associated with increased C_{18}-ceramide and CerS1 expression (Senkal et al., 2007, 2006). Interestingly, in GEM/DOX-treated HNSCC xenografts, downregulation of CerS6 expression and C_{16}-ceramide was also observed.

Based on these preclinical data, a phase II clinical trial was designed to test the hypothesis that treatment with a gemcitabine (GEM)/doxorubicin (DOX) will be efficacious via reconstitution of C_{18}-ceramide signaling in HNSCC patients for whom first-line platinum-based therapy failed (Saddoughi et al., 2011). Response rates in the refractory HNSCC patient population were encouraging compared to other clinical trials studying a similar population of patients. In particular, five patients had progressive disease (PD), one had complete response (CR), three exhibited partial response (PR), and eight had stable disease (SD). The median progression-free survival was 1.6 months (95% CI, 1.4, 4.2) with a median survival of 5.6 months (95% CI, 3.8, 18.2) (Saddoughi et al., 2011). To understand ceramide's role as a potential novel serum biomarker of chemotherapy response, serum ceramide was measured in patients after every two cycles of chemotherapy. Interestingly, CR/PR and SD patients had a significant increase in serum C_{18}-ceramide over the course of treatment when compared to PD patients (Saddoughi et al., 2011). This was the first time that a clinical trial highlighted the potential role of ceramide as a novel serum biomarker for monitoring chemotherapy response in patients.

Drug resistance is a major obstacle in cancer therapeutics. Recent studies indicate that one mechanism of drug resistance is the alteration of ceramide accumulation. In fact, ceramide is highly metabolized into GlcCer due to an increase in glucosylceramide synthase (GCS) activity and/or expression in some cancer cells (Gouaze-Andersson et al., 2007). This phenomenon has been implicated in development of drug resistance in various cancer cell types, especially in breast cancer cells (Gouaze et al., 2005). Although the role of GCS in the development of drug resistance has been challenged in some cancer models, a mechanistic link between GCS and P-glycoprotein (P-gp), an ABC transporter implicated in drug resistance, has been recently revealed. The knockdown of GCS expression with siRNA significantly inhibits the expression of *MDR1*, a gene that encodes for P-gp, and reverses drug resistance (Gouaze-Andersson & Cabot, 2006). These data are consistent with an earlier study, which showed that SDZ PSC 833, an inhibitor of P-gp, inhibits GCS and alters GlcCer levels (Goulding, Giuliano, & Cabot, 2000). Additionally, increased accumulation of GlcCer is found in cells overexpressing P-gp. Interestingly, several members of the ABC transporter family are implicated in the translocation of phospholipids and sphingolipids across the lipid bilayer, and P-gp has been proposed as a specific transporter for GlcCer that translocates this molecule across the Golgi to deliver it for the synthesis of neutral glycosphingolipids (De Rosa, Sillence, Ackerley, & Lingwood, 2004). Thus, P-gp and GCS appear to function in the same pathway of ceramide/GlcCer metabolism, and this may provide an important link for the function of GCS in drug resistance. These results also suggest that inhibitors of GCS may be useful in preventing chemotherapy resistance (De Rosa et al., 2008, 2004). Overall, ceramide plays an important clinical role in multiple areas of cancer pathogenesis making it an exciting avenue to study.

5. CERAMIDE/SPHINGOSINE-BASED ANTICANCER THERAPEUTICS

Pro-cell death function of ceramide suggests that ceramide mimetics or analogues may open doors to new therapies to battle cancer (Table 2.2). Thus, finding ways to increase ceramide either by exogenous treatment or by elevating endogenous ceramide in cancer cells is desired. Conversely, S1P has been shown to promote cancer pathogenesis; thus, suppression of its generation/accumulation could suppress tumor growth as an alternative treatment strategy.

Table 2.2 Sphingolipid-based anticancer therapeutics

Compound	Mechanism of action/ target	Cancer type
Ceramidoids (Pyr-ceramides)	Mitochondrial targeting	HNSCC, lung, breast
4,6,-Diene-ceramide	Ceramide analogue	Breast
C_{16}-serinol	Ceramide analogue	Neuroblastoma
B13 and its derivatives	A-CDase inhibitor	Prostate, colon, HNSCC
D-MAPP	A-, N-CDase inhibitor	Squamous cell carcinoma
PPMP, PPPP	GCS inhibitors	Solid tumors
DMS	SK inhibitor	Leukemia, colon, breast
Anti-S1P-mAb	Binds S1P	Solid tumors
Pegylated liposomes with ceramide (nanoparticles)	Improved delivery	Breast
Vincristine in SM-liposomes (sphingosomes)	Improved delivery	Leukemia (ALL)
Safingol (L-t-dihydro-sphingosine)	SK inhibitor	Solid tumors
FTY-720	Myriocin analogue	Prostate, breast, CML, ALL

One common approach to promote apoptosis in cancer cells is the use of exogenous ceramide analogues or mimetic as therapeutic agents (Table 2.2). There have been recent improvements in the design and delivery of these ceramides. For example, varied chain pyridinium ceramides have been synthesized with increased water solubility and cell membrane permeability. The positive charge of the pyridinium ring in these structures allows targeting and accumulation of these ceramide analogues mainly into mitochondria and, to a lesser extent, into the nucleus of cancer cells (Senkal et al., 2006). Studies suggest that cancer cells acquire a more negative charge in their subcellular structures (especially mitochondria) (Modica-Napolitano & Aprille, 2001); therefore, Pyr-Cer can preferentially target cancer cells with minimum toxicity to normal cells. In fact, data indicate that L-t-C_6-Pyr-Cer and D-e-C_{16}-Pyr-Cer preferentially accumulate in mitochondria- and nuclei-enriched fractions in several human cancer cells *in vitro*, and this was consistent with the higher accumulation of the compound in the HNSCC tumor site, compared to the liver and intestines *in vivo* (Novgorodov et al., 2005; Senkal et al., 2006). The accumulation of Pyr-Cer in the mitochondria dramatically altered the structures and functions of mitochondria. This resulted in a decrease of the mitochondrial membrane potential, release of mitochondrial cytochrome C, activation of caspase-3 and caspase-9, and causing apoptotic cell death (Novgorodov et al., 2005; Senkal et al., 2006). Structural analogues of ceramide have also been synthesized and had promising anticancer effects (Bieberich, Kawaguchi, & Yu, 2000; Crawford, Bittman, Chun, Byun, & Bowen, 2003).

Alternatively, to improve cellular delivery and targeting, researchers have attempted to deliver exogenous ceramide in pegylated (polyethylglycol) liposomes. Data suggest that this delivery method of ceramide increased growth inhibitory effects in human breast cancer cells, via enhanced accumulation of ceramide (Stover & Kester, 2003). The liposomal delivery of exogenous natural ceramide also resulted in the inhibition of phosphorylated Akt and stimulated the activity of caspase-3/7 more effectively than nonliposomal ceramide. *In vivo* therapeutic efficacy of the pegylated ceramide for the delivery of exogenous ceramide, which resulted in slower tumor growth in murine models of breast cancer, was also demonstrated (Stover, Sharma, Robertson, & Kester, 2005). Additionally, encapsulated vincristine in SM-liposomes, also called sphingosomes, has improved efficacy compared to the conventional drug used in animal models of adult acute lymphocytic leukemia (ALL). Excitingly, sphingosomal vincristine is now in clinical phase II trials for treatment of patients with recurrent and refractory adult ALL (Thomas et al., 2006). Polymeric nanoparticle delivery systems are being explored as an alternative route to improve the delivery of ceramide for the treatment of cancer cells to overcome resistance (van Vlerken, Duan, Seiden, & Amiji, 2007).

Increasing intracellular ceramide concentrations has been suggested as an effective method to regulate cancer cell growth. To this end, certain chemotherapeutic agents such as daunorubicin, camptothecin, fludarabine, etoposide, and gemcitabine increase ceramide generation through the *de novo* ceramide pathway or via activation of SMases (Ogretmen & Hannun, 2004). Recently, it was shown that a Sonic-Hedgehog inhibitor cyclopamine induces apoptosis in medulloblastoma and head and neck cancer cells via induction of n-Smase-2/ceramide by nitric oxide generation, independently of Sonic-Hedgehog inhibition (Meyers-Needham, Lewis, et al., 2012; Meyers-

Needham, Ponnusamy, et al., 2012). Moreover, vorinostat and sorafenib in combination were shown to increase CD95 activation in gastrointestinal tumor cells via inducing the Ca(2+)/*de novo* ceramide/PP2A/reactive oxygen species-dependent signaling pathway (Park et al., 2010). Alternatively, another approach to increase endogenous ceramide is to inhibit enzymes that regulate ceramide clearance, which leads to increased antiproliferative responses in various cancer cells. Theoretically, combining these two therapeutic strategies could lead to promising outcomes. In fact, combining the chemotherapeutic agent gemcitabine with SM synergistically inhibits pancreatic tumor growth *in vivo* (Modrak, Cardillo, Newsome, Goldenberg, & Gold, 2004). The therapeutic effects of doxorubicin were also enhanced when used in combination with SM in various human cancer cell lines (Veldman et al., 2005). Mechanistically, SM was shown to increase the cellular uptake of doxorubicin via altering cell membrane permeability, leading to increased accumulation and bioavailability of the drug in these cells (Veldman et al., 2005).

Small molecule inhibitors of important metabolic enzymes involved in the sphingolipid pathway have been used to induce the accumulation of ceramide in some cancers. For example, B13, an inhibitor of acid CDase, was used in a metastatic colon cancer mouse model (Selzner et al., 2001) and a prostate cancer xenograph model (Samsel et al., 2004). In both cases, B13 caused the accumulation of ceramide and resulted in prevention of tumor growth.

FTY720 (Gilenya; Fingolimod), a fungal metabolite produced by *Isaria sinclairii*, is a myriocin analogue, was approved by the FDA for treatment of patients with refractory multiple sclerosis, and has gained a lot of attention as a possible anticancer agent (Brinkmann et al., 2010; Liu et al., 2010, 2008; Neviani et al., 2007). Mechanistically, FTY720 acts as an anti-MS drug by regulating sphingosine-1-phosphate receptor signaling and suppresses autoimmunity upon its phosphorylation by SK2 (Billich et al., 2003; Paugh, Payne, Barbour, Milstien, & Spiegel, 2003) to produce (S)-P-FTY720, the active enantiomer. Of the five S1P receptors, P-FTY720 interacts with $SIP_{1,3,4,5}$ at 0.3–3.1 nM, but does not interact with $S1P_2$. Once P-FTY720 binds to $S1P_1$ receptor, it leads to internalization and degradation of the receptor preventing lymphocyte egress (Matloubian, 2004).

One suggested anticancer mechanism of FTY720 is through activation of PP2A (Neviani et al., 2007). Activation of PP2A by FTY720 has been demonstrated in ALL, chronic lymphocytic leukemia, and c-KIT+ cancers (Liu et al., 2008; Roberts et al., 2010). Another proposed mechanism for

FTY720 anticancer effects is through inhibition or degradation of SK1. In prostate cancer cells, SK1 inhibition by FTY720 was considered an essential part of the drug mechanism since overexpression of SK1 had protective effects against FTY720 (Lim et al., 2011; Pchejetski et al., 2010; Tonelli et al., 2010). It was also shown that FTY720 alters ceramide metabolism (Berdyshev et al., 2009; Lahiri et al., 2009); however, whether FTY720-mediated cell death involves ceramide signaling is unclear.

The anticancer role of dihydroceramide has been recently explored. This was highlighted when prostate and lung cancer cells were treated with g-tocopherol, the main dietary form of vitamin E, which inhibited cell proliferation and induced apoptosis, with concomitant accumulation of dihydroceramides (Jiang, Wong, Fyrst, Saba, & Ames, 2004). Additionally, fenretinide, which was initially thought to increase ceramide generation, has been reported by various independent groups to elevate dihydroceramides, possibly via the inhibition of DES (Kraveka et al., 2007; Rahmaniyan, Curley, Obeid, Hannun, & Kraveka, 2011). In fact, recent data revealed that accumulation of endogenous dihydroceramides via downregulation of DES results in Rb-dependent growth arrest in human neuroblastoma cells (Kraveka et al., 2007). Previously, dihydroceramides were thought to be biologically inactive molecules; however, these data suggest that, when generated in cancer cells, they might be important in the regulation of cancer cell growth and/or survival.

These preliminary approaches point to the feasibility of developing more active and perhaps more selective analogues of ceramide, which can be tested in clinical trials for the treatment of patients with cancer. They also raise the possibility that several enzymes of ceramide clearance (SM synthases, CDases, CK, and GCS) may serve as novel therapeutic targets (Ogretmen & Hannun, 2001, 2004). In fact, several independent studies showed that downregulation of acid CDase, or inhibition of its activity, induces apoptosis and inhibits tumor growth (Ogretmen & Hannun, 2001, 2004). Similar data were also reported for the downregulation of CK, which inhibited cellular proliferation and enhanced apoptosis induced by serum starvation in A549 human lung cancer cells (Mitra et al., 2007).

6. CONCLUSIONS

Over the past two decades, the field of sphingolipid research has expanded immensely. The importance of sphingolipids, in particular ceramide, is well established in cancer. Previously, ceramide was classified as a "pro-cell death molecule"; however, it is now evident that not all ceramides are created equally, and that ceramides with different fatty-acid chain lengths exhibit distinct biological functions in a context-dependent manner. Although mechanisms that regulate distinct functions of endogenous ceramides are unknown, it is possible that ceramide's bioactive properties stem from its specific binding partners in cells. Depending on the subcellular localization, ceramide may interact with different proteins leading to the activation/inhibition of various downstream responses, such as binding of ceramide to I2PP2A activates tumor suppressor PP2A signaling in human lung cancer cells. Collectively, this information suggests that exogenous treatment or reactivation of antiproliferative ceramide signaling may be a novel therapeutic option for the treatment of cancer. To this end, recently discovered small molecule activators of CerS1, or inhibitors of CerS6 (Schiffmann et al., 2012), might be useful for tumor suppression. As the details of ceramide signaling and metabolism in cancer are dissected, new ceramide-based therapeutic options for the treatment of this disease will be developed.

ACKNOWLEDGMENTS

This work was supported by research grants CA088932, DE016572, and CA097132 to B. O. S. A. S was supported by funds from Wachovia Foundation and Abney Foundation, and NIH F30 ES019464.

REFERENCES

Alesse, E., Zazzeroni, F., Angelucci, A., Giannini, G., Di Marcotullio, L., & Gulino, A. (1998). The growth arrest and downregulation of c-myc transcription induced by ceramide are related events dependent on p21 induction, Rb underphosphorylation and E2F sequestering. *Cell Death and Differentiation*, *5*, 381–389.

Alvarez, S. E., Harikumar, K. B., Hait, N. C., Allegood, J., Strub, G. M., Kim, E. Y., et al. (2010). Sphingosine-1-phosphate is a missing cofactor for the E3 ubiquitin ligase TRAF2. *Nature*, *465*, 1084–1088.

Bauer, R., Voelzmann, A., Breiden, B., Schepers, U., Farwanah, H., Hahn, I., et al. (2009). Schlank, a member of the ceramide synthase family controls growth and body fat in Drosophila. *The EMBO Journal*, *28*, 3706–3716.

Ben-David, O., Pewzner-Jung, Y., Brenner, O., Laviad, E. L., Kogot-Levin, A., Weissberg, I., et al. (2011). Encephalopathy caused by ablation of very long acyl chain ceramide synthesis may be largely due to reduced galactosylceramide levels. *The Journal of Biological Chemistry*, *2011*(34), 30022–30033.

Berdyshev, E. V., Gorshkova, I., Skobeleva, A., Bittman, R., Lu, X., Dudek, S. M., et al. (2009). FTY720 inhibits ceramide synthases and up-regulates dihydrosphingosine 1-phosphate formation in human lung endothelial cells. *The Journal of Biological Chemistry*, *284*, 5467–5477.

Bieberich, E., Kawaguchi, T., & Yu, R. K. (2000). N-acylated serinol is a novel ceramide mimic inducing apoptosis in neuroblastoma cells. *The Journal of Biological Chemistry*, *275*, 177–181.

Billich, A., Bornancin, F., Devay, P., Mechtcheriakova, D., Urtz, N., & Baumruker, T. (2003). Phosphorylation of the immunomodulatory drug FTY720 by sphingosine kinases. *The Journal of Biological Chemistry, 278*, 47408–47415.

Bourbon, N. A., Yun, J., & Kester, M. (2000). Ceramide directly activates protein kinase C zeta to regulate a stress-activated protein kinase signaling complex. *The Journal of Biological Chemistry, 275*, 35617–35623.

Brinkmann, V., Billich, A., Baumruker, T., Heining, P., Schmouder, R., Francis, G., et al. (2010). Fingolimod (FTY720): Discovery and development of an oral drug to treat multiple sclerosis. *Nature Reviews. Drug Discovery, 9*, 883–897.

Chalfant, C. E., Kishikawa, K., Mumby, M. C., Kamibayashi, C., Bielawska, A., & Hannun, Y. A. (1999). Long chain ceramides activate protein phosphatase-1 and protein phosphatase-2A. Activation is stereospecific and regulated by phosphatidic acid. *The Journal of Biological Chemistry, 274*, 20313–20317.

Chalfant, C. E., Ogretmen, B., Galadari, S., Kroesen, B. J., Pettus, B. J., & Hannun, Y. A. (2001). FAS activation induces dephosphorylation of SR proteins; dependence on the de novo generation of ceramide and activation of protein phosphatase 1. *The Journal of Biological Chemistry, 276*, 44848–44855.

Crawford, K. W., Bittman, R., Chun, J., Byun, H. S., & Bowen, W. D. (2003). Novel ceramide analogues display selective cytotoxicity in drug-resistant breast tumor cell lines compared to normal breast epithelial cells. *Cellular and Molecular Biology (Noisy-le-Grand, France), 49*, 1017–1023.

D'Angelo, G., Polishchuk, E., Di Tullio, G., Santoro, M., Di Campli, A., Godi, A., et al. (2007). Glycosphingolipid synthesis requires FAPP2 transfer of glucosylceramide. *Nature, 449*, 62–67.

De Rosa, M. F., Ackerley, C., Wang, B., Ito, S., Clarke, D. M., & Lingwood, C. (2008). Inhibition of multidrug resistance by adamantylgb3, a globotriaosylceramide analog. *The Journal of Biological Chemistry, 283*, 4501–4511.

De Rosa, M. F., Sillence, D., Ackerley, C., & Lingwood, C. (2004). Role of multiple drug resistance protein 1 in neutral but not acidic glycosphingolipid biosynthesis. *The Journal of Biological Chemistry, 279*, 7867–7876.

Deng, X., Deng, X., Yin, X., Allan, R., Lu, D. D., Maurer, C. W., et al. (2010). Ceramide biogenesis is required for radiation-induced apoptosis in the germ line of *C. elegans*. *Science, 322*, 110–115.

Erez-Roman, R., Pienik, R., & Futerman, A. H. (2010). Increased ceramide synthase 2 and 6 mRNA levels in breast cancer tissues and correlation with sphingosine kinase expression. *Biochemical and Biophysical Research Communications, 391*, 219–223.

Gault, C. R., Obeid, L. M., & Hannun, Y. A. (2010). An overview of sphingolipid metabolism: From synthesis to breakdown. *Advances in Experimental Medicine and Biology, 688*, 1–23.

Gouaze, V., Liu, Y. Y., Prickett, C. S., Yu, J. Y., Giuliano, A. E., & Cabot, M. C. (2005). Glucosylceramide synthase blockade down-regulates P-glycoprotein and resensitizes multidrug-resistant breast cancer cells to anticancer drugs. *Cancer Research, 65*, 3861–3867.

Gouaze-Andersson, V., & Cabot, M. C. (2006). Glycosphingolipids and drug resistance. *Biochimica et Biophysica Acta, 1758*, 2096–2103.

Gouaze-Andersson, V., Yu, J. Y., Kreitenberg, A. J., Bielawska, A., Giuliano, A. E., & Cabot, M. C. (2007). Ceramide and glucosylceramide upregulate expression of the multidrug resistance gene MDR1 in cancer cells. *Biochimica et Biophysica Acta, 1771*, 1407–1417.

Goulding, C. W., Giuliano, A. E., & Cabot, M. C. (2000). SDZ PSC 833 the drug resistance modulator activates cellular ceramide formation by a pathway independent of P-glycoprotein. *Cancer Letters, 149*, 143–151.

Guillas, I., Kirchman, P. A., Chuard, R., Pfefferli, M., Jiang, J. C., Jazwinski, S. M., et al. (2001). C26-CoA-dependent ceramide synthesis of *Saccharomyces cerevisiae* is operated by Lag1p and Lac1p. *The EMBO Journal, 20*, 2655–2665.

Hait, N. C., Allegood, J., Maceyka, M., Strub, G. M., Harikumar, K. B., Singh, S. K., et al. (2009). Regulation of histone acetylation in the nucleus by sphingosine-1-phosphate. *Science, 325*, 1254–1257.

Hanada, K., Kumagai, K., Tomishige, N., & Yamaji, T. (2009). CERT-mediated trafficking of ceramide. *Biochimica et Biophysica Acta, 1791*, 684–691.

Hanada, K., Kumagai, K., Yasuda, S., Miura, Y., Kawano, M., Fukasawa, M., et al. (2003). Molecular machinery for non-vesicular trafficking of ceramide. *Nature, 426*, 803–809.

Heinrich, M., Neumeyer, J., Jakob, M., Hallas, C., Tchikov, V., Winoto-Morbach, S., et al. (2004). Cathepsin D links TNF-induced acid sphingomyelinase to Bid-mediated caspase-9 and -3 activation. *Cell Death and Differentiation, 11*, 550–563.

Heinrich, M., Wickel, M., Schneider-Brachert, W., Sandberg, C., Gahr, J., Schwandner, R., et al. (1999). Cathepsin D targeted by acid sphingomyelinase-derived ceramide. *The EMBO Journal, 18*, 5252–5263.

Huwiler, A., Brunner, J., Hummel, R., Vervoordeldonk, M., Stabel, S., van den Bosch, H., et al. (1996). Ceramide-binding and activation defines protein kinase c-Raf as a ceramide-activated protein kinase. *Proceedings of the National Academy of Sciences of the United States of America, 93*, 6959–6963.

Imgrund, S., Hartmann, D., Farwanah, H., Eckhardt, M., Sandhoff, R., Degen, J., et al. (2009). Adult ceramide synthase 2 (CERS2)-deficient mice exhibit myelin sheath defects, cerebellar degeneration, and hepatocarcinomas. *The Journal of Biological Chemistry, 284*, 33549–33560.

Jazwinski, S. M., & Conzelmann, A. (2002). LAG1 puts the focus on ceramide signaling. *The International Journal of Biochemistry & Cell Biology, 34*, 1491–1495.

Jennemann, R., Rabionet, M., Gorgas, K., Epstein, S., Dalpke, A., Rothermel, U., et al. (2012). Loss of ceramide synthase 3 causes lethal skin barrier disruption. *Human Molecular Genetics, 21*, 586–608.

Jiang, Q., Wong, J., Fyrst, H., Saba, J. D., & Ames, B. N. (2004). Gamma-tocopherol or combinations of vitamin E forms induce cell death in human prostate cancer cells by interrupting sphingolipid synthesis. *Proceedings of the National Academy of Sciences of the United States of America, 101*, 17825–17830.

Karahatay, S., Thomas, K., Koybasi, S., Senkal, C. E., Elojeimy, S., Liu, X., et al. (2007). Clinical relevance of ceramide metabolism in the pathogenesis of human head and neck squamous cell carcinoma (HNSCC): Attenuation of C(18)-ceramide in HNSCC tumors correlates with lymphovascular invasion and nodal metastasis. *Cancer Letters, 256*, 101–111.

Katayose, Y., Li, M., Al-Murrani, S. W., Shenolikar, S., & Damuni, Z. (2000). Protein phosphatase 2A inhibitors, I(1)(PP2A) and I(2)(PP2A), associate with and modify the substrate specificity of protein phosphatase 1. *The Journal of Biological Chemistry, 275*, 9209–9214.

Kitatani, K., Idkowiak-Baldys, J., & Hannun, Y. A. (2008). The sphingolipid salvage pathway in ceramide metabolism and signaling. *Cellular Signalling, 20*, 1010–1018.

Koybasi, S., Senkal, C. E., Sundararaj, K., Spassieva, S., Bielawski, J., Osta, W., et al. (2004). Defects in cell growth regulation by C18:0-ceramide and longevity assurance gene 1 in human head and neck squamous cell carcinomas. *The Journal of Biological Chemistry, 279*, 44311–44319.

Kraveka, J. M., Li, L., Szulc, Z. M., Bielawski, J., Ogretmen, B., Hannun, Y. A., et al. (2007). Involvement of dihydroceramide desaturase in cell cycle progression in human neuroblastoma cells. *The Journal of Biological Chemistry, 282*, 16718–16728.

Lahiri, S., Park, H., Laviad, E. L., Lu, X., Bittman, R., & Futerman, A. H. (2009). Ceramide synthesis is modulated by the sphingosine analog FTY720 via a mixture of uncompetitive and noncompetitive inhibition in an Acyl-CoA chain length-dependent manner. *The Journal of Biological Chemistry*, *284*, 16090–16098.

Lim, K. G., Tonelli, F., Li, Z., Lu, X., Bittman, R., Pyne, S., et al. (2011). FTY720 analogues as sphingosine kinase 1 inhibitors: Enzyme inhibition kinetics, allosterism, proteasomal degradation, and actin rearrangement in MCF-7 breast cancer cells. *The Journal of Biological Chemistry*, *286*, 18633–18640.

Liu, Q., Alinari, L., Chen, C. S., Yan, F., Dalton, J. T., Lapalombella, R., et al. (2010). FTY720 shows promising *in vitro* and *in vivo* preclinical activity by downmodulating Cyclin D1 and phospho-Akt in mantle cell lymphoma. *Clinical Cancer Research*, *16*, 3182–3192.

Liu, Q., Zhao, X., Frissora, F., Ma, Y., Santhanam, R., Jarjoura, D., et al. (2008). FTY720 demonstrates promising preclinical activity for chronic lymphocytic leukemia and lymphoblastic leukemia/lymphoma. *Blood*, *111*, 275–284.

Matloubian, M., Lo, C. G., Cinamon, G., Lesneski, M. J., Xu, Y., Brinkmann, V., et al. (2004). Lymphocyte egress from thymus and peripheral lymphoid organs is dependent on S1P receptor 1. *Nature*, *427*, 355–360.

Menuz, V., Howell, K. S., Gentina, S., Epstein, S., Riezman, I., Fornallaz-Mulhauser, M., et al. (2009). Protection of *C. elegans* from anoxia by HYL-2 ceramide synthase. *Science*, *324*, 381–384.

Mesika, A., Ben-Dor, S., Laviad, E. L., & Futerman, A. H. (2007). A new functional motif in Hox domain-containing ceramide synthases: Identification of a novel region flanking the Hox and TLC domains essential for activity. *The Journal of Biological Chemistry*, *282*, 27366–27373.

Meyers-Needham, M., Lewis, J. A., Gencer, S., Sentelle, R. D., Saddoughi, S. A., Clarke, C. J., et al. (2012). Off-target function of the sonic-hedgehog inhibitor cyclopamine in mediating apoptosis via nitric oxide-dependent neutral sphingomyelinase 2/ceramide induction. *Molecular Cancer Therapeutics*, *11*, 1092–1102.

Meyers-Needham, M., Ponnusamy, S., Gencer, S., Jiang, W., Thomas, R. J., Senkal, C. E., et al. (2012). Concerted functions of HDAC1 and microRNA-574-5p repress alternatively spliced ceramide synthase 1 expression in human cancer cells. *EMBO Molecular Medicine*, *4*, 78–92.

Mitra, P., Maceyka, M., Payne, S. G., Lamour, N., Milstien, S., Chalfant, C. E., et al. (2007). Ceramide kinase regulates growth and survival of A549 human lung adenocarcinoma cells. *FEBS Letters*, *581*, 735–740.

Modica-Napolitano, J. S., & Aprille, J. R. (2001). Delocalized lipophilic cations selectively target the mitochondria of carcinoma cells. *Advanced Drug Delivery Reviews*, *49*, 63–70.

Modrak, D. E., Cardillo, T. M., Newsome, G. A., Goldenberg, D. M., & Gold, D. V. (2004). Synergistic interaction between sphingomyelin and gemcitabine potentiates ceramide-mediated apoptosis in pancreatic cancer. *Cancer Research*, *64*, 8405–8410.

Mukhopadhyay, A., Saddoughi, S. A., Song, P., Sultan, I., Ponnusamy, S., Senkal, C. E., et al. (2009). Direct interaction between the inhibitor 2 and ceramide via sphingolipid-protein binding is involved in the regulation of protein phosphatase 2A activity and signaling. *The FASEB Journal*, *23*, 751–763.

Muto, S., Senda, M., Akai, Y., Sato, L., Suzuki, T., Nagai, R., et al. (2007). Relationship between the structure of SET/TAF-Ibeta/INHAT and its histone chaperone activity. *Proceedings of the National Academy of Sciences of the United States of America*, *104*, 4285–4290.

Neviani, P., Santhanam, R., Oaks, J. J., Eiring, A. M., Notari, M., Blaser, B. W., et al. (2007). FTY720, a new alternative for treating blast crisis chronic myelogenous leukemia and

Philadelphia chromosome-positive acute lymphocytic leukemia. *The Journal of Clinical Investigation, 117*, 2408–2421.

Novgorodov, S. A., Szulc, Z. M., Luberto, C., Jones, J. A., Bielawski, J., Bielawska, A., et al. (2005). Positively charged ceramide is a potent inducer of mitochondrial permeabilization. *The Journal of Biological Chemistry, 280*, 16096–16105.

Ogretmen, B., & Hannun, Y. A. (2001). Updates on functions of ceramide in chemotherapy-induced cell death and in multidrug resistance. *Drug Resistance Updates, 4*, 368–377.

Ogretmen, B., & Hannun, Y. A. (2004). Biologically active sphingolipids in cancer pathogenesis and treatment. *Nature Reviews. Cancer, 4*, 604–616.

Ogretmen, B., Pettus, B. J., Rossi, M. J., Wood, R., Usta, J., Szulc, Z., et al. (2002). Biochemical mechanisms of the generation of endogenous long chain ceramide in response to exogenous short chain ceramide in the A549 human lung adenocarcinoma cell line. Role for endogenous ceramide in mediating the action of exogenous ceramide. *The Journal of Biological Chemistry, 277*, 12960–12969.

Park, M. A., Mitchell, C., Zhang, G., Yacoub, A., Allegood, J., Häussinger, D., et al. (2010). Vorinostat and sorafenib increase CD95 activation in gastrointestinal tumor cells through a Ca(2+)-de novo ceramide-PP2A-reactive oxygen species-dependent signaling pathway. *Cancer Research, 70*(15), 6313–6324.

Paugh, S. W., Payne, S. G., Barbour, S. E., Milstien, S., & Spiegel, S. (2003). The immunosuppressant FTY720 is phosphorylated by sphingosine kinase type 2. *FEBS Letters, 554*, 189–193.

Pchejetski, D., Bohler, T., Brizuela, L., Sauer, L., Doumerc, N., Golzio, M., et al. (2010). FTY720 (fingolimod) sensitizes prostate cancer cells to radiotherapy by inhibition of sphingosine kinase-1. *Cancer Research, 70*, 8651–8661.

Pettus, B. J., Bielawska, A., Subramanian, P., Wijesinghe, D. S., Maceyka, M., Leslie, C. C., et al. (2004). Ceramide 1-phosphate is a direct activator of cytosolic phospholipase A2. *The Journal of Biological Chemistry, 279*, 11320–11326.

Pewzner-Jung, Y., Ben-Dor, S., & Futerman, A. H. (2006). When do Lasses (longevity assurance genes) become CerS (ceramide synthases)?: Insights into the regulation of ceramide synthesis. *The Journal of Biological Chemistry, 281*(35), 25001–25005.

Pewzner-Jung, Y., Park, H., Laviad, E. L., Silva, L. C., Lahiri, S., Stiban, J., et al. (2010). A critical role for ceramide synthase 2 in liver homeostasis: I. Alterations in lipid metabolic pathways. *The Journal of Biological Chemistry, 285*, 10902–10910.

Ponnusamy, S., Meyers-Needham, M., Senkal, C. E., Saddoughi, S. A., Sentelle, D., Selvam, S. P., et al. (2010). Sphingolipids and cancer: Ceramide and sphingosine-1-phosphate in the regulation of cell death and drug resistance. *Future Oncology, 6*, 1603–1624.

Rahmaniyan, M., Curley, R. W., Jr., Obeid, L. M., Hannun, Y. A., & Kraveka, J. M. (2011). Identification of dihydroceramide desaturase as a direct *in vitro* target for fenretinide. *The Journal of Biological Chemistry, 286*, 24754–24764.

Roberts, K. G., Smith, A. M., McDougall, F., Carpenter, H., Horan, M., Neviani, P., et al. (2010). Essential requirement for PP2A inhibition by the oncogenic receptor c-KIT suggests PP2A reactivation as a strategy to treat c-KIT+ cancers. *Cancer Research, 70*, 5438–5447.

Saddoughi, S. A., Garrett-Mayer, E., Chaudhary, U., O'Brien, P. E., Afrin, L. B., Day, T. A., et al. (2011). Results of a phase II trial of gemcitabine plus doxorubicin in patients with recurrent head and neck cancers: Serum C-ceramide as a novel biomarker for monitoring response. *Clinical Cancer Research, 17*, 6097–6105.

Salas, A., Ponnusamy, S., Senkal, C. E., Meyers-Needham, M., Selvam, S. P., Saddoughi, S. A., et al. (2011). Sphingosine kinase-1 and sphingosine 1-phosphate receptor 2 mediate Bcr-Abl1 stability and drug resistance by modulation of protein phosphatase 2A. *Blood, 117*, 5941–5952.

Samsel, L., Zaidel, G., Drumgoole, H. M., Jelovac, D., Drachenberg, C., Rhee, J. G., et al. (2004). The ceramide analog, B13, induces apoptosis in prostate cancer cell lines and inhibits tumor growth in prostate cancer xenografts. *The Prostate, 58*, 382–393.

Schiffmann, S., Hartmann, D., Fuchs, S., Birod, K., Ferreiròs, N., Schreiber, Y., et al. (2012). Inhibitors of specific ceramide synthases. *Biochimie, 94*(2), 558–565.

Schiffmann, S., Sandner, J., Birod, K., Wobst, I., Angioni, C., Ruckhaberle, E., et al. (2009). Ceramide synthases and ceramide levels are increased in breast cancer tissue. *Carcinogenesis, 30*, 745–752.

Selzner, M., Bielawska, A., Morse, M. A., Rudiger, H. A., Sindram, D., Hannun, Y. A., et al. (2001). Induction of apoptotic cell death and prevention of tumor growth by ceramide analogues in metastatic human colon cancer. *Cancer Research, 61*, 1233–1240.

Senkal, C. E., Ponnusamy, S., Bielawski, J., Hannun, Y. A., & Ogretmen, B. (2010). Antiapoptotic roles of ceramide-synthase-6-generated C16-ceramide via selective regulation of the ATF6/CHOP arm of ER-stress-response pathways. *The FASEB Journal, 24*, 296–308.

Senkal, C. E., Ponnusamy, S., Manevich, Y., Meyers-Needham, M., Saddoughi, S. A., Mukhopadyay, A., et al. (2011). Alteration of ceramide synthase 6/C16-ceramide induces activating transcription factor 6-mediated endoplasmic reticulum (ER) stress and apoptosis via perturbation of cellular Ca2+ and ER/Golgi membrane network. *The Journal of Biological Chemistry, 286*, 42446–42458.

Senkal, C. E., Ponnusamy, S., Rossi, M. J., Bialewski, J., Sinha, D., Jiang, J. C., et al. (2007). Role of human longevity assurance gene 1 and C18-ceramide in chemotherapy-induced cell death in human head and neck squamous cell carcinomas. *Molecular Cancer Therapeutics, 6*, 712–722.

Senkal, C. E., Ponnusamy, S., Rossi, M. J., Sundararaj, K., Szulc, Z., Bielawski, J., et al. (2006). Potent antitumor activity of a novel cationic pyridinium-ceramide alone or in combination with gemcitabine against human head and neck squamous cell carcinomas *in vitro* and *in vivo*. *The Journal of Pharmacology and Experimental Therapeutics, 317*, 1188–1199.

Sentelle, R. D., Senkal, C. E., Jiang, W., Ponnusamy, S., Gencer, S., Panneer Selvam, S., et al. (2012). Ceramide targets autophagosomes to mitochondria and induces lethal mitophagy. *Nature Chemical Biology*, Epub ahead of print.

Stahelin, R. V., Subramanian, P., Vora, M., Cho, W., & Chalfant, C. E. (2007). Ceramide-1-phosphate binds group IVA cytosolic phospholipase a2 via a novel site in the C2 domain. *The Journal of Biological Chemistry, 282*, 20467–20474.

Stover, T., & Kester, M. (2003). Liposomal delivery enhances short-chain ceramide-induced apoptosis of breast cancer cells. *The Journal of Pharmacology and Experimental Therapeutics, 307*, 468–475.

Stover, T. C., Sharma, A., Robertson, G. P., & Kester, M. (2005). Systemic delivery of liposomal short-chain ceramide limits solid tumor growth in murine models of breast adenocarcinoma. *Clinical Cancer Research, 11*, 3465–3474.

Sultan, I., Senkal, C. E., Ponnusamy, S., Bielawski, J., Szulc, Z., Bielawska, A., et al. (2006). Regulation of the sphingosine-recycling pathway for ceramide generation by oxidative stress, and its role in controlling c-Myc/Max function. *The Biochemical Journal, 393*(Pt. 2), 513–521.

Swanton, C., Marani, M., Pardo, O., Warne, P. H., Kelly, G., Sahai, E., et al. (2007). Regulators of mitotic arrest and ceramide metabolism are determinants of sensitivity to paclitaxel and other chemotherapeutic drugs. *Cancer Cell, 11*, 498–512.

Switzer, C. H., Cheng, R. Y., Vitek, T. M., Christensen, D. J., Wink, D. A., & Vitek, M. P. (2011). Targeting SET/I(2)PP2A oncoprotein functions as a multi-pathway strategy for cancer therapy. *Oncogene, 30*, 2504–2513.

Thomas, D. A., Sarris, A. H., Cortes, J., Faderl, S., O'Brien, S., Giles, F. J., et al. (2006). Phase II study of sphingosomal vincristine in patients with recurrent or refractory adult acute lymphocytic leukemia. *Cancer, 106*, 120–127.

Tidhar, R., Ben-Dor, S., Wang, E., Kelly, S., Merrill, A. H., Jr., & Futerman, A. H. (2012). Acyl chain specificity of ceramide synthases is determined within a region of 150 residues in the Tram-Lag-CLN8 (TLC) domain. *The Journal of Biological Chemistry, 287*(5), 3197–3206.

Tonelli, F., Lim, K. G., Loveridge, C., Long, J., Pitson, S. M., Tigyi, G., et al. (2010). FTY720 and (S)-FTY720 vinylphosphonate inhibit sphingosine kinase 1 and promote its proteasomal degradation in human pulmonary artery smooth muscle, breast cancer and androgen-independent prostate cancer cells. *Cellular Signalling, 22*, 1536–1542.

van Vlerken, L. E., Duan, Z., Seiden, M. V., & Amiji, M. M. (2007). Modulation of intracellular ceramide using polymeric nanoparticles to overcome multidrug resistance in cancer. *Cancer Research, 67*, 4843–4850.

Veldman, R. J., Koning, G. A., van Hell, A., Zerp, S., Vink, S. R., Storm, G., et al. (2005). Coformulated N-octanoyl-glucosylceramide improves cellular delivery and cytotoxicity of liposomal doxorubicin. *The Journal of Pharmacology and Experimental Therapeutics, 315*, 704–710.

Venkataraman, K., Riebeling, C., Bodennec, J., Riezman, H., Allegood, J. C., Sullards, M. C., et al. (2002). Upstream of growth and differentiation factor 1 (uog1), a mammalian homolog of the yeast longevity assurance gene 1 (LAG1), regulates N-stearoyl-sphinganine (C18-(dihydro)ceramide) synthesis in a fumonisin B1-independent manner in mammalian cells. *The Journal of Biological Chemistry, 277*, 35642–35649.

Wooten, L. G., & Ogretmen, B. (2005). Sp1/Sp3-dependent regulation of human telomerase reverse transcriptase promoter activity by the bioactive sphingolipid ceramide. *The Journal of Biological Chemistry, 280*, 28867–28876.

Wooten-Blanks, L. G., Song, P., Senkal, C. E., & Ogretmen, B. (2007). Mechanisms of ceramide-mediated repression of the human telomerase reverse transcriptase promoter via deacetylation of Sp3 by histone deacetylase 1. *The FASEB Journal, 21*, 3386–3397.

Yeh, E., Cunningham, M., Arnold, H., Chasse, D., Monteith, T., Ivaldi, G., et al. (2004). A signalling pathway controlling c-Myc degradation that impacts oncogenic transformation of human cells. *Nature Cell Biology, 6*, 308–318.

Yin, X., Zafrullah, M., Lee, H., Haimovitz-Friedman, A., Fuks, Z., & Kolesnick, R. (2009). A ceramide-binding C1 domain mediates kinase suppressor of ras membrane translocation. *Cellular Physiology and Biochemistry, 24*, 219–230.

Zhang, Y., Yao, B., Delikat, S., Bayoumy, S., Lin, X. H., Basu, S., et al. (1997). Kinase suppressor of Ras is ceramide-activated protein kinase. *Cell, 89*, 63–72.

Zhao, L., Spassieva, S. D., Jucius, T. J., Shultz, L. D., Shick, H. E., Macklin, W. B., et al. (2011). A deficiency of ceramide biosynthesis causes cerebellar purkinje cell neurodegeneration and lipofuscin accumulation. *PLoS Genetics*, 7, e1002063.

CHAPTER THREE

Ceramide Glycosylation Catalyzed by Glucosylceramide Synthase and Cancer Drug Resistance

Yong-Yu Liu*,[1], Ronald A. Hill*, Yu-Teh Li†
*Department of Basic Pharmaceutical Sciences, University of Louisiana at Monroe, Monroe, Louisiana, USA
†Department of Biochemistry and Molecular Biology, Tulane University School of Medicine, New Orleans, Louisiana, USA
[1]Corresponding author: e-mail address: yliu@ulm.edu

Contents

Abstract

Glucosylceramide synthase (GCS), converting ceramide to glucosylceramide, catalyzes the first reaction of ceramide glycosylation in sphingolipid metabolism. This glycosylation by GCS is a critical step regulating the modulation of cellular activities by controlling ceramide and glycosphingolipids (GSLs). An increase of ceramide in response to stresses, such as chemotherapy, drives cells to proliferation arrest and apoptosis or autophagy; however, ceramide glycosylation promptly eliminates ceramide and consequently, these induced processes, thus protecting cancer cells. Further, persistently

Advances in Cancer Research, Volume 117
ISSN 0065-230X
http://dx.doi.org/10.1016/B978-0-12-394274-6.00003-0

enhanced ceramide glycosylation can increase GSLs, participating in selecting cancer cells to drug resistance. GCS is overexpressed in diverse drug-resistant cancer cells and in tumors of breast, colon, and leukemia that display poor response to chemotherapy. As ceramide glycosylation by GCS is a rate-limiting step in GSL synthesis, inhibition of GCS sensitizes cancer cells to anticancer drugs and eradicates cancer stem cells. Mechanistic studies indicate that uncoupling ceramide glycosylation can modulate gene expression, decreasing MDR1 through the cSrc/β-catenin pathway and restoring p53 expression via RNA splicing. These studies not only expand our knowledge in understanding how ceramide glycosylation affects cancer cells but also provide novel therapeutic approaches for targeting refractory tumors.

1. INTRODUCTION

Sphingolipids are mainly present in eukaryote membranes and are lipids sharing similar structures that consist of sphinganine linked to a fatty acid (Hannun & Obeid, 2008; Merrill, 2011). Ceramide is the simplest in structure, and other complex sphingolipids possess additional hydrophilic domains, such as phosphate, phosphorylcholine, and sugar moieties attached to their sphingoid bases. C_{18}-ceramide is one of the common molecular species of ceramide in most cells (Fig. 3.1), though the generic term "ceramide" designates a family of more than 50 distinct molecular species that are synthesized by six ceramide synthases (CerS1-6, also known as the longevity assurance gene products, LASS1-6) (Pewzner-Jung, Ben-Dor, & Futerman, 2006; Rabionet et al., 2008).

One fate of ceramide in cells is glycosylation. In the first step leading to a multiplicity of glycosylated species, glucose or galactose becomes attached to the 1-hydroxy group of ceramide, yielding a simple glycosphingolipid (GSL), glucosylceramide (GlcCer), or galactosylceramide (GalCer), respectively (Fig. 3.1). From these, more complex GSLs, such as lactosylceramide (LacCer), globotriaosylceramide (Gb3), monosialoganglioside (GM3), and others can be synthesized by incorporation of additional sugar residues (Hannun & Obeid, 2008; Yu, Nakatani, & Yanagisawa, 2009).

Ceramide and GSLs are important biological molecules in cellular processes of cancer progression, and key modulators of the outcome of cancer treatments. Besides providing structural integrity in membranes, ceramide and GSLs play critical roles in modulating cellular signaling and gene expression (Hakomori, 2010; Hannun & Obeid, 2008; Patwardhan & Liu, 2011). Through these, they alter various aspects of cell functions, notably including apoptosis, proliferation, autophagy, endocytosis, transport, migration, senescence, and inflammation.

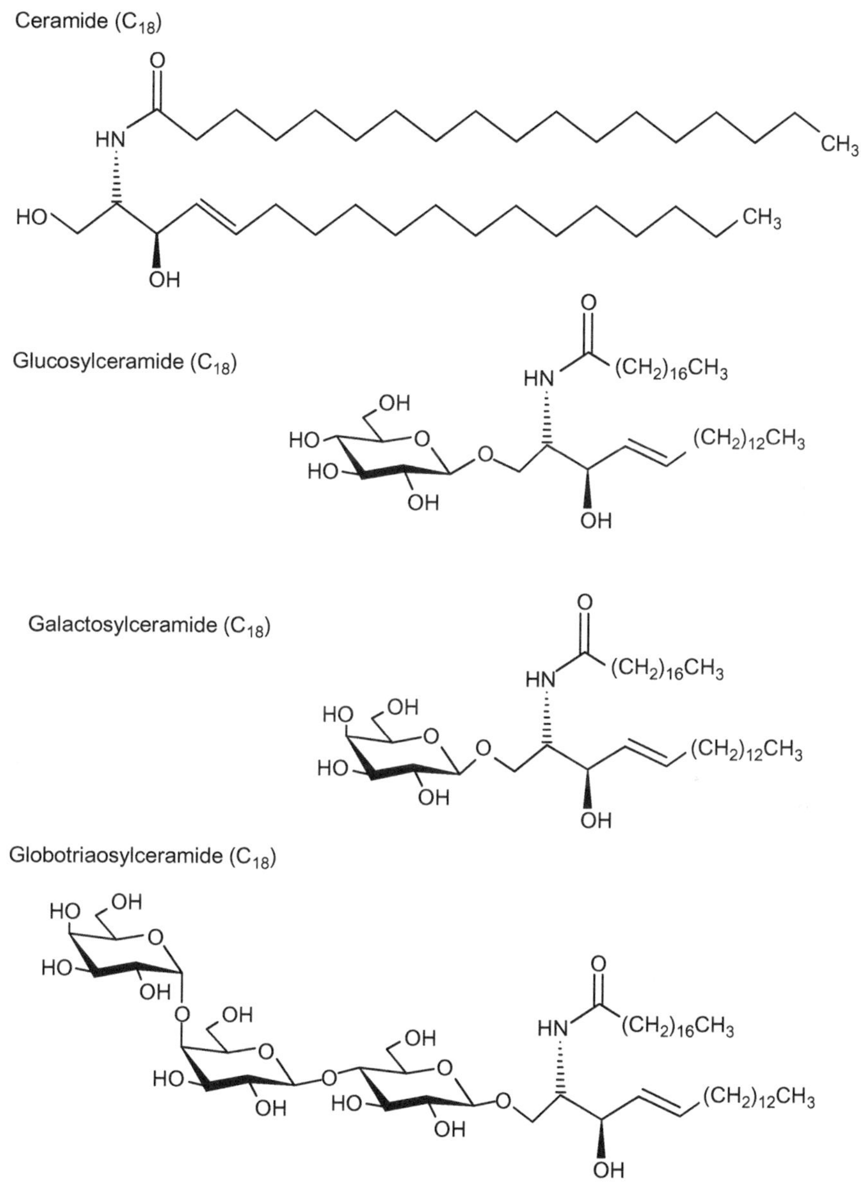

Figure 3.1 Basic structures of ceramide, glucosylceramide, galactosylceramide, and globotriaosylceramide. In mammals, one of the common ceramide species is C_{18}-ceramide, which has a sphinganine chain length of 18 carbon atoms, with an *E* double bond between C4 and C5, and a C18 fatty acid acylating its C-2 amino group. Glucose or galactose is attached to the 1-hydroxy group of ceramide to form glucosylceramide or galactosylceramide. A series of glycosylations transfer various sugar units to the glucose moiety of glucosylceramide to generate different GSLs, such as globotriaosylceramide.

These sphingolipid-modulated processes in turn are crucial in tumorigenesis, cancer progression, and the efficacies of cancer therapies (Ogretmen, 2006; Ogretmen & Hannun, 2001, 2004; Patwardhan & Liu, 2011; Senchenkov, Litvak, & Cabot, 2001). The balance between ceramide and GlcCer or other GSLs can induce cells to undergo malignant growth, or rescue cancerous cells to normal. The rate-limiting enzymes in ceramide glycosylation, particularly glucosylceramide synthase (GCS), actively participate in the cell biology of cancer progression by shifting reactions to generate metabolites in favor of cancer (Hakomori, 2010; Liu, Han, Giuliano, & Cabot, 2001; Ogretmen & Hannun, 2004; Patwardhan & Liu, 2011). Ceramide can modulate cellular processes directly through interactions with effectors, such as in ceramide-induced mitochondria activation to initiate apoptosis (Chipuk et al., 2012; Hannun & Obeid, 2008; von Haefen et al., 2002). GSLs mainly form lipid rafts, or GSL-enriched microdomains (GEMs), in the plasma membrane, thus supporting or modulating definite signaling cascades (Hannun & Obeid, 2008; Patwardhan & Liu, 2011; Sonnino, Prinetti, Mauri, Chigorno, & Tettamanti, 2006). Several comprehensive reviews have summarized the progress on dysregulated sphingolipids and cancers (Ogretmen & Hannun, 2004; Pyne & Pyne, 2010). Here, we address evidence showing that cancer drug resistance is attributed to ceramide glycosylation. Glucosylation is one critical step controlling ceramide levels, and also the synthesis of GSLs in cells responding to stresses such as chemotherapy or radiation therapy. As an increase of ceramide after treatments initiates processes of proliferation arrest, apoptosis, and autophagy, this ceramide glucosylation can promptly arrest these cellular processes, and thereby protect cancer cells. Further, persistently enhanced ceramide glycosylation can facilitate cancer progression by modulating the expression of genes involved in tumor metastasis, altering the status of cancer stem cells (CSCs), and facilitating drug resistance (Modrak, Gold, & Goldenberg, 2006; Ogretmen, Kraveka, et al., 2001; Ogretmen, McCauley, & Safa, 1998; Ogretmen et al., 2002; Ogretmen & Safa, 1996, 1997; Ogretmen, Schady, et al., 2001; Patwardhan & Liu, 2011). To face the challenge of understanding how ceramide glycosylation by GCS confers drug resistance in cells, we examine these findings with relation to ABC transporters, CSCs, and p53 mutations. We also consider the treatment of drug-resistant cancers through the inhibition of GCS-mediated processes.

2. CERAMIDE GLYCOSYLATION AND GEMs

Ceramide is mainly generated in the endoplasmic reticulum (ER), but its glycosylation is primarily conducted by GCS in the Golgi apparatus of mammalian cells. Via a cascade of enzymatic reactions, more than 3000

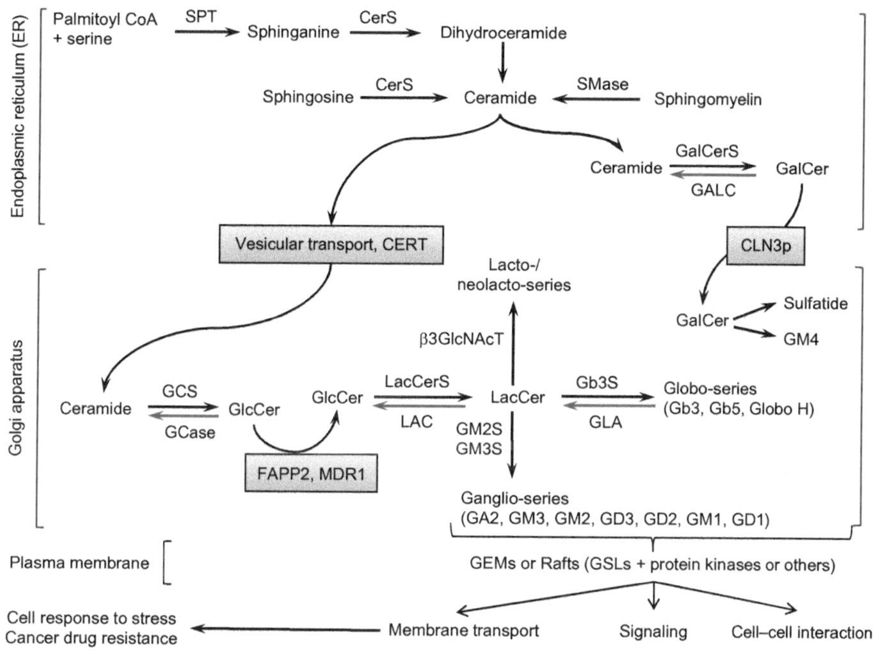

Figure 3.2 Glycosphingolipid biosynthesis and its cellular functions. SPT, serine-palmitoyl transferase; CerS, ceramide synthase; CERT, ceramide transporter; GCS, glucosylceramide synthase; GalCerS, galactosylceramide synthase; GALC, galactocerebrosidase (β-galactosidase); LacCerS, lactosylceramide synthase; Gb3S, globotriaosylceramide synthase; GCase, glucocerebrosidase (β-glucosidase); GLA, α-galactosidase A; GM2S, GM2 synthase; GM3S, GM3 synthase; GEMs, GSL-enriched microdomains.

different GSLs can be generated in the ER. In addition to variations in the activities of enzymes involved in glycosylation, the transport of ceramide from the ER to the Golgi, and of GSLs from the Golgi to other membranes, also modulates the distribution of GSLs in the membrane microdomains, thus altering cellular processes (Gault, Obeid, & Hannun, 2010; Fig. 3.2).

2.1. Ceramide synthase and its translocation from the ER to Golgi

In mammalian cells, ceramide is synthesized predominantly by the *de novo* pathway from serine and palmitoyl-CoA in the ER and ER-associated membrane (Hannun & Obeid, 2008; Merrill, 2011). Ceramide can also be produced from sphingomyelin breakdown catalyzed by sphingomyelinases (SMases) in the inner leaflet of the plasma membrane (neutral SMase) or the outer leaflet of lysosomal membrane (acid SMase) (Hannun & Obeid, 2008; Kolesnick, Haimovitz-Friedman, & Fuks, 1994; Fig. 3.2). The cells employ two major mechanisms to mobilize ceramide, either ceramide transfer (CERT) or vesicular transport (Gault et al., 2010; Halter et al., 2007; Yamaji, Kumagai, Tomishige, & Hanada, 2008). CERT is a cytosolic protein that transfers ceramide from the ER to the Golgi, where it can be modified into sphingomyelin, and possibly GSLs, given that a sphingomyelin synthase (SMS1) and GCS have both been localized biochemically to the *cis*-medial Golgi (Futerman & Pagano, 1991). The CERT protein is composed of at least three functional domains that determine its function: pleckstrin homology (PH) domain, FFAT domain, and START domain, sequentially from the N-terminus to the C-terminus (Kudo et al., 2008). The PH domain is able to recognize phosphatidylinositol 4-monophosphate (PI4P) on acceptor Golgi membranes, thereby allowing for directed transport to the Golgi. The FFAT domain is thought to enable binding to ER-resident VAMP-associated proteins so that CERT can only accept ceramide from the ER (Derre, Swiss, & Agaisse, 2011; Hanada, Kumagai, Tomishige, & Yamaji, 2009). The START domain provides a hydrophobic pocket enabling ceramide transport through the aqueous environment of the cytoplasm for delivery to the Golgi. *In vitro* studies have shown that phosphorylation of CERT at multiple serine residues, by casein kinase I or others, results in inhibitory interaction between the START and PH domains that inactivates the PI4P binding and CERT (Kumagai, Kawano, Shinkai-Ouchi, Nishijima, & Hanada, 2007). It is as of yet unclear if such phosphorylation or oligomerization that is known to occur with this protein, constitute general mechanism by which cellular stresses can inactivate CERT (Charruyer et al., 2008).

CERT displays a preference for ceramide species with acyl chains less than C22. Although CERT still transfers C_{22}- and $C_{24:1}$-ceramide, it does so with only 40% of the efficiency of shorter chain species (Kudo et al., 2008; Kumagai et al., 2007). In addition, CERT shows minimal to no transfer of $C_{24:0}$-ceramide. Ceramide transported by CERT is preferentially incorporated into sphingomyelin rather than into GSLs (Hanada et al., 2003; Kudo et al., 2010). Because CERT has preference for specific ceramide chain lengths, this may at least partially govern which forms of ceramide are preferentially utilized for sphingomyelin synthesis versus which ceramide species are preferred for GSL synthesis.

An alternative pathway for the transport of ceramide species to the Golgi is coat protein dependent, and makes use of vesicular transport (Watson & Stephens, 2005). The principal driving force behind the formation of vesicular carriers is the multi-subunit coat protein complex (COPII); however, little is known about how this pathway is regulated with respect to ceramide transport. Vesicular transport is thought to be the major pathway delivering ceramide to the *cis*-Golgi for GSL synthesis (Gault et al., 2010; Fig. 3.2).

2.2. Ceramide galactosylation catalyzed by galactosylceramide synthase and the synthesis of sulfatide and GM4

Human galactosylceramide synthase (GalCerS), also known as UDP-galactose:ceramide galactosyltransferase (UGT8 or CGT) (EC2.4.2.62), transfers the galactose residue from UDP-galactose to ceramide at the 1-hydroxyl moiety, thereby producing GalCer (Schulte & Stoffel, 1993; Stahl, Jurevics, Morell, Suzuki, & Popko, 1994; Fig. 3.1). GalCerS (61.1 kDa), encoded by human *UGT8* (2906 bp, accession# NM_001128174), is an ER transmembrane protein, and has its catalytic site facing the lumen of the ER (Bosio, Binczek, & Stoffel, 1996; Kapitonov & Yu, 1997; Sprong et al., 1998). GalCerS is structurally related to the UDP-glucuronyltransferases, enzymes critical to type II biotransformation of xenobiotics, and to porphyrin metabolism (Stahl et al., 1994). GalCerS has a limited tissue distribution, and its expression is detected primarily in Schwann cells, oligodendrocytes, kidneys, testes, intestine, and milk (Bouhours & Bouhours, 1979; Vos, Lopes-Cardozo, & Gadella, 1994). In the central nervous system, GalCer and its subsequent metabolites (sulfatide and GM4) are highly enriched in myelin. *UGT8* knockout mice display a tremor phenotype, severe motor weakness due to loss of nerve conduction, male infertility, and premature death (Fujimoto et al., 2000). Interestingly, the neuronal phenotype in mice lacking *UGT8* can be rescued by expression of an oligodendrocyte-specific *UGT8* gene; this suggests that GalCer is extremely important for oligodendrocyte function (Zoller, Bussow, Gieselmann, & Eckhardt, 2005). GalCer is a precursor for sulfatide (3-sulfogalactosylceramide) and GM4, both of which are synthesized in the Golgi. CLN3 (neuronal ceroid-lipofuscinosis 3) protein, which has five transmembrane domains, including a GalCer lipid raft-binding domain, is crucial for lysosomal function, and is responsible for the transport of GalCer from ER/Golgi to lipid rafts in membrane (Persaud-Sawin, McNamara, Rylova, Vandongen, & Boustany, 2004; Rusyn, Mousallem, Persaud-Sawin, Miller, & Boustany, 2008; Fig. 3.2). Sulfatides are synthesized from GalCer via sulfation by GalCer sulfotransferase, which transfers a sulfate residue from an activated sulfate donor, 3′-phosphoadenosine-5′-phosphosulfate (Benjamins, Hadden, & Skoff, 1982; Honke et al., 1997). Sulfatides are enriched in myelin, and many of the known myelination defects may be due to deficiency of sulfatide production. Evidence for this assertion comes from the observation that mice deficient in GalCer sulfotransferase have major defects in myelination, although their pathology is less severe than in an outright *UGT8* knockout mouse (Marcus et al., 2006). Following its biosynthesis, a fraction of GalCer reaches the lumen of the Golgi and is reacted with cytidine-5′-monophospho-*N*-acetylneuraminic acid by the action of sialyltransferase to form *N*-acetylneuraminyl-GalCer (GM4; sialyl-GalCer) (Shanker & Pieringer, 1983; Fig. 3.2).

It has been reported that GalCerS expression levels are strongly associated with histological typing in human oligodendrogliomas and astrocytomas; GalCerS can be used as molecular marker in combination with those of other myelin proteins (MBP, CNP, PLP) to distinguish these tumors (Popko et al., 2002). Transcriptome profiling of prostate cancer cell lines showed that cells with metastatic properties express much higher GalCerS mRNA levels in comparison with nonmetastatic cells (Oudes et al., 2005). GalCerS is one of the six genes whose elevated expression levels are correlated with increasing risk of lung metastasis in breast cancer patients (Landemaine et al., 2008). Dziecedil Giel et al. have further reported that expression of GalCer is higher in breast tumors metastasized to the lung than in matched primary tumors, and that increased amounts of GalCerS enzyme in cancerous tissue are associated with the progression to a more malignant phenotype (Dziecedil Giel et al., 2010). The pronounced expression of GalCerS and GalCer appears only in those breast cancer cell lines observed to form metastases in a nude mice model.

2.3. GCS and GlcCer translocation

Human GlcCer synthase (GCS) (EC2.4.1.80), also known as UDP-glucose: ceramide glucosyltransferase (GlcT-1), transfers glucose from UDP-glucose to ceramide, thereby producing GlcCer (Basu, Kaufman, & Roseman, 1968; Ichikawa, Sakiyama, Suzuki, Hidari, & Hirabayashi, 1996; Shukla & Radin, 1990; Fig. 3.1). GCS (44.9 kDa), encoded by human *UGCG* (1730 bp; accession# BC038711), is a transmembrane protein present on the *cis*-Golgi, and has its catalytic site facing the cytosol, where newly produced GlcCer can be recognized by the four-phosphate adaptor protein 2 (FAPP2; Fig. 3.2; D'Angelo et al., 2007; Ichikawa et al., 1996; Jeckel, Karrenbauer, Burger, van Meer, & Wieland, 1992). Unlike GalCer, GlcCer is a precursor for more than 3000 GSLs, the majority of all GSLs that can be produced by mammalian cells, and GCS is the first rate-limiting enzyme in the synthesis of these GSLs (Merrill, 2011; Radin, 1994). Ceramide substrate is transported by vesicles from the ER, or by CERT (see Section 2.1), and GlcCer synthesized on a cytosolic surface of the Golgi is then translocated across the Golgi membrane for higher GSL synthesis in the *trans*-Golgi (Fig. 3.2; D'Angelo et al., 2007; Halter et al., 2007). Studies with rat liver Golgi membrane have found that transbilayer movement of spin-labeled GlcCer is rapid, saturable, and inhibited by protease treatment, suggesting the membranes contain a GlcCer flippase (Buton et al., 2002). MDR1 has been shown to have flippase activity (De Rosa, Sillence, Ackerley, & Lingwood, 2004; Eckford & Sharom, 2005).

FAPP2 is a cytosolic protein consisting of an N-terminal PH domain recognizing the Golgi marker, PI4P, followed by a central proline-rich region, and a glycolipid transfer protein (GLTP)-like domain toward the C-terminus (Halter et al., 2007). The FAPP2 has transfer activity for GlcCer both *in vitro* and in cells (D'Angelo et al., 2007; Halter et al., 2007). Knocking down FAPP2 by RNAi reduces the conversion of GlcCer to LacCer, and to downstream higher-order GSLs. It has been suggested that FAPP2 functions directly in the formation of apical carriers in the *trans*-Golgi network (TGN). Evidence suggesting that FAPP2 regulates membrane transport from the Golgi by its glycolipid transfer function has also been brought forward. D'Angelo et al. favor a transfer of GlcCer from the *cis*-Golgi to the *trans*-Golgi (D'Angelo et al., 2007), whereas Halter et al. suggest that FAPP2 takes GlcCer from the *trans*-Golgi membrane to the ER (Halter et al., 2007). FAPP2 is a dimeric protein that has the capability to form tubules from membrane sheets (an activity that is dependent on the PI4P binding activity of the PH domain of FAPP2). Cao et al. report that FAPP2 exerts membrane tubulating activity, by binding the small GTPase, Arf1, to induce membrane deformations leading to tubulation at the TGN (Cao et al., 2009). The function of the GLTP domain in FAPP2 remains unclear, but it could be involved in transferring GlcCer to the cellular site where GlcCer can be translocated across the membrane, either at the TGN or in the ER, to function as a precursor luminally for complex GSL synthesis, as suggested previously (D'Angelo et al., 2007; Halter et al., 2007). It is also possible that FAPP2 functions as a sensor for regulating glycolipid levels in the cell. The presence of GlcCer on the cytoplasmic sides of the TGN membrane could serve as a signal for FAPP2 to bind. It would do so by coincidence, binding to PI4P, Arf1, and potentially other factors. This ensemble would contribute to the formation and tubulation of transport carriers, which exit from the TGN to deliver both protein and glycolipid cargos to the cell surface. A feedback mechanism would limit GlcCer translocation from the cytosolic to the luminal leaflet when LacCer and other downstream

GSLs accumulate in the luminal leaflet of the TGN (Cao et al., 2009). Such a function would be in keeping with the proposition that lipid transfer proteins in general could function as biosensors regulating lipid levels in the cell (D'Angelo, Rega, & De Matteis, 2012; Mattjus, 2009).

2.4. The synthesis of complex GSLs and the formation of membrane microdomains

In mammals, GlcCer and GalCer are initial monohexosylceramides for the synthesis of complex GSLs, but almost all the high-order GSLs are produced from GlcCer following additional reactions catalyzed by glycosyltransferases (Fig. 3.2; Merrill, 2011). Human LacCer synthase (LacCerS), also known as UDP-galactose:glucosylceramide β-1→4-galactosyltransferase, is encoded mainly by *β4GalT-V* (3931 bp; accession# AF097159) or *β4GalT-VI*. LacCerS transfers galactose from UDP-galactose to GlcCer to produce LacCer in the Golgi (Fig. 3.2; Takizawa et al., 1999). β4GalT-V is also implicated in the synthesis of N-glycans of cell-surface glycoproteins. Some of the factors reported to regulate LacCerS include growth factors, cytokines, lipids, lipoproteins, and hemodynamic factors, such as fluid shear stress (Chatterjee, Kolmakova, & Rajesh, 2008).

LacCer is the precursor for synthesis of ganglio-series, globo-series, lacto-series, and neolacto-series GSLs (Fig. 3.1; Merrill, 2011; Sandhoff & Kolter, 2003). For the ganglio-series GSLs, the enzyme responsible for the first neutral metabolite, GalNAcβ1→4Galβ1→4Glcβ1Cer (GA2, also called asialo-GM2), is GM2 synthase (β4GalNAcT, β1→4-*N*-acetylgalactosylaminyltransferase) or GM2/GD2 synthase because it additionally converts gangliosides GM3 to GM2, GD3 to GD2, and so forth (Furukawa & Takamiya, 2002). LacCer is sialylated to ganglioside GM3 by ST3Gal-V (SAT-I, CMP-*N*-acetyl-neuraminate:lactosylceramide α2→3-sialyltransferase, also known as GM3 synthase) (Yu, Bieberich, Xia, & Zeng, 2004). Biosynthesis of the Gb3 (Galα1→4Galβ1→4Glcβ1Cer), the initial step for the globo-series GSLs, is catalyzed by Gb3 synthase (α1→4-galactosyltransferase, α1→4GalT) (Kojima et al., 2000). Gb3 is converted to Gb4 by Gb4 synthase (β3GalNAcT). Next in this series is Gb5, synthesized by the action of β3GalT-V. Gb5 is also known as the stage-specific embryonic antigen-3 (SSEA-3), a frequently used stem cell marker (Yu & Yanagisawa, 2006; Zhou, Henion, Jungalwala, Berger, & Hennet, 2000). Biosynthesis of lacto-/neolacto-series GSLs begins with the formation of GlcNAcβ1→3Galβ1→4Glcβ1Cer (also referred to as Lc3, or amino-ceramide trihexoside, amino-CTH) by β-1 → 3-*N*-acetylglucosaminyltransferase (UDP-*N*-acetylglucosamine: β-galactose β1 → 3-*N*-acetylglucosaminyltransferase, amino-CTH synthase) (Togayachi et al., 2001).

GSLs synthesized in the Golgi are clustered with sphingolipids and other membrane components to form GEMs, or lipid rafts, and glycosynapses found within cell membranes (Gupta & Surolia, 2010; Hakomori, 2010; Sonnino et al., 2006). GSLs are inclined toward formation of lipid-ordered phases in membranes, both with and without cholesterol; they are therefore prime players in microdomain formation. Lipid rafts are small, heterogeneous, and dynamic domains enriched with GSLs, sphingolipids, cholesterol, and glycosylphosphatidylinositol-anchored proteins or other proteins (tetraspanins, caveolins, growth factor receptors, integrins) (Hakomori, 2010; Hancock, 2006). These specialized membrane microdomains profoundly influence membrane organization, and are known to compartmentalize cellular processes by serving as organizing centers for the assembly of signaling molecules, influencing membrane fluidity and membrane protein trafficking, and regulating neurotransmission and receptor trafficking (Lingwood & Simons, 2010; Simons & Ikonen, 1997). Lipid rafts modulate membrane transport, signal transduction, and cell–cell interactions, thus modulating cell responses to stress, and playing key roles in cellular development of drug resistance (Lingwood & Simons, 2010; Liu, Gupta, et al., 2010).

3. CERAMIDE GLYCOSYLATION AND CANCER DRUG RESISTANCE

Drug resistance is a characteristic detected in 40–80% of solid tumors, and constitutes a serious barrier to successful treatment of cancer patients. Although resistance is often intrinsic to cancers, as therapy becomes more and more effective, the incidence of acquired resistance are increased (Gottesman, 2002). Due to genetic instability and survival responses to stress, cancer cells develop multiple mechanisms to evade drug toxicity (Dean, Fojo, & Bates, 2005; Liu, 2011; Senchenkov et al., 2001). A growing body of evidence indicates that upregulating ceramide glycosylation catalyzed by GCS is one of these (Liu, Gupta, et al., 2010; Liu et al., 2001; Reynolds, Maurer, & Kolesnick, 2004; Senchenkov et al., 2001), and that this upregulation, through its resultant ability to increase levels of GlcCer, and of globo-series GSLs and others, modulates drug transport, reduces cell apoptosis, favors proliferation, and promotes enrichment of tumors with cells having pronounced drug resistance.

3.1. GCS and drug resistance

The weight of evidence suggests that GCS overexpression is a primary effector of acquired drug resistance of cancer cells. This assertion has been proven primarily by using drug-resistant cell models, as cell lines selected by stepwise exposure to drugs are a cornerstone for investigating molecular mechanisms underlying cellular resistance (Calcagno et al., 2010; Fairchild et al., 1987; Rogan, Hamilton, Young, Klecker, & Ozols, 1984). Lavie et al. reported that GlcCer was accumulated in NCI/ADR-RES ovarian cancer and KB-V-1 cervical cancer cells, indicating a correlation between cellular drug resistance and alterations in GlcCer metabolism (Lavie, Cao, Bursten, Giuliano, & Cabot, 1996). Subsequently, excessive GCS, which is responsible for GlcCer production, has been detected as a cause of drug resistance in more than 14 different cancer cell lines of human breast, ovarian, colon, and cervical cancers, and leukemia (Baran, Bielawski, Gunduz, & Ogretmen, 2011; Chai et al., 2011; Itoh et al., 2003; Liu et al., 2001, 2008; Song, Zang, Zhang, Cao, & Yang, 2012; van Vlerken, Duan, Seiden, & Amiji, 2007; Xie et al., 2008; Zhang et al., 2009). These multidrug-resistant cells, selected by diverse agents (doxorubicin, paclitaxel, vinblastine, imatinib), overexpress GCS at levels two- to fourfold higher than their sensitive counterparts (Table 3.1). Additionally, GCS mRNA levels are significantly increased in drug-resistant HL-60/VCR and MeWo-Etol cells, as compared with their drug-sensitive counterparts (HL-60, MeWo) (Gouaze et al., 2004). Further, silencing GCS expression (using siRNA or antisense oligonucleotide) or inhibition of GCS (using PDMP or Genz-123346) sensitizes these resistant cells, up to 100-fold, to more than 20 anticancer agents of diverse types including doxorubicin (Dox), paclitaxel (Tax), cisplatin (CDDP), vinblastine (Vin), and imatinib (Ima) (Table 3.1). Further, introduction of the GCS gene into drug-sensitive cells confers cellular resistance to doxorubicin, tumor-necrosis factor-α, daunorubicin and C_6-ceramide in human MCF-7 breast cancer, A549 lung cancer, and HL-60 leukemia cells (Itoh et al., 2003; Liu, Han, Giuliano, & Cabot, 1999; Liu, Han, Giuliano, Hansen, & Cabot, 2000; Liu, Han, Giuliano, Ichikawa, et al., 1999; Ogretmen, Kraveka, et al., 2001). It has also been noted that GCS could not protect such cell lines as mouse melanoma GM95 and human T-lymphoblastoid Jurkat J16 cells against Dox or CD95 (Tepper, Diks, van Blitterswijk, & Borst, 2000; Veldman et al., 2003), indicating that drug resistance is an outcome of sophisticated cellular processes, and is cell-type or cancer dependent. Enhancing cellular ceramide via GCS inhibition (Liu et al., 2001; Patwardhan et al.,

Table 3.1 GCS and acquired drug resistance in cancer cells

Cell lines	Acquired resistance (fold)	GCS/GlcCer level (fold)	Other markers	Increased sensitivity (fold)
Breast cancer				
MCF-7	1	1		
MCF-7 P500	55 (Dox)	3	MDR1	35 (oligo)
MCF-7/ Dox	22 (Dox)	3	MDR1, CS	20 (oligo)
MCF-7/ ADM	192 (Dox)	2	MDR1	8 (shRNA)
Ovarian cancer				
OVCA8	1	1		
NCI/ ADR-RES	33 (Dox)	4	MDR1, mp53	36 (oligo)
A2780	1	1		
A2780-AD	194 (Dox)	3	MDR1	4 (oligo)
SKOV3	1			
SKOV3TR	>100 (Tax)		MDR1	100 (nano-Cer)
Colon cancer				
SW620	1	1		
SW620AD	121 (Dox)	4	MDR1	62 (oligo)
HCT-8	1	1		
HCT-8/ VCR	4.5 (cisplatin)	1.33	MDR1	2 (shRNA)
HCT-15	n/a	n/a	MDR1	>30 (Genz-123346)
$HT29G^{+}$	1			
$HT29^{col}$	25 (Col)	2	MRP1	
Cervical cancer				
KB-3-1	1	1		
KB-A1	121 (Dox)	4	MDR1	76 (siRNA)
KB-V1	213 (Vin)	2	MDR1	17 (Genz-123346)

Continued

Table 3.1 GCS and acquired drug resistance in cancer cells—cont'd

Cell lines	Acquired resistance (fold)	GCS/GlcCer level (fold)	Other markers	Increased sensitivity (fold)
Leukemia				
K562	1	1		
K562/IMA	19 (imatinib)	2	n/a	4 (PDMP)
K562/A02	50 (Dox)	4	MDR1, Bcl-2	4.5 (PDMP)
HL-60	1	1		
HL-60/ADR	16 (Dox)	2–3	MDR1	
Glioblastoma				
T98G	1	1		
TMZ-R	6.1 (TMZ)	2	MGMT	2 (PPMP/NB-DGJ)
PCL-R	15 (Tax)	2	MDR1	2 (PPMP/NB-DGJ)

Cited from published reports (Baran et al., 2011; Chai et al., 2011; Gouaze et al., 2004; Itoh et al., 2003; Liu et al., 2001, 2008; Song et al., 2012; van Vlerken et al., 2007; Xie et al., 2008; Zhang et al., 2009). GCS levels have been reported in as mRNA or protein levels. Acquired resistance was evaluated in cell viability assay or apoptosis markers. Resistance to a particular anticancer drug or the identity of an agent used to inhibit GCS is indicated in parentheses. Dox, doxorubicin; Vin, vinblastine; Col, colchicine; Tax, paclitaxel; TMZ, temozolomide; MGMT, O^6-methylguanine-DNA methyltransferase; shRNA, small hairpin RNA; siRNA, small interfering RNA; PDMP, PPMP, NB-DGJ, see main text.

2009; Weiss, Hettmer, Smith, & Ladisch, 2003), or delivery of ceramide by using polymeric nanoparticles (van Vlerken et al., 2007), overcomes multidrug resistance, providing further evidence that GCS-mediated abolishment of ceramide-induced apoptosis is one mechanism underlying acquired drug resistance in these resistant cells.

GCS is expressed at diverse levels in normal tissues and cells; however, it has been found that *increased expression of GCS* (rather than any particular absolute expression level) is correlated to the progression of breast cancer, urinary cancer, ovarian cancer, and leukemia (Liu, Patwardhan, Xie, et al., 2011). Itoh et al. reported that decreases of cellular ceramide concentrations serve as an indicator of chemoresistance in leukemia (Itoh et al., 2003). Expression levels of GCS, as well as of sphingomyelin synthase (which converts ceramide to sphingomyelin), were found to be twofold higher in chemoresistant leukemia ($n=14$) than in chemosensitive leukemia ($n=9$). This finding has been corroborated by another study, wherein it was found that GCS mRNA was elevated by twofold in leukemia patients who displayed nonresponse to chemotherapy ($n=30$), as compared to the complete response group ($n=35$) (Xie et al., 2008). In the same study, overexpression of GCS was also accompanied with increased MDR1 and Bcl-2 expression levels in leukemia that was unresponsive to chemotherapy. Additionally, retrospective analyses of microarray data in clinical trials indicate that elevated GCS expression is associated with ER-positive breast cancer and with poor response to paclitaxel in breast cancer patients (Juul et al., 2010; Ruckhaberle et al., 2009). Other groups reported that, based on microarray data, upregulated GCS expression is a genetic signature for the progression and metastasis of renal cell cancer (Jones et al., 2005), and is associated with lymphatic metastasis in penile carcinoma (Kroon et al., 2008). In bladder cancers, GCS overexpression is associated with lymph node metastasis; the overall 5-year survival and disease-free survival rates are reduced to 75% (45.1 vs. 60.3 months and 27.3 vs. 36.2 months, respectively) in patients with tumors exhibiting high levels of GCS (Sun, Zhang, Zhang, et al., 2012). The levels of GCS mRNA or protein are elevated by fourfold in approximately 80% of metastatic breast tumors in state III ($n=5/7$) or lymph node-positive ($n=7/8$) and ER-positive tumors ($n=7/9$) (Liu, Patwardhan, Xie, et al., 2011).

Again reflecting cancer-type dependence and/or not-yet-understood levels of complexity, however, Tomioka et al. recently showed by whole-genome array screening that 9q31 (includes *UGCG*, which encodes GCS) is one of the four loci deleted in primary gastric cancers ($n=56$). If any one of these four loci was deleted, the prognosis of the patient was significantly worse ($P=0.0019$) (Tomioka et al., 2010).

3.2. GSLs and ABC transporters

Our understanding of how cancer cells acquire drug resistance during the course of chemotherapy remains incomplete. It is understood, though, that in addition to inducing cell death, ceramide generated in cells exposed to anticancer drugs actively participates in modulating gene expression (Patwardhan & Liu, 2011). Possibly, certain molecular species of ceramides modulate the expression of genes contributing to drug resistance of cancers.

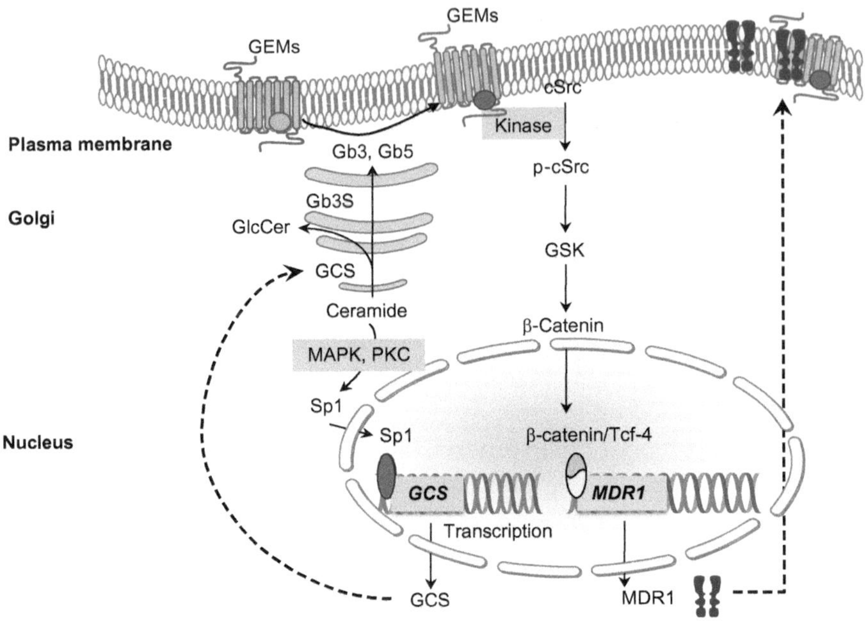

Figure 3.3 Cells exposed to drugs upregulate drug-resistant genes via actions of ceramide and GSLs. Ceramide, generated by *de novo* synthesis in response to stresses, transactivates GCS expression, possibly by way of the MAPK or PKC cascades and the Sp1 transcription factor; globo-series GSLs (Gb3, Gb5) interact with lipids/protein on GEMs and activate the cSrc–GSK cascade, consequently increasing the recruitment of β-catenin/Tcf-4 to upregulate MDR1. MAPK, mitogen-activated protein kinase; GEMs, GSL-enriched microdomains; GSK, glycogen synthase kinase-3. (For color version of this figure, the reader is referred to the online version of this chapter.)

We at least know that, although doxorubicin (0.5 μM, a concentration below its IC_{50}) elicits increases in cellular ceramide levels in MCF-7 cells, it also leads activation of the GCS promoter via the transcription factor Sp1 to amplify GCS expression in a positive feedback response (Fig. 3.3; Liu et al., 2008; Uchida et al., 2004). Further, disruption of ceramide synthesis by the CerS inhibitor fumonisin B1 (FB1) prevents the transactivation of GCS expression by doxorubicin; exogenous C_6-ceramide (5 μM, a concentration below its IC_{50}) or SMase can mimic doxorubicin's ability to induce GCS expression (Liu et al., 2008). Either stepwise selection under treatment with cell-permeable ceramide (C_8-ceramide), or direct GlcCer treatments, can lead cancer cells to develop resistance to anticancer drugs (Gouaze-Andersson et al., 2007). Alterations of sphingolipids and GSLs in cancer cells exposed to less-toxic concentrations of anticancer drugs may also favor cancer cell survival.

GCS overexpression is frequently correlated with MDR1 levels in drug-resistant cells and tumors. Our recent work has shown that globo-series GSLs, synthesized downstream of ceramide glycosylation by GCS, upregulate MDR1 expression via activation of cSrc signaling and TCF4/β-catenin recruitment on the MDR1 gene promoter (Fig. 3.3; Liu, Gupta, et al., 2010). Increases or decreases of various GSLs will alter lipid–lipid or lipid–protein interactions and affect the action of protein kinases (cSrc kinases) in GEMs of the plasma membrane. As noted above, doxorubicin treatment increases ceramide generation via the *de novo* synthesis pathway, but also transactivates GCS expression via the Sp1 transcription factor; the consequent increase in concentrations of certain globo-series GSLs (Gb3, Gb5) in turn activates cSrc kinases, increases nuclear β-catenin by diminishing its degradation after phosphorylation, and transactivates MDR1 gene expression. In this way, sphingolipids (ceramide, globo-series GSLs) upregulate GCS and MDR1 expressions in response to anticancer drugs, and thereby confer cell resistance by preventing ceramide-induced apoptosis and MDR1-mediated drug efflux (Fig. 3.3; Liu, Gupta, et al., 2010; Liu et al., 2008). In accord with these findings, we also found that MBO-asGCS, which silences GCS in the nanomolar range, reverses cell resistance by suppression of MDR1 in drug-resistant cells and tumors (Liu, Gupta, et al., 2010; Patwardhan et al., 2009).

3.3. GSLs and CSCs

Sphingolipids play crucial roles in determining stem cell fate, including self-renewal, proliferation, and differentiation; accordingly, sphingolipid modulators can potentially be developed as therapeutic agents to eliminate CSCs (Bieberich, 2004; Hakomori, 2008; Yu, Suzuki, & Yanagisawa, 2010). The cell-surface GSLs globopentaosylceramide (Gb5), and monosialyl-Gb5 (MSGb5) are also known as SSEA-3 and SSEA-4, in recognition of their usefulness as markers on human ES cells (Bieberich, 2004; Pera & Tam, 2010; Stewart et al., 2006). Several studies have shown that alterations of GSLs, as compared with non-stem cells, are associated CSCs. As an example, SSEA-3 and Globo H are markers for a subpopulation of CSCs in breast cancer patients (Chang et al., 2008). Breast CSCs with CD55 are highly resistant to ceramide- or serum deprivation-induced apoptosis; and exposure to ceramide (nano-liposomal C_6-ceramide, 3 μM) prevents premature human ES cell differentiation and maintains pluripotent stem cell populations *in vitro* (Salli et al., 2009; Xu et al., 2007). Addition of serum (10% fetal bovine serum, 24 h) or inhibition of STAT3 phosphorylation with WP1193 (5 μM, 24 h) significantly decreases the numbers of human GCS11 glioblastoma CSCs ($CD133^{+}$), accompanied with decreased GCS expression (He, H'Ng, Leong, Hutmacher, & Melendez, 2010). On the other hand, deoxycholate promotes the survival of mouse breast CSCs ($CD44^{+}$/Flk-1^{+}) by reducing ceramide levels (Krishnamurthy, Wang, Rokhfeld, & Bieberich, 2008). CSCs isolated from drug-resistant breast cancer cells (MCF-7/Dox) highly express GCS and other stem cell markers (Calcagno et al., 2010); silencing of GCS by using MBO-asGCS (100 nM for 6 days) significantly decreases the CSC numbers in human MCF-7/Dox cells (Gupta, Zhang, & Liu, 2011). Our unpublished works indicate that extent of ceramide glycosylation by GCS, and the levels of globo-series GSLs, are significantly higher in breast CSCs than in non-CSCs or in $ABCG2^{+}$ bone marrow stem cells. Collectively, these observations suggest that ceramide glycosylation by GCS plays one of the key mechanistic roles in maintaining CSCs in their dedifferentiated state.

3.4. Ceramide glycosylation and the expression of p53 mutants

p53 is a key tumor suppressor in preventing tumorigenesis and cancer progression; however, mutant p53, detected in over 50% cancers, promotes tumor progression and resistance to therapies, and such mutants have become the most common prognostic indicator for both tumor recurrence and cancer death (Brosh & Rotter, 2009; Hollstein, Sidransky, Vogelstein, & Harris, 1991; Olivier et al., 2006). The majority of p53 mutants in human cancers abrogate their transactivation effects to p53-responsive genes, such as $p21^{Waf1/Cip1}$, *PUMA*, and *BAX*. Moreover, the mutants confer a dominant-negative activity over the remaining wild-type allele, and also gain new oncogenic properties (Brosh & Rotter, 2009; Bullock & Fersht, 2001). Restoration of wild-type p53 function has succeeded in bringing about regression of tumors, and in fact, this represents a promising approach for treating cancers (Brosh & Rotter, 2009; Ventura et al., 2007).

Disrupting ceramide glycosylation is a new approach to target mutant p53 for cancer treatments (Liu, 2011). Various strategies have been developed to reconstitute p53 functions in suppressing tumor progression and improving treatments. These mainly include replacing wild-type p53 by gene therapy, augmenting wild-type p53 by inhibiting MDM2/MDMX-mediated p53 degradation, and converting mutant p53 into a wild-type mimic by altering its protein conformation (Brosh & Rotter, 2009; Chen, Wang, & El-Deiry,

2010; Wiman, 2010). A recent advance, in addition to these approaches, was the discovery that suppression of GCS could restore wild-type p53 expression and induce p53-dependent apoptosis in p53-mutant cancer cells (Liu, Patwardhan, Bhinge, et al., 2011). Human NCI/ADR-RES and OVCAR-8 cancer cells dominantly express p53 mutants, with a deletion of seven and six amino acids (encoded by exon 5), respectively, within the DNA-binding domain of p53 (Liu, Patwardhan, Bhinge, et al., 2011; Ogretmen & Safa, 1997). Silencing of GCS with MBO-asGCS substantially increases the levels of phosphorylated p53, and of the products of p53-responsive genes, including p21, BAX, and PUMA, consequently directing "mutant p53" cells to apoptosis. Conversely, inhibition of ceramide synthase with FB1 prevents p53 restoration induced by MBO-asGCS, while addition of exogenous C_6-ceramide reactivates p53 function in p53-mutant cells (Liu, Patwardhan, Bhinge, et al., 2011). Further, assessment of hnRNA shows that the wild-type p53 hnRNA is transcribed in both wild-type and mutant p53 cell lines, although the latter only expresses mutant mRNA and protein, suggesting that silencing GCS may restore p53 at the level of posttranscriptional processing. This study, as proof of concept, indicates that dysfunctional regulation of transcription and associated posttranscriptional processes is an important cause of p53 mutants in cancer cells.

4. TARGETING CERAMIDE GLYCOSYLATION TO REVERSE DRUG RESISTANCE

Based on the body of work reviewed above, GCS, a rate-liming enzyme of ceramide glycosylation, constitutes a new therapeutic target for reversing drug resistance. Blocking ceramide glycosylation by the inhibition of GCS in drug-resistant cancers can result in increased levels of ceramide and decreased levels of GSLs, thus sensitizing cancer cells to chemotherapy. Compounds drawn from several groups of small molecules have been used to inhibit GCS activity, and gene-based agents, including oligonucleotides and siRNA, have shown promise in specifically targeting cancer cells that overexpress GCS (Fig. 3.4).

4.1. GCS inhibitors

Ceramide glucosylation can be blocked by GCS inhibitors (Abe, Wild, Lee, & Shayman, 2001). Two classes of GCS inhibitors have been described, including analogs of D-*threo*-1-phenyl-2-decanoylamino-3-morpholino-propanol (PDMP) and a group of imino sugars (Shayman, Lee, Abe, & Shu, 2000).

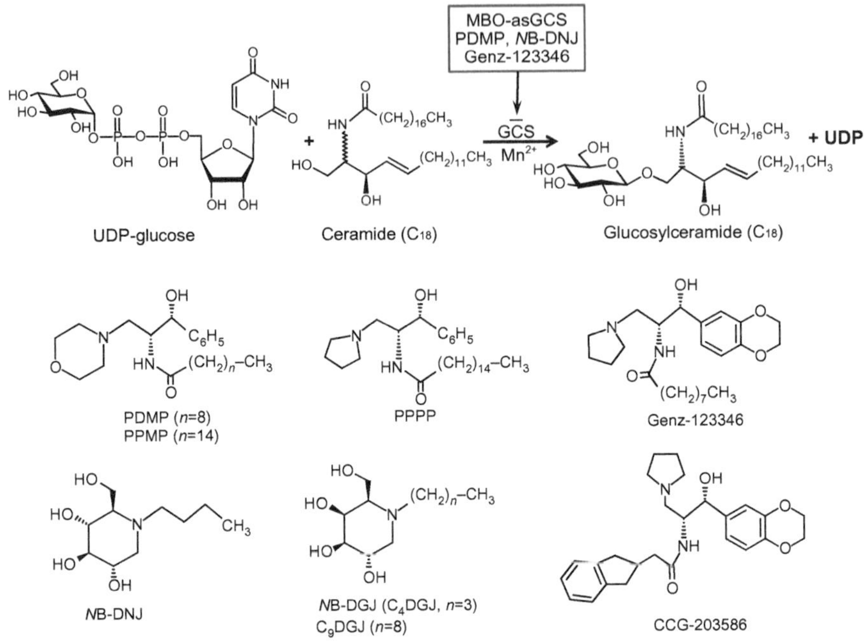

Figure 3.4 Targeting ceramide glycosylation by GCS.

PDMP is the parent compound of "P" drugs, including PPMP, PPPP, and Genz-123346 (Fig. 3.4; Chai et al., 2011; McEachern et al., 2007). The sensitizing effects of PDMP, PPMP, and PPPP have been observed in several types of cancer cells (di Bartolomeo & Spinedi, 2001; Lavie et al., 1997; Nicholson, Quinn, Kellett, & Warr, 1999; Sietsma et al., 2000) and in tumor-bearing mice (Huang et al., 2011). Unfortunately, pharmacological interpretation with PDMP is confounded because, in addition to inhibiting GCS, PDMP reportedly can inhibit other enzymes involved in GSL metabolism (Lee, Abe, & Shayman, 1999; Liour & Yu, 2002), and can also affect calcium homeostasis and membrane fluidity (Kok et al., 1998). PPPP is at least 10-fold more potent than PDMP or PPMP, and pharmacological interpretation is cleaner because PPPP can much more selectively inhibit human GCS (Hillig, Warnecke, & Heinz, 2005; Lee et al., 1999; Liour & Yu, 2002). Genz-123346 is another novel analog of PDMP also having improved selectivity for inhibiting GCS (Chai et al., 2011; McEachern et al., 2007). Genz-123346 can sensitize drug-resistant KB-V-1 cells to vinblastine, and its chemosensitizing activity appears to be mediated primarily through suppression of MDR1 function (Chai et al., 2011). A new compound, CCG-203586, has been found to inhibit GCS at low-nanomolar concentrations, with little to no direct inhibition of MDR1 (Larsen et al., 2012). This compound may be a useful tool for clarifying the mechanistic association between GCS and MDR1 (see above) in modulating drug resistance of cancer cells.

The imino sugar, *N*-butyl-deoxynojirimycin (*N*B-DNJ, also known as miglustat, Zavesca®, OGT 918) is a reversible and competitive inhibitor of ceramide, but not UDP-glucose, in the reaction catalyzed by GCS (Fig. 3.4; Butters, Dwek, & Platt, 2005). *N*B-DNJ inhibits GCS with modest potency ($K_i \approx 5\ \mu M$) and inhibits α-galactosidase only at much higher concentrations ($K_i > 100\ \mu M$) (Butters et al., 2005). OGT 2378, C_4DGJ (*N*-butyl-deoxygalactonojirimycin or lucerastat, also known as OGB-1) and C_9DGJ (*N*-nonyl-deoxygalactonojirimycin, also known as OGB-2) are more selective GCS inhibitors in this class (Norris-Cervetto, Callaghan, Platt, Dwek, & Butters, 2004; Weiss et al., 2003). Genz-529468 is a new and more potent imino sugar-based inhibitor of GCS (Nietupski et al., 2012). *N*B-DNJ inhibition of GCS and, consequently, ganglioside synthesis delayed murine melanoma onset (Guerrera & Ladisch, 2003; Weiss et al., 2003). C_9DGJ can sensitize glioblastoma TMZ-R and PCL-R cells to paclitaxel or temozolomide (Giussani et al., 2012), and colon carcinoma HCT-15 cells to vinblastine (Chai et al., 2011). C_4DGJ and C_9DGJ sensitize chronic lymphatic leukemia patients that overexpress MDR1 (Gerrard, Butters, Ganeshaguru, & Mehta, 2009). It was also observed, however, that C_4DGJ and C_9DGJ could not reverse drug resistance in NCI/ADR–RES ovarian cancer cells and MES-SA/DX-5 uterine sarcoma cells (Norris-Cervetto et al., 2004). The applications of this group of inhibitors are limited by their micromolar-level inhibitory activity, and a generally low specificity against GCS (Larsen et al., 2012), although newer analogs may solve this problem (Richards et al., 2012).

The medical safety and therapeutic efficacy of GCS inhibitors, including *N*B-DNJ and "P" drugs, have been tested in clinical trials of Fabry disease, HIV, diabetes, and type 1 Gaucher disease (Butters et al., 2005; Larsen et al., 2012); however, their chemosensitizing effects on tumors have not yet been tested in clinical trials. Several therapeutic agents, including tamoxifen, mifepristone, cyclosporine A, and arsenic trioxide, are also known to nonspecifically inhibit GCS and to sensitize cancer cells to anticancer agents (Dbaibo et al., 2007; Senchenkov et al., 2001).

4.2. Agents silencing GCS

Silencing of the GCS gene, which directly proves GCS to be a cause of drug resistance, also offers a specific approach for sensitizing tumors that poorly respond to chemotherapy due to GCS overexpression (Fig. 3.4). Antisense gene transfection (full length) demonstrates the concept that suppression of GCS can sensitize drug-resistant cells, such as NCI/ADR-RES, to anthracyclines, *Vinca* alkaloids, taxanes, and other anticancer drugs (Liu et al., 2001, 2000). Oligonucleotides (20-mer) that target the open reading frame of GCS (ORF 18-37), as in a phosphorothioate DNA (Liu et al., 2004, 2008), or a 2′-*O*-methyl RNA with phosphorothioate DNA (mixed-backbone oligonucleotide, MBO-asGCS) (Liu, Gupta, et al., 2010; Patwardhan et al., 2009), sensitize drug cytotoxicity in resistant human NCI/ADR–RES, A2780AD, KB-A1, SW620/AD, and murine EMT6/AR1 cancer cells. RNA interference by treatment with siRNA duplex (Gouaze et al., 2005) or vector-mediated transfection of short hairpin DNA (pSUPER-GCSshRNA) (Liu, Xie, et al., 2010; Sun, Zhang, Gao, et al., 2010; Sun et al., 2006; Zhang et al., 2009, 2011) sensitizes drug-resistant human breast cancer cells and leukemia cells; however, these siRNA agents have not yet been tested *in vivo*. MBO-asGCS directly administered in animal models has relatively higher uptake by tumors than other tissues, and little nonspecific toxicity to normal tissues (Patwardhan et al., 2009).

5. PROSPECTIVE

Instead of pinpointing a single target, current studies in this field provide compelling evidence that ceramide glycosylation is highly associated with cancer drug resistance, particularly acquired resistance. As a rate-limiting enzyme in ceramide glycosylation, GCS is essential for many cellular processes in normal physiological as well as pathological conditions. Sensitizing cancer cells, and not normal tissues, to chemotherapy requires that we gain further understanding as to how ceramide glycosylation by GCS specifically alters cancerous processes, including drug transport, induced apoptosis, mutant expression of tumor suppressors, and CSC formation. Identifying the roles of particular species of ceramide and GSLs (notably GlcCer, LacCer, Gb3, Gb5) in these processes, and their effects on lipid–lipid and lipid–protein interactions should help us discover approaches to reverse drug resistance, while sparing normal cells and tissues of adverse effects. Elucidating details of GCS protein structure and its conformational changes when interacting with inhibitors will assist in designing and synthesizing more potent and more selective GCS inhibitors. Finally, clinical studies designed specifically to assess the association of GCS and the responses of cancer patients to chemotherapy would pave the way for subsequent trials investigating the seemingly great promise of GCS inhibitors or modulating agents for reversing drug resistance in cancer patients.

ACKNOWLEDGMENTS

This work was supported by grants from the Mizutani Foundation for Glycoscience, the National Center for Research Resources (5P20RR016456-11), and the National Institute of General Medical Sciences (8 P20 GM103424-11) from the National Institutes of Health.

REFERENCES

Abe, A., Wild, S. R., Lee, W. L., & Shayman, J. A. (2001). Agents for the treatment of glycosphingolipid storage disorders. *Current Drug Metabolism, 2*, 331–338.

Baran, Y., Bielawski, J., Gunduz, U., & Ogretmen, B. (2011). Targeting glucosylceramide synthase sensitizes imatinib-resistant chronic myeloid leukemia cells via endogenous ceramide accumulation. *Journal of Cancer Research and Clinical Oncology, 137*, 1535–1544.

Basu, S., Kaufman, B., & Roseman, S. (1968). Enzymatic synthesis of ceramide-glucose and ceramide-lactose by glycosyltransferases from embryonic chicken brain. *The Journal of Biological Chemistry, 243*, 5802–5804.

Benjamins, J. A., Hadden, T., & Skoff, R. P. (1982). Cerebroside sulfotransferase in Golgi-enriched fractions from rat brain. *Journal of Neurochemistry, 38*, 233–241.

Bieberich, E. (2004). Integration of glycosphingolipid metabolism and cell-fate decisions in cancer and stem cells: Review and hypothesis. *Glycoconjugate Journal, 21*, 315–327.

Bosio, A., Binczek, E., & Stoffel, W. (1996). Molecular cloning and characterization of the mouse CGT gene encoding UDP-galactose ceramide-galactosyltransferase (cerebroside synthetase). *Genomics, 35*, 223–226.

Bouhours, J. F., & Bouhours, D. (1979). Galactosylceramide is the major cerebroside of human milk fat globule membrane. *Biochemical and Biophysical Research Communications, 88*, 1217–1222.

Brosh, R., & Rotter, V. (2009). When mutants gain new powers: News from the mutant p53 field. *Nature Reviews. Cancer, 9*, 701–713.

Bullock, A. N., & Fersht, A. R. (2001). Rescuing the function of mutant p53. *Nature Reviews. Cancer, 1*, 68–76.

Buton, X., Herve, P., Kubelt, J., Tannert, A., Burger, K. N., Fellmann, P., et al. (2002). Transbilayer movement of monohexosylsphingolipids in endoplasmic reticulum and Golgi membranes. *Biochemistry, 41*, 13106–13115.

Butters, T. D., Dwek, R. A., & Platt, F. M. (2005). Imino sugar inhibitors for treating the lysosomal glycosphingolipidoses. *Glycobiology, 15*, 43R–52R.

Calcagno, A. M., Salcido, C. D., Gillet, J. P., Wu, C. P., Fostel, J. M., Mumau, M. D., et al. (2010). Prolonged drug selection of breast cancer cells and enrichment of cancer stem cell characteristics. *Journal of the National Cancer Institute, 102*, 1637–1652.

Cao, X., Coskun, U., Rossle, M., Buschhorn, S. B., Grzybek, M., Dafforn, T. R., et al. (2009). Golgi protein FAPP2 tubulates membranes. *Proceedings of the National Academy of Sciences of the United States of America, 106*, 21121–21125.

Chai, L., McLaren, R. P., Byrne, A., Chuang, W. L., Huang, Y., Dufault, M. R., et al. (2011). The chemosensitizing activity of inhibitors of glucosylceramide synthase is mediated primarily through modulation of P-gp function. *International Journal of Oncology, 38*, 701–711.

Chang, W. W., Lee, C. H., Lee, P., Lin, J., Hsu, C. W., Hung, J. T., et al. (2008). Expression of Globo H and SSEA3 in breast cancer stem cells and the involvement of fucosyl transferases 1 and 2 in Globo H synthesis. *Proceedings of the National Academy of Sciences of the United States of America, 105*, 11667–11672.

Charruyer, A., Bell, S. M., Kawano, M., Douangpanya, S., Yen, T. Y., Macher, B. A., et al. (2008). Decreased ceramide transport protein (CERT) function alters sphingomyelin production following UVB irradiation. *The Journal of Biological Chemistry, 283*, 16682–16692.

Chatterjee, S., Kolmakova, A., & Rajesh, M. (2008). Regulation of lactosylceramide synthase (glucosylceramide beta1–>4 galactosyltransferase); implication as a drug target. *Current Drug Targets*, *9*, 272–281.

Chen, F., Wang, W., & El-Deiry, W. S. (2010). Current strategies to target p53 in cancer. *Biochemical Pharmacology*, *80*, 724–730.

Chipuk, J. E., McStay, G. P., Bharti, A., Kuwana, T., Clarke, C. J., Siskind, L. J., et al. (2012). Sphingolipid metabolism cooperates with BAK and BAX to promote the mitochondrial pathway of apoptosis. *Cell*, *148*, 988–1000.

D'Angelo, G., Polishchuk, E., Di Tullio, G., Santoro, M., Di Campli, A., Godi, A., et al. (2007). Glycosphingolipid synthesis requires FAPP2 transfer of glucosylceramide. *Nature*, *449*, 62–67.

D'Angelo, G., Rega, L. R., & De Matteis, M. A. (2012). Connecting vesicular transport with lipid synthesis: FAPP2. *Biochimica et Biophysica Acta*, *1821*, 1089–1095.

Dbaibo, G. S., Kfoury, Y., Darwiche, N., Panjarian, S., Kozhaya, L., Nasr, R., et al. (2007). Arsenic trioxide induces accumulation of cytotoxic levels of ceramide in acute promyelocytic leukemia and adult T-cell leukemia/lymphoma cells through de novo ceramide synthesis and inhibition of glucosylceramide synthase activity. *Haematologica*, *92*, 753–762.

Dean, M., Fojo, T., & Bates, S. (2005). Tumour stem cells and drug resistance. *Nature Reviews. Cancer*, *5*, 275–284.

De Rosa, M. F., Sillence, D., Ackerley, C., & Lingwood, C. (2004). Role of multiple drug resistance protein 1 in neutral but not acidic glycosphingolipid biosynthesis. *The Journal of Biological Chemistry*, *279*, 7867–7876.

Derre, I., Swiss, R., & Agaisse, H. (2011). The lipid transfer protein CERT interacts with the Chlamydia inclusion protein IncD and participates to ER-Chlamydia inclusion membrane contact sites. *PLoS Pathogens*, 7, e1002092.

di Bartolomeo, S., & Spinedi, A. (2001). Differential chemosensitizing effect of two glucosylceramide synthase inhibitors in hepatoma cells. *Biochemical and Biophysical Research Communications*, *288*, 269–274.

Dziecedil Giel, P., Owczarek, T., Plazuk, E., Gomulkiewicz, A., Majchrzak, M., Podhorska-Okolow, M., et al. (2010). Ceramide galactosyltransferase (UGT8) is a molecular marker of breast cancer malignancy and lung metastases. *British Journal of Cancer*, *103*, 524–531.

Eckford, P. D., & Sharom, F. J. (2005). The reconstituted P-glycoprotein multidrug transporter is a flippase for glucosylceramide and other simple glycosphingolipids. *The Biochemical Journal*, *389*, 517–526.

Fairchild, C. R., Ivy, S. P., Kao-Shan, C. S., Whang-Peng, J., Rosen, N., Israel, M. A., et al. (1987). Isolation of amplified and overexpressed DNA sequences from adriamycin-resistant human breast cancer cells. *Cancer Research*, *47*, 5141–5148.

Fujimoto, H., Tadano-Aritomi, K., Tokumasu, A., Ito, K., Hikita, T., Suzuki, K., et al. (2000). Requirement of seminolipid in spermatogenesis revealed by UDP-galactose: Ceramide galactosyltransferase-deficient mice. *The Journal of Biological Chemistry*, *275*, 22623–22626.

Furukawa, K., & Takamiya, K. (2002). Beta1,4-N-acetylgalactosaminyltransferase—GM2/GD2 synthase: A key enzyme to control the synthesis of brain-enriched complex gangliosides. *Biochimica et Biophysica Acta*, *1573*, 356–362.

Futerman, A. H., & Pagano, R. E. (1991). Determination of the intracellular sites and topology of glucosylceramide synthesis in rat liver. *The Biochemical Journal*, *280*(Pt 2), 295–302.

Gault, C. R., Obeid, L. M., & Hannun, Y. A. (2010). An overview of sphingolipid metabolism: From synthesis to breakdown. *Advances in Experimental Medicine and Biology*, *688*, 1–23.

Gerrard, G., Butters, T. D., Ganeshaguru, K., & Mehta, A. B. (2009). Glucosylceramide synthase inhibitors sensitise CLL cells to cytotoxic agents without reversing P-gp functional activity. *European Journal of Pharmacology*, *609*, 34–39.

Giussani, P., Bassi, R., Anelli, V., Brioschi, L., De Zen, F., Riccitelli, E., et al. (2012). Glucosylceramide synthase protects glioblastoma cells against autophagic and apoptotic death induced by temozolomide and Paclitaxel. *Cancer Investigation, 30*, 27–37.
Gottesman, M. M. (2002). Mechanisms of cancer drug resistance. *Annual Review of Medicine, 53*, 615–627.
Gouaze, V., Liu, Y. Y., Prickett, C. S., Yu, J. Y., Giuliano, A. E., & Cabot, M. C. (2005). Glucosylceramide synthase blockade down-regulates P-glycoprotein and resensitizes multidrug-resistant breast cancer cells to anticancer drugs. *Cancer Research, 65*, 3861–3867.
Gouaze, V., Yu, J. Y., Bleicher, R. J., Han, T. Y., Liu, Y. Y., Wang, H., et al. (2004). Overexpression of glucosylceramide synthase and P-glycoprotein in cancer cells selected for resistance to natural product chemotherapy. *Molecular Cancer Therapeutics, 3*, 633–639.
Gouaze-Andersson, V., Yu, J. Y., Kreitenberg, A. J., Bielawska, A., Giuliano, A. E., & Cabot, M. C. (2007). Ceramide and glucosylceramide upregulate expression of the multidrug resistance gene MDR1 in cancer cells. *Biochimica et Biophysica Acta, 1771*, 1407–1417.
Guerrera, M., & Ladisch, S. (2003). N-butyldeoxynojirimycin inhibits murine melanoma cell ganglioside metabolism and delays tumor onset. *Cancer Letters, 201*, 31–40.
Gupta, G., & Surolia, A. (2010). Glycosphingolipids in microdomain formation and their spatial organization. *FEBS Letters, 584*, 1634–1641.
Gupta, V., Zhang, Q. J., & Liu, Y. Y. (2011). Evaluation of anticancer agents using flow cytometry analysis of cancer stem cells. *Methods in Molecular Biology, 716*, 179–191.
Hakomori, S. I. (2008). Structure and function of glycosphingolipids and sphingolipids: Recollections and future trends. *Biochimica et Biophysica Acta, 1780*, 325–346.
Hakomori, S. I. (2010). Glycosynaptic microdomains controlling tumor cell phenotype through alteration of cell growth, adhesion, and motility. *FEBS Letters, 584*, 1901–1906.
Halter, D., Neumann, S., van Dijk, S. M., Wolthoorn, J., de Maziere, A. M., Vieira, O. V., et al. (2007). Pre- and post-Golgi translocation of glucosylceramide in glycosphingolipid synthesis. *The Journal of Cell Biology, 179*, 101–115.
Hanada, K., Kumagai, K., Tomishige, N., & Yamaji, T. (2009). CERT-mediated trafficking of ceramide. *Biochimica et Biophysica Acta, 1791*, 684–691.
Hanada, K., Kumagai, K., Yasuda, S., Miura, Y., Kawano, M., Fukasawa, M., et al. (2003). Molecular machinery for non-vesicular trafficking of ceramide. *Nature, 426*, 803–809.
Hancock, J. F. (2006). Lipid rafts: Contentious only from simplistic standpoints. *Nature Reviews. Molecular Cell Biology, 7*, 456–462.
Hannun, Y. A., & Obeid, L. M. (2008). Principles of bioactive lipid signalling: Lessons from sphingolipids. *Nature Reviews. Molecular Cell Biology, 9*, 139–150.
He, X., H'Ng, S. C., Leong, D. T., Hutmacher, D. W., & Melendez, A. J. (2010). Sphingosine-1-phosphate mediates proliferation maintaining the multipotency of human adult bone marrow and adipose tissue-derived stem cells. *Journal of Molecular Cell Biology, 2*, 199–208.
Hillig, I., Warnecke, D., & Heinz, E. (2005). An inhibitor of glucosylceramide synthase inhibits the human enzyme, but not enzymes from other organisms. *Bioscience, Biotechnology, and Biochemistry, 69*, 1782–1785.
Hollstein, M., Sidransky, D., Vogelstein, B., & Harris, C. C. (1991). p53 mutations in human cancers. *Science, 253*, 49–53.
Honke, K., Tsuda, M., Hirahara, Y., Ishii, A., Makita, A., & Wada, Y. (1997). Molecular cloning and expression of cDNA encoding human 3′-phosphoadenylylsulfate: galactosylceramide 3′-sulfotransferase. *The Journal of Biological Chemistry, 272*, 4864–4868.
Huang, W. C., Tsai, C. C., Chen, C. L., Chen, T. Y., Chen, Y. P., Lin, Y. S., et al. (2011). Glucosylceramide synthase inhibitor PDMP sensitizes chronic myeloid leukemia T315I mutant to Bcr-Abl inhibitor and cooperatively induces glycogen synthase kinase-3-regulated apoptosis. *The FASEB Journal, 25*, 3661–3673.

Ichikawa, S., Sakiyama, H., Suzuki, G., Hidari, K. I., & Hirabayashi, Y. (1996). Expression cloning of a cDNA for human ceramide glucosyltransferase that catalyzes the first glycosylation step of glycosphingolipid synthesis. *Proceedings of the National Academy of Sciences of the United States of America, 93*, 4638–4643.

Itoh, M., Kitano, T., Watanabe, M., Kondo, T., Yabu, T., Taguchi, Y., et al. (2003). Possible role of ceramide as an indicator of chemoresistance: Decrease of the ceramide content via activation of glucosylceramide synthase and sphingomyelin synthase in chemoresistant leukemia. *Clinical Cancer Research, 9*, 415–423.

Jeckel, D., Karrenbauer, A., Burger, K. N., van Meer, G., & Wieland, F. (1992). Glucosylceramide is synthesized at the cytosolic surface of various Golgi subfractions. *The Journal of Cell Biology, 117*, 259–267.

Jones, J., Otu, H., Spentzos, D., Kolia, S., Inan, M., Beecken, W. D., et al. (2005). Gene signatures of progression and metastasis in renal cell cancer. *Clinical Cancer Research, 11*, 5730–5739.

Juul, N., Szallasi, Z., Eklund, A. C., Li, Q., Burrell, R. A., Gerlinger, M., et al. (2010). Assessment of an RNA interference screen-derived mitotic and ceramide pathway metagene as a predictor of response to neoadjuvant paclitaxel for primary triple-negative breast cancer: A retrospective analysis of five clinical trials. *The Lancet Oncology, 11*, 358–365.

Kapitonov, D., & Yu, R. K. (1997). Cloning, characterization, and expression of human ceramide galactosyltransferase cDNA. *Biochemical and Biophysical Research Communications, 232*, 449–453.

Kojima, Y., Fukumoto, S., Furukawa, K., Okajima, T., Wiels, J., Yokoyama, K., et al. (2000). Molecular cloning of globotriaosylceramide/CD77 synthase, a glycosyltransferase that initiates the synthesis of globo series glycosphingolipids. *The Journal of Biological Chemistry, 275*, 15152–15156.

Kok, J. W., Babia, T., Filipeanu, C. M., Nelemans, A., Egea, G., & Hoekstra, D. (1998). PDMP blocks brefeldin A-induced retrograde membrane transport from golgi to ER: Evidence for involvement of calcium homeostasis and dissociation from sphingolipid metabolism. *The Journal of Cell Biology, 142*, 25–38.

Kolesnick, R. N., Haimovitz-Friedman, A., & Fuks, Z. (1994). The sphingomyelin signal transduction pathway mediates apoptosis for tumor necrosis factor, Fas, and ionizing radiation. *Biochemistry and Cell Biology, 72*, 471–474.

Krishnamurthy, K., Wang, G., Rokhfeld, D., & Bieberich, E. (2008). Deoxycholate promotes survival of breast cancer cells by reducing the level of pro-apoptotic ceramide. *Breast Cancer Research, 10*, R106.

Kroon, B. K., Leijte, J. A., van Boven, H., Wessels, L. F., Velds, A., Horenblas, S., et al. (2008). Microarray gene-expression profiling to predict lymph node metastasis in penile carcinoma. *BJU International, 102*, 510–515.

Kudo, N., Kumagai, K., Matsubara, R., Kobayashi, S., Hanada, K., Wakatsuki, S., et al. (2010). Crystal structures of the CERT START domain with inhibitors provide insights into the mechanism of ceramide transfer. *Journal of Molecular Biology, 396*, 245–251.

Kudo, N., Kumagai, K., Tomishige, N., Yamaji, T., Wakatsuki, S., Nishijima, M., et al. (2008). Structural basis for specific lipid recognition by CERT responsible for nonvesicular trafficking of ceramide. *Proceedings of the National Academy of Sciences of the United States of America, 105*, 488–493.

Kumagai, K., Kawano, M., Shinkai-Ouchi, F., Nishijima, M., & Hanada, K. (2007). Interorganelle trafficking of ceramide is regulated by phosphorylation-dependent cooperativity between the PH and START domains of CERT. *The Journal of Biological Chemistry, 282*, 17758–17766.

Landemaine, T., Jackson, A., Bellahcene, A., Rucci, N., Sin, S., Abad, B. M., et al. (2008). A six-gene signature predicting breast cancer lung metastasis. *Cancer Research, 68*, 6092–6099.

Larsen, S. D., Wilson, M. W., Abe, A., Shu, L., George, C. H., Kirchhoff, P., et al. (2012). Property-based design of a glucosylceramide synthase inhibitor that reduces glucosylceramide in the brain. *Journal of Lipid Research, 53*, 282–291.

Lavie, Y., Cao, H., Bursten, S. L., Giuliano, A. E., & Cabot, M. C. (1996). Accumulation of glucosylceramides in multidrug-resistant cancer cells. *The Journal of Biological Chemistry, 271*, 19530–19536.

Lavie, Y., Cao, H., Volner, A., Lucci, A., Han, T. Y., Geffen, V., et al. (1997). Agents that reverse multidrug resistance, tamoxifen, verapamil, and cyclosporin A, block glycosphingolipid metabolism by inhibiting ceramide glycosylation in human cancer cells. *The Journal of Biological Chemistry, 272*, 1682–1687.

Lee, L., Abe, A., & Shayman, J. A. (1999). Improved inhibitors of glucosylceramide synthase. *The Journal of Biological Chemistry, 274*, 14662–14669.

Lingwood, D., & Simons, K. (2010). Lipid rafts as a membrane-organizing principle. *Science, 327*, 46–50.

Liour, S. S., & Yu, R. K. (2002). Differential effects of three inhibitors of glycosphingolipid biosynthesis on neuronal differentiation of embryonal carcinoma stem cells. *Neurochemical Research, 27*, 1507–1512.

Liu, Y. Y. (2011). Resuscitating wild-type p53 expression by disrupting ceramide glycosylation: A novel approach to target mutant p53 tumors. *Cancer Research, 71*, 6295–6299.

Liu, Y. Y., Gupta, V., Patwardhan, G. A., Bhinge, K., Zhao, Y., Bao, J., et al. (2010). Glucosylceramide synthase upregulates MDR1 expression in the regulation of cancer drug resistance through cSrc and beta-catenin signaling. *Molecular Cancer, 9*, 145.

Liu, Y. Y., Han, T. Y., Giuliano, A. E., & Cabot, M. C. (1999). Expression of glucosylceramide synthase, converting ceramide to glucosylceramide, confers adriamycin resistance in human breast cancer cells. *The Journal of Biological Chemistry, 274*, 1140–1146.

Liu, Y. Y., Han, T. Y., Giuliano, A. E., & Cabot, M. C. (2001). Ceramide glycosylation potentiates cellular multidrug resistance. *The FASEB Journal, 15*, 719–730.

Liu, Y. Y., Han, T. Y., Giuliano, A. E., Hansen, N., & Cabot, M. C. (2000). Uncoupling ceramide glycosylation by transfection of glucosylceramide synthase antisense reverses adriamycin resistance. *The Journal of Biological Chemistry, 275*, 7138–7143.

Liu, Y. Y., Han, T. Y., Giuliano, A. E., Ichikawa, S., Hirabayashi, Y., & Cabot, M. C. (1999). Glycosylation of ceramide potentiates cellular resistance to tumor necrosis factor-alpha-induced apoptosis. *Experimental Cell Research, 252*, 464–470.

Liu, Y. Y., Han, T. Y., Yu, J. Y., Bitterman, A., Le, A., Giuliano, A. E., et al. (2004). Oligonucleotides blocking glucosylceramide synthase expression selectively reverse drug resistance in cancer cells. *Journal of Lipid Research, 45*, 933–940.

Liu, Y. Y., Patwardhan, G. A., Bhinge, K., Gupta, V., Gu, X., & Jazwinski, S. M. (2011). Suppression of glucosylceramide synthase restores p53-dependent apoptosis in mutant p53 cancer cells. *Cancer Research, 71*, 2276–2285.

Liu, Y. Y., Patwardhan, G. A., Xie, P., Gu, X., Giuliano, A. E., & Cabot, M. C. (2011). Glucosylceramide synthase, a factor in modulating drug resistance, is overexpressed in metastatic breast carcinoma. *International Journal of Oncology, 39*, 425–431.

Liu, Y., Xie, K. M., Yang, G. Q., Bai, X. M., Shi, Y. P., Mu, H. J., et al. (2010). GCS induces multidrug resistance by regulating apoptosis-related genes in K562/AO2 cell line. *Cancer Chemotherapy and Pharmacology, 66*, 433–439.

Liu, Y. Y., Yu, J. Y., Yin, D., Patwardhan, G. A., Gupta, V., Hirabayashi, Y., et al. (2008). A role for ceramide in driving cancer cell resistance to doxorubicin. *The FASEB Journal, 22*, 2541–2551.

Marcus, J., Honigbaum, S., Shroff, S., Honke, K., Rosenbluth, J., & Dupree, J. L. (2006). Sulfatide is essential for the maintenance of CNS myelin and axon structure. *Glia, 53*, 372–381.

Mattjus, P. (2009). Glycolipid transfer proteins and membrane interaction. *Biochimica et Biophysica Acta, 1788*, 267–272.
McEachern, K. A., Fung, J., Komarnitsky, S., Siegel, C. S., Chuang, W. L., Hutto, E., et al. (2007). A specific and potent inhibitor of glucosylceramide synthase for substrate inhibition therapy of Gaucher disease. *Molecular Genetics and Metabolism, 91*, 259–267.
Merrill, Jr. A. H., (2011). Sphingolipid and glycosphingolipid metabolic pathways in the era of sphingolipidomics. *Chemical Reviews, 111*, 6387–6422.
Modrak, D. E., Gold, D. V., & Goldenberg, D. M. (2006). Sphingolipid targets in cancer therapy. *Molecular Cancer Therapeutics, 5*, 200–208.
Nicholson, K. M., Quinn, D. M., Kellett, G. L., & Warr, J. R. (1999). Preferential killing of multidrug-resistant KB cells by inhibitors of glucosylceramide synthase. *British Journal of Cancer, 81*, 423–430.
Nietupski, J. B., Pacheco, J. J., Chuang, W. L., Maratea, K., Li, L., Foley, J., et al. (2012). Iminosugar-based inhibitors of glucosylceramide synthase prolong survival but paradoxically increase brain glucosylceramide levels in Niemann-Pick C mice. *Molecular Genetics and Metabolism, 105*, 621–628.
Norris-Cervetto, E., Callaghan, R., Platt, F. M., Dwek, R. A., & Butters, T. D. (2004). Inhibition of glucosylceramide synthase does not reverse drug resistance in cancer cells. *The Journal of Biological Chemistry, 279*, 40412–40418.
Ogretmen, B. (2006). Sphingolipids in cancer: Regulation of pathogenesis and therapy. *FEBS Letters, 580*, 5467–5476.
Ogretmen, B., & Hannun, Y. A. (2001). Updates on functions of ceramide in chemotherapy-induced cell death and in multidrug resistance. *Drug Resistance Updates, 4*, 368–377.
Ogretmen, B., & Hannun, Y. A. (2004). Biologically active sphingolipids in cancer pathogenesis and treatment. *Nature Reviews. Cancer, 4*, 604–616.
Ogretmen, B., Kraveka, J. M., Schady, D., Usta, J., Hannun, Y. A., & Obeid, L. M. (2001). Molecular mechanisms of ceramide-mediated telomerase inhibition in the A549 human lung adenocarcinoma cell line. *The Journal of Biological Chemistry, 276*, 32506–32514.
Ogretmen, B., McCauley, M. D., & Safa, A. R. (1998). Molecular mechanisms of loss of beta 2-microglobulin expression in drug-resistant breast cancer sublines and its involvement in drug resistance. *Biochemistry, 37*, 11679–11691.
Ogretmen, B., Pettus, B. J., Rossi, M. J., Wood, R., Usta, J., Szulc, Z., et al. (2002). Biochemical mechanisms of the generation of endogenous long chain ceramide in response to exogenous short chain ceramide in the A549 human lung adenocarcinoma cell line. Role for endogenous ceramide in mediating the action of exogenous ceramide. *The Journal of Biological Chemistry, 277*, 12960–12969.
Ogretmen, B., & Safa, A. R. (1996). Down-regulation of apoptosis-related bcl-2 but not bcl-xL or bax proteins in multidrug-resistant MCF-7/Adr human breast cancer cells. *International Journal of Cancer, 67*, 608–614.
Ogretmen, B., & Safa, A. R. (1997). Expression of the mutated p53 tumor suppressor protein and its molecular and biochemical characterization in multidrug resistant MCF-7/Adr human breast cancer cells. *Oncogene, 14*, 499–506.
Ogretmen, B., Schady, D., Usta, J., Wood, R., Kraveka, J. M., Luberto, C., et al. (2001). Role of ceramide in mediating the inhibition of telomerase activity in A549 human lung adenocarcinoma cells. *The Journal of Biological Chemistry, 276*, 24901–24910.
Olivier, M., Langerod, A., Carrieri, P., Bergh, J., Klaar, S., Eyfjord, J., et al. (2006). The clinical value of somatic TP53 gene mutations in 1,794 patients with breast cancer. *Clinical Cancer Research, 12*, 1157–1167.
Oudes, A. J., Roach, J. C., Walashek, L. S., Eichner, L. J., True, L. D., Vessella, R. L., et al. (2005). Application of Affymetrix array and Massively Parallel Signature Sequencing for identification of genes involved in prostate cancer progression. *BMC Cancer, 5*, 86.

Patwardhan, G. A., & Liu, Y. Y. (2011). Sphingolipids and expression regulation of genes in cancer. *Progress in Lipid Research, 50*, 104–114.
Patwardhan, G. A., Zhang, Q. J., Yin, D., Gupta, V., Bao, J., Senkal, C. E., et al. (2009). A new mixed-backbone oligonucleotide against glucosylceramide synthase sensitizes multidrug-resistant tumors to apoptosis. *PLoS One, 4*, e6938.
Pera, M. F., & Tam, P. P. (2010). Extrinsic regulation of pluripotent stem cells. *Nature, 465*, 713–720.
Persaud-Sawin, D. A., McNamara, J. O., 2nd, Rylova, S., Vandongen, A., & Boustany, R. M. (2004). A galactosylceramide binding domain is involved in trafficking of CLN3 from Golgi to rafts via recycling endosomes. *Pediatric Research, 56*, 449–463.
Pewzner-Jung, Y., Ben-Dor, S., & Futerman, A. H. (2006). When do Lasses (longevity assurance genes) become CerS (ceramide synthases)?: Insights into the regulation of ceramide synthesis. *The Journal of Biological Chemistry, 281*, 25001–25005.
Popko, B., Pearl, D. K., Walker, D. M., Comas, T. C., Baerwald, K. D., Burger, P. C., et al. (2002). Molecular markers that identify human astrocytomas and oligodendrogliomas. *Journal of Neuropathology and Experimental Neurology, 61*, 329–338.
Pyne, N. J., & Pyne, S. (2010). Sphingosine 1-phosphate and cancer. *Nature Reviews. Cancer, 10*, 489–503.
Rabionet, M., van der Spoel, A. C., Chuang, C. C., von Tumpling-Radosta, B., Litjens, M., Bouwmeester, D., et al. (2008). Male germ cells require polyenoic sphingolipids with complex glycosylation for completion of meiosis: A link to ceramide synthase-3. *The Journal of Biological Chemistry, 283*, 13357–13369.
Radin, N. S. (1994). Glucosylceramide in the nervous system—A mini-review. *Neurochemical Research, 19*, 533–540.
Reynolds, C. P., Maurer, B. J., & Kolesnick, R. N. (2004). Ceramide synthesis and metabolism as a target for cancer therapy. *Cancer Letters, 206*, 169–180.
Richards, S., Larson, C. J., Koltun, E. S., Hanel, A., Chan, V., Nachtigall, J., et al. (2012). Discovery and characterization of an inhibitor of glucosylceramide synthase. *Journal of Medicinal Chemistry, 55*, 4322–4335.
Rogan, A. M., Hamilton, T. C., Young, R. C., Klecker, R. W., Jr., & Ozols, R. F. (1984). Reversal of adriamycin resistance by verapamil in human ovarian cancer. *Science, 224*, 994–996.
Ruckhaberle, E., Karn, T., Hanker, L., Gatje, R., Metzler, D., Holtrich, U., et al. (2009). Prognostic relevance of glucosylceramide synthase (GCS) expression in breast cancer. *Journal of Cancer Research and Clinical Oncology, 135*, 81–90.
Rusyn, E., Mousallem, T., Persaud-Sawin, D. A., Miller, S., & Boustany, R. M. (2008). CLN3p impacts galactosylceramide transport, raft morphology, and lipid content. *Pediatric Research, 63*, 625–631.
Salli, U., Fox, T. E., Carkaci-Salli, N., Sharma, A., Robertson, G. P., Kester, M., et al. (2009). Propagation of undifferentiated human embryonic stem cells with nanoliposomal ceramide. *Stem Cells and Development, 18*, 55–65.
Sandhoff, K., & Kolter, T. (2003). Biosynthesis and degradation of mammalian glycosphingolipids. *Philosophical Transactions of the Royal Society of London. Series B, Biological Sciences, 358*, 847–861.
Schulte, S., & Stoffel, W. (1993). Ceramide UDPgalactosyltransferase from myelinating rat brain: Purification, cloning, and expression. *Proceedings of the National Academy of Sciences of the United States of America, 90*, 10265–10269.
Senchenkov, A., Litvak, D. A., & Cabot, M. C. (2001). Targeting ceramide metabolism—A strategy for overcoming drug resistance. *Journal of the National Cancer Institute, 93*, 347–357.
Shanker, G., & Pieringer, R. A. (1983). Effect of thyroid hormone on the synthesis of sialosyl galactosylceramide (GM4) in myelinogenic cultures of cells dissociated from embryonic mouse brain. *Brain Research, 282*, 169–174.

Shayman, J. A., Lee, L., Abe, A., & Shu, L. (2000). Inhibitors of glucosylceramide synthase. *Methods in Enzymology, 311*, 373–387.

Shukla, G. S., & Radin, N. S. (1990). Glucosyceramide synthase of mouse kidney: Further characterization with an improved assay method. *Archives of Biochemistry and Biophysics, 283*, 372–378.

Sietsma, H., Veldman, R. J., Kolk, D., Ausema, B., Nijhof, W., Kamps, W., et al. (2000). 1-phenyl-2-decanoylamino-3-morpholino-1-propanol chemosensitizes neuroblastoma cells for taxol and vincristine. *Clinical Cancer Research, 6*, 942–948.

Simons, K., & Ikonen, E. (1997). Functional rafts in cell membranes. *Nature, 387*, 569–572.

Song, M., Zang, W., Zhang, B., Cao, J., & Yang, G. (2012). GCS overexpression is associated with multidrug resistance of human HCT-8 colon cancer cells. *Journal of Experimental & Clinical Cancer Research, 31*, 23.

Sonnino, S., Prinetti, A., Mauri, L., Chigorno, V., & Tettamanti, G. (2006). Dynamic and structural properties of sphingolipids as driving forces for the formation of membrane domains. *Chemical Reviews, 106*, 2111–2125.

Sprong, H., Kruithof, B., Leijendekker, R., Slot, J. W., van Meer, G., & van der Sluijs, P. (1998). UDP-galactose:ceramide galactosyltransferase is a class I integral membrane protein of the endoplasmic reticulum. *The Journal of Biological Chemistry, 273*, 25880–25888.

Stahl, N., Jurevics, H., Morell, P., Suzuki, K., & Popko, B. (1994). Isolation, characterization, and expression of cDNA clones that encode rat UDP-galactose: Ceramide galactosyltransferase. *Journal of Neuroscience Research, 38*, 234–242.

Stewart, M. H., Bosse, M., Chadwick, K., Menendez, P., Bendall, S. C., & Bhatia, M. (2006). Clonal isolation of hESCs reveals heterogeneity within the pluripotent stem cell compartment. *Nature Methods, 3*, 807–815.

Sun, Y., Zhang, T., Gao, P., Meng, B., Gao, Y., Wang, X., et al. (2010). Targeting glucosylceramide synthase downregulates expression of the multidrug resistance gene MDR1 and sensitizes breast carcinoma cells to anticancer drugs. *Breast Cancer Research and Treatment, 121*, 591–599.

Sun, C. C., Zhang, Z., Zhang, S. Y., Li, J., Li, Z. L., & Kong, C. Z. (2012). Up-regulation of glucosylceramide synthase in urinary bladder neoplasms. *Urologic Oncology, 30*, 444–449.

Sun, Y. L., Zhou, G. Y., Li, K. N., Gao, P., Zhang, Q. H., Zhen, J. H., et al. (2006). Suppression of glucosylceramide synthase by RNA interference reverses multidrug resistance in human breast cancer cells. *Neoplasma, 53*, 1–8.

Takizawa, M., Nomura, T., Wakisaka, E., Yoshizuka, N., Aoki, J., Arai, H., et al. (1999). cDNA cloning and expression of human lactosylceramide synthase. *Biochimica et Biophysica Acta, 1438*, 301–304.

Tepper, A. D., Diks, S. H., van Blitterswijk, W. J., & Borst, J. (2000). Glucosylceramide synthase does not attenuate the ceramide pool accumulating during apoptosis induced by CD95 or anti-cancer regimens. *The Journal of Biological Chemistry, 275*, 34810–34817.

Togayachi, A., Akashima, T., Ookubo, R., Kudo, T., Nishihara, S., Iwasaki, H., et al. (2001). Molecular cloning and characterization of UDP-GlcNAc:lactosylceramide beta 1,3-N-acetylglucosaminyltransferase (beta 3Gn-T5), an essential enzyme for the expression of HNK-1 and Lewis X epitopes on glycolipids. *The Journal of Biological Chemistry, 276*, 22032–22040.

Tomioka, N., Morita, K., Kobayashi, N., Tada, M., Itoh, T., Saitoh, S., et al. (2010). Array comparative genomic hybridization analysis revealed four genomic prognostic biomarkers for primary gastric cancers. *Cancer Genetics and Cytogenetics, 201*, 6–14.

Uchida, Y., Itoh, M., Taguchi, Y., Yamaoka, S., Umehara, H., Ichikawa, S., et al. (2004). Ceramide reduction and transcriptional up-regulation of glucosylceramide synthase through doxorubicin-activated Sp1 in drug-resistant HL-60/ADR cells. *Cancer Research, 64*, 6271–6279.

van Vlerken, L. E., Duan, Z., Seiden, M. V., & Amiji, M. M. (2007). Modulation of intracellular ceramide using polymeric nanoparticles to overcome multidrug resistance in cancer. *Cancer Research, 67*, 4843–4850.
Veldman, R. J., Mita, A., Cuvillier, O., Garcia, V., Klappe, K., Medin, J. A., et al. (2003). The absence of functional glucosylceramide synthase does not sensitize melanoma cells for anticancer drugs. *The FASEB Journal, 17*, 1144–1146.
Ventura, A., Kirsch, D. G., McLaughlin, M. E., Tuveson, D. A., Grimm, J., Lintault, L., et al. (2007). Restoration of p53 function leads to tumour regression in vivo. *Nature, 445*, 661–665.
von Haefen, C., Wieder, T., Gillissen, B., Starck, L., Graupner, V., Dorken, B., et al. (2002). Ceramide induces mitochondrial activation and apoptosis via a Bax-dependent pathway in human carcinoma cells. *Oncogene, 21*, 4009–4019.
Vos, J. P., Lopes-Cardozo, M., & Gadella, B. M. (1994). Metabolic and functional aspects of sulfogalactolipids. *Biochimica et Biophysica Acta, 1211*, 125–149.
Watson, P., & Stephens, D. J. (2005). ER-to-Golgi transport: Form and formation of vesicular and tubular carriers. *Biochimica et Biophysica Acta, 1744*, 304–315.
Weiss, M., Hettmer, S., Smith, P., & Ladisch, S. (2003). Inhibition of melanoma tumor growth by a novel inhibitor of glucosylceramide synthase. *Cancer Research, 63*, 3654–3658.
Wiman, K. G. (2010). Pharmacological reactivation of mutant p53: From protein structure to the cancer patient. *Oncogene, 29*, 4245–4252.
Xie, P., Shen, Y. F., Shi, Y. P., Ge, S. M., Gu, Z. H., Wang, J., et al. (2008). Overexpression of glucosylceramide synthase in associated with multidrug resistance of leukemia cells. *Leukemia Research, 32*, 475–480.
Xu, J. X., Morii, E., Liu, Y., Nakamichi, N., Ikeda, J., Kimura, H., et al. (2007). High tolerance to apoptotic stimuli induced by serum depletion and ceramide in side-population cells: High expression of CD55 as a novel character for side-population. *Experimental Cell Research, 313*, 1877–1885.
Yamaji, T., Kumagai, K., Tomishige, N., & Hanada, K. (2008). Two sphingolipid transfer proteins, CERT and FAPP2: Their roles in sphingolipid metabolism. *IUBMB Life, 60*, 511–518.
Yu, R. K., Bieberich, E., Xia, T., & Zeng, G. (2004). Regulation of ganglioside biosynthesis in the nervous system. *Journal of Lipid Research, 45*, 783–793.
Yu, R. K., Nakatani, Y., & Yanagisawa, M. (2009). The role of glycosphingolipid metabolism in the developing brain. *Journal of Lipid Research, 50*(Suppl), S440–S445.
Yu, R. K., Suzuki, Y., & Yanagisawa, M. (2010). Membrane glycolipids in stem cells. *FEBS Letters, 584*, 1694–1699.
Yu, R. K., & Yanagisawa, M. (2006). Glycobiology of neural stem cells. *CNS & Neurological Disorders Drug Targets, 5*, 415–423.
Zhang, X., Li, J., Qiu, Z., Gao, P., Wu, X., & Zhou, G. (2009). Co-suppression of MDR1 (multidrug resistance 1) and GCS (glucosylceramide synthase) restores sensitivity to multidrug resistance breast cancer cells by RNA interference (RNAi). *Cancer Biology & Therapy, 8*, 1117–1121.
Zhang, Y. Y., Xie, K. M., Yang, G. Q., Mu, H. J., Yin, Y., Zhang, B., et al. (2011). The effect of glucosylceramide synthase on P-glycoprotein function in K562/AO2 leukemia drug-resistance cell line. *International Journal of Hematology, 93*, 361–367.
Zhou, D., Henion, T. R., Jungalwala, F. B., Berger, E. G., & Hennet, T. (2000). The beta 1,3-galactosyltransferase beta 3GalT-V is a stage-specific embryonic antigen-3 (SSEA-3) synthase. *The Journal of Biological Chemistry, 275*, 22631–22634.
Zoller, I., Bussow, H., Gieselmann, V., & Eckhardt, M. (2005). Oligodendrocyte-specific ceramide galactosyltransferase (CGT) expression phenotypically rescues CGT-deficient mice and demonstrates that CGT activity does not limit brain galactosylceramide level. *Glia, 52*, 190–198.

CHAPTER FOUR

Use of Acid Sphingomyelinase for Cancer Therapy

Radoslav Savić, Edward H. Schuchman[1]
Department of Genetics and Genomic Sciences, Mount Sinai School of Medicine, New York, USA
[1]Corresponding author: e-mail address: edward.schuchman@mssm.edu

Contents

Abstract

Acid sphingomyelinase (ASM) is a lipid hydrolase that cleaves the sphingolipid, sphingomyelin, into ceramide. Mutations in the ASM gene (*SMPD1*) result in the rare lysosomal storage disorder, Niemann–Pick disease (NPD). In addition to its role in NPD, over the past two decades, the importance of sphingolipids, and ASM in particular, in normal physiology and the pathophysiology of numerous common diseases also has become known. For example, altered sphingolipid metabolism occurs in many cancers, generally reducing the levels of the pro-apoptotic lipid, ceramide, and/or elevating the levels of the proliferative lipid, sphingosine-1-phosphate (S1P). These changes likely contribute to the tumorigenicity and/or metastatic capacity of the cancer. In addition, many cancer therapies induce ceramide-mediated death, and cancer cells have evolved novel mechanisms to overcome this effect. In the present review, we discuss sphingolipid metabolism in cancer, and specifically the potential for pharmacological modulation using ASM. Of note, recombinant human ASM (rhASM) has been produced for human use and is being evaluated as a treatment for NPD. Thus, its use for cancer therapy could be rapidly evaluated in the clinic after appropriate animal model studies

Advances in Cancer Research, Volume 117
ISSN 0065-230X
http://dx.doi.org/10.1016/B978-0-12-394274-6.00004-2

have been completed. As this enzyme was initially studied in the context of NPD, we start with a brief overview of the history of ASM and NPD, followed by a discussion of the role of ASM in cancer biology, and then summarize emerging preclinical efficacy studies using rhASM as an adjunct in the treatment of solid tumors.

1. INTRODUCTION

1.1. Niemann–Pick disease, discovery of ASM, and creation of ASM knockout mice

In the early 1900s, a German physician, Albert Niemann, published the first description of a previously unknown disease. The publication was a case study of an infant presenting with hepatosplenomegaly, nervous system impairment, and death at 18 months of age (Niemann, 1914). A fellow German, Ludwig Pick, in the 1920s, studied the pathological abnormalities in patients with this new disease and recognized that it was a lipid storage disorder. He also identified adult individuals with similar lipid abnormalities (Pick, 1927). In honor of Drs. Niemann and Pick, an eponym "Niemann–Pick disease" (NPD) was used henceforth to describe such patients. Studies soon followed showing that the majority of the accumulated lipid in NPD was sphingomyelin (SM) (Klenk, 1934) and greater than 10-fold increases in the amounts of SM were observed in the organs of deceased NPD patients. SM was most prominently accumulated in the liver and the spleen (Menten & Welton, 1946). The main clinical distinction among different types of NPD includes the very early onset of severe neurodegeneration (Type A NPD), compared to other patients in which the central nervous system (CNS) is spared and who present primarily with reticuloendothelial system (RES) disease (Type B NPD) (Crocker, 1961; Crocker & Farber, 1958). There is also another form of NPD in which SM accumulates, and it was originally thought that these patients had the same metabolic abnormality as those with Types A and B NPD. However, it is now known that this form of the disease (Type C NPD) is caused by mutations in two distinct genes involved in cholesterol metabolism (Carstea et al., 1997; Naureckiene et al., 2000). Type C NPD individuals suffer from progressive neurological degeneration, and despite the distinct genetic abnormalities, share many of the same lipid and other pathological abnormalities with Type A and B NPD.

In 1966, the loss of activity of a specific enzyme, acid sphingomyelinase (ASM) (E.C. #3.1.4.12; SM phosphodiesterase), was identified in NPD patients (Brady, Kanfer, Mock, & Fredrickson, 1966; Kanfer, Young, Shapiro, & Brady, 1966). Data from humans showed that patients with Type A NPD had almost no residual enzymatic activity and died in infancy, whereas patients with primarily RES abnormalities (Type B) had measurable residual activity of the enzyme and were largely free from neurological defects, surviving into childhood or early adulthood (Schneider & Kennedy, 1967). Cells from patients with Type C NPD also had low ASM activity, but this could be recovered by incubation in media lacking cholesterol.

The cDNA and the gene-encoding ASM (*SMPD1*) were cloned in 1991 and 1992 (Schuchman, Levran, Pereira, & Desnick, 1992; Schuchman, Suchi, Takahashi, Sandhoff, & Desnick, 1991), and the gene location was mapped to the short arm of chromosome 11 (11p15.1) (Schuchman et al., 1992). A catalytically active, full-length, ASM cDNA and two shorter inactive splice variants were identified. The human gene was small (~6 kb), and the coding region was divided into six exons (da Veiga Pereira, Desnick, Adler, Disteche, & Schuchman, 1991; Schuchman et al., 1991). The full-length cDNA predicted a full-length protein of 629 amino acids with a characteristic signal peptide sequence. Six N-glycosylation sites were identified, and it was subsequently found that five of the sites were utilized (Ferlinz et al., 1997). Several polymorphic amino acid changes also were identified, along with a polymorphic size variant within the signal peptide region (Wan & Schuchman, 1995).

Because of its acidic pH optimum and the lipid storage pattern within NPD lysosomes, ASM was considered a lysosomal protein. Subsequent biosynthetic and trafficking studies in cultured skin fibroblasts confirmed that the majority of the enzyme was targeted to lysosomes via mannose-6-phosphate (M6P) residues on the N-linked oligosaccharides (as in other lysosomal proteins), although secreted and other forms also were identified (Hurwitz, Ferlinz, Vielhaber, Moczall, & Sandhoff, 1994). For example, in the late 1980s, a zinc-activated, secreted form of ASM was found in serum, and it was subsequently shown that this enzyme was derived from the *SMPD1* gene (Schissel, Schuchman, Williams, & Tabas, 1996; Spence, Byers, Palmer, & Cook, 1989). It should also be noted that there are several other (nonacidic) sphingomyelinases that have been identified in mammals, and at least four additional genes that encode these proteins (see Table 4.1). These enzymes and genes will not be discussed further in this review.

An important step toward the development of enzyme replacement therapy for NPD was the overexpression of the full-length human ASM cDNA in Chinese Hamster ovary (CHO) cells, which led to the selective secretion of the recombinant human ASM (rhASM) into the culture media, facilitating its

Table 4.1 Different types of mammalian sphingomyelinases

Sphingomyelinase type	Primary location	*In vitro* optimal pH	Other locations	Responsible gene
Acid	Lysosome	~4.5	Plasma membrane, serum	*SMPD1*
Neutral	Plasma membrane	~7	Golgi, ER, mitochondria	*SMPD2-4*
Alkaline	Intestine	~9	Bile	*ENPP7*

Bernardo et al. (2000), Duan et al. (1996), and Goni and Alonso (2002).

large-scale purification (He et al., 1999). A further step was the construction of ASM knockout mice (ASMKO), which proved to be an invaluable tool not only for the studies of NPD pathology and treatment (Horinouchi et al., 1995; Otterbach & Stoffel, 1995) but also sphingolipid signaling (Schuchman, 2010).

1.2. Production of recombinant ASM and clinical trials of ERT in NPD

The ASMKO mice were constructed by two groups in 1995 (Horinouchi et al., 1995; Otterbach & Stoffel, 1995), although only one has been studied in detail (Horinouchi et al., 1995). These animals had no residual ASM activity in the liver, heart, spleen, kidney, brain, or lung and exhibited massive accumulation of SM in the liver and brain (Horinouchi et al., 1995; Otterbach & Stoffel, 1995). To a much lesser degree, there was also accumulation of ceramide in the ASMKO tissues (presumably due to SM breakdown by other sphingomyelinases), and most recently, high-level accumulation of sphingosine has been found as well (unpublished observation, E.H.S). The ASMKO mice have proven to be an excellent model in which to test the potential efficacy of hrASM treatment for alleviating the pathology of NPD. Such an enzyme replacement treatment approach is available today for 6 of the more than 50 lysosomal storage disorders, and under development for several others (Hemsley & Hopwood, 2011; Lachmann, 2011).

As noted above, the development of rhASM for the treatment of NPD started by using CHO cells to overexpress the full-length human enzyme and purify it from the culture media (He et al., 1999). The human recombinant enzyme (rhASM) had a peak activity at pH 4.5, which was Zn^{2+} dependent, and successfully degraded SM in fibroblasts from Type A NPD patients in culture (He et al., 1999). Following an intravenous injection of rhASM, the biodistribution of the enzyme in ASMKO mice favored the liver (>85%), followed by the spleen. Based on these results, and the fact that the liver and spleen are major sites of pathology in NPD, preclinical efficacy studies using rhASM were then carried out in the ASMKO mice (Miranda et al., 2000). Up to 10 mg/kg of rhASM was injected systemically (tail vein) into mice every other day, and marked reduction in SM levels, as well as improvement in histopathology, was observed in nonneurological tissues after only 2 weeks of treatment. Reversal of disease pathology was not observed in the brain, and the neurological deficits and short lifespan of the treated ASMKO mice remained (Miranda et al., 2000). The conclusions of this study clearly demonstrated that systemic administration of rhASM might be used to treat Type B, but not Type A, NPD patients. Such results are most certainly driven by the pharmacokinetic distribution of the enzyme, which heavily favors the monocyte/macrophage system, and does not cross the blood–brain barrier (Miranda et al., 2000).

The Food and Drug Administration (FDA) agency granted orphan drug status for use of rhASM in treatment of Type B NPD in 2000, and 1 year later, the European Medicines Agency (EMA) granted the same status in Europe (EMA, 2001; FDA, 2003). An open-label, nonrandomized, single-center, Phase I dose-safety trial in adults with ASM deficiency was initiated in 2007 and completed in 2009 (ClinicalTrials.gov Identifier: NCT00410566). A dose range of 0.03–1.0 mg/kg of rhASM was evaluated in a total of 11 patients (McGovern et al., 2009). The maximum tolerated starting dose for NPD was concluded to be 0.6 mg/kg, with a possibility of within-patient escalation. Of note, at doses at or above 0.3 mg/kg, several biomarker changes were observed in the serum, including the elevation of ceramide, several acute-phase reactive proteins including CRP, and bilirubin (at 0.6 and 1 mg/kg) (McGovern et al., 2009). These changes were not observed in normal mice and other animals at doses up to 30 mg/kg, suggesting that they were occurring as a result of the massive SM storage in NPD.

2. INVOLVEMENT OF ASM IN STRESS RESPONSE, APOPTOSIS, AND CANCER CELL RESISTANCE

2.1. ASM can remodel cell membranes and induce cell signaling

In an important paper by Tabas and colleagues, it was demonstrated that lysosomal and secreted ("zinc-activated") ASM were encoded by the same gene (Schissel et al., 1996). Soon after, the zinc requirement was further verified by showing that both forms of the enzyme require zinc—the lysosomal form had high levels of zinc already bound, while the secreted form had less and therefore required zinc to obtain full *in vitro* activity (Schissel, Keesler, Schuchman, Williams, & Tabas, 1998). Importantly, it was also found that the activity of lysosomal and secreted ASM was not exclusive to acidic pH, and that significant cleavage of SM occurred at a neutral pH as well (Schissel, Jiang, et al., 1998; Schissel, Keesler, et al., 1998). The latter point is particularly important in the context of cellular signaling, as it suggested that ASM could be involved in restructuring of the membrane and generation of ceramide at the cell surface. Key properties of the major mammalian sphingomyelinases are given in Table 4.1.

Cancer cells have long been recognized to have altered plasma membranes that allow them to lose contact growth inhibition, become more loosely attached to culture dishes, enhance transport of nutrients, and express tumor-specific antigens (Hakomori, 1973). The plasma membranes of most mammalian cells contain ~50% lipids by mass, with SM as one of the major sphingolipids in the exoplasmic leaflet of the plasma membrane (Bretscher, 1973; Koval & Pagano, 1991). SM is critical for the formation of lipid rafts and determination of plasma membrane fluidity (Simons & Ikonen, 1997). Diverse stress stimuli can elicit translocation of ASM to the plasma membrane, where ASM cleaves the SM into ceramide and phosphorylcholine (Grassme et al., 2001). The latter action increases the plasma membrane fluidity and affects the distribution of cholesterol in cells, such as, for example, displacement of cholesterol from membrane rafts by ceramide (Megha & London, 2004; Slotte, Hedstrom, Rannstrom, & Ekman, 1989). Hydrophobic ceramide molecules associate into ceramide-rich microdomains, strongly stabilizing the membrane rafts (Holopainen, Subramanian, & Kinnunen, 1998; Xu et al., 2001), which can lead to activation of signaling cascades through facilitating protein–protein interactions, including dimerization of receptor proteins, and lowering the activation threshold for cell signaling pathways (Grassme, Riethmuller, & Gulbins, 2007). Consequently, ASM-induced membrane remodeling is critical in the initiation of downstream signaling cascades (Hannun & Bell, 1989).

Some of the direct targets of ceramide include protein kinases such as PKC zeta. For example, NIH 3T3 fibroblasts treated with TNFα, which is known to promote ceramide generation (Kim, Linardic, Obeid, & Hannun, 1991), have increased phosphorylation and activity of PKC zeta (Muller et al., 1995). Catalytically active PKC zeta also has been observed to move into the nucleus, along with concomitant increases in ceramide, in HL-60 cells stimulated with all-*trans* retinoic acid, suggesting a role in the regulation of transcription (Bertolaso et al., 1998). On the other hand, interleukin 1B-stimulated activation of SM signaling (Mathias et al., 1993) initiates ceramide accumulation that can activate a proto-oncogene serine/threonine kinase c-Raf, resulting in activation of mitogen-activated protein kinases and expression of genes involved in the inflammatory response (Huwiler et al., 1996). Endogenous and exogenous ceramide also can lead to inhibition of telomerase activity via concurrent decrease in telomerase reverse transcriptase (hTERT) and c-Myc protein levels (Ogretmen et al., 2001). Activation of stress-activated protein kinase JNK by ceramide is also well documented (Huang, Ma, Ding, Bowden, & Dong, 1997). Importantly, these and other studies demonstrate that the effects of ASM-generated ceramide most prominently involve apoptotic cell death in response to a variety of cell stress stimuli.

2.2. Early apoptosis and ASM

Forty years ago, a seminal review describing programmed cell death and distinguishing it from necrosis was published (Kerr, Wyllie, & Currie, 1972). Unlike necrotic cells, which swell and rupture, apoptotic cells shrink and get cleared by phagocytes without spilling their content into the surrounding milieu. The term apoptosis was proposed by Kerr and colleagues in 1972 and has been used ever since (Kerr et al., 1972). The early events in apoptotic cell death include membrane blebbing and inversion of phosphatidylserine from the inner to outer leaflet of the plasma membrane (Fadok, Bratton, Frasch, Warner, & Henson, 1998), which provide signals important in clearance of apoptotic cells by phagocytosis. Changes in the structural integrity of membranes can lead to programmed cell death as well. For example, ceramides are extremely hydrophobic lipids with the predominant species (*N*-(octadecanoyl)-sphing-4-enine) having a predicted octanol water partition coefficient of +13.9 (Avdeef, 2001; Pubchem, 2012). Based on such physical–chemical properties, ceramides can interact with organelle membranes including the mitochondria, creating pores which increase the leakage of the mitochondrial proteins, including cytochrome *c*, and propagation of apoptosis by activation of cellular caspases (Siskind, 2005). Increased permeability of mitochondrial membranes is often an early event in apoptosis (Jacotot et al., 1999), and exposure of cells to UV light or oxidative stress has been shown to activate sphingomyelinases (Komatsu et al., 2001), resulting in conformational changes at the mitochondrial membrane and, eventually, apoptotic cell death (Kashkar, Wiegmann, Yazdanpanah, Haubert, & Kronke, 2005).

The role of ASM in apoptosis was first suggested by Kolesnick and colleagues using cells and mice lacking normal ASM activity (Santana et al., 1996). Lymphoblasts from two NPD patients, one with <3% of normal ASM activity and another with <1% of normal ASM activity, were used for *in vitro* studies, and ASMKO mice were used *in vivo*. Irradiation of lymphoblasts from NPD patients did not induce measurable apoptosis, as determined by morphological changes of the cell nuclei. A decrease in SM and concomitant increase in ceramide were only observed in normal lymphoblasts. Lymphoblasts from NPD patients retained elevated levels of SM without an increase in ceramide following the irradiation. The radiation-resistant phenotype also was observed in lungs of the ASMKO mice, all together providing strong evidence for the requirement of an active ASM in irradiation-induced apoptosis (Santana et al., 1996). In addition, using fibroblasts and lymphoblasts from Type A NPD patients, UV light exposure failed to induce conformational change in Bax protein in the ASM-deficient cells, without detectable release of cytochrome *c* or caspase-9 processing (Kashkar et al., 2005). However, transfecting the lymphoblasts with the ASM cDNA led to the exact opposite effects and induction of apoptosis. These data revealed an important role of active ASM in the initiation of early apoptosis in response to stress stimuli such as UV irradiation.

As noted above, the contribution of ceramide to cellular signaling includes interactions with diverse targets in specific cellular compartments. In mitochondria, ceramide generated by neutral sphingomyelinase activates protein phosphatases 1 and 2A, which leads to inactivation of antiapoptotic Bcl2 protein, cytochrome *c* release, and apoptosis (Chalfant, Szulc, Roddy, Bielawska, & Hannun, 2004; Jacotot et al., 1999; Ruvolo, Deng, Ito, Carr, & May, 1999). Activation of protein phosphatase 1 also can lead to dephosphorylation of a family of serine/arginine-rich proteins that regulate splicing in the nucleus (Valcarcel & Green, 1996), resulting in alternative splicing of Bcl-2-related protein Bcl-X into pro-apoptotic or antiapoptotic isoforms (Akgul, Moulding, & Edwards, 2004; Chalfant et al., 2001; Ogretmen & Hannun, 2004). Tumor necrosis factor alpha-induced activation of ASM can activate the endolysosomal protease cathepsin D, a known target of ceramide, which can then activate the pro-apoptotic Bid protein followed by activation of initiator caspase-9 and executor caspase-3, resulting in apoptotic cell death (Heinrich et al., 2004). It is clear that ceramide, a product of ASM activity, has diverse targets and effects on cell growth, survival, and apoptosis (Lin, Chen, & Lin, 2006; Ogretmen & Hannun, 2004).

Of note, a metabolite of ceramide with pro-survival functions is sphingosine-1-phosphate (S1P). Unlike the tumor-suppressor lipid ceramide, S1P represses ceramide-mediated cell death and has a tumorigenic role (Cuvillier et al., 1996; Pyne & Pyne, 2010), as well as the ability to induce angiogenesis (Lee et al., 1999). Indeed, increased density of blood vessels was observed in MCF-7 breast cancer xenograft tumors overexpressing sphingosine kinase 1, the enzyme responsible for producing S1P (Nava, Hobson, Murthy, Milstien, & Spiegel, 2002). The ability of S1P to induce the development of new blood vessels from already existing endothelium is enhanced by its synergistic interaction with known angiogenic factors, including VEGF (Licht, Tsirulnikov, Reuveni, Yarnitzky, & Ben-Sasson, 2003). Therefore, the balance of different products of SM metabolism in cells plays an important role in determining the cell fate (Fig. 4.1).

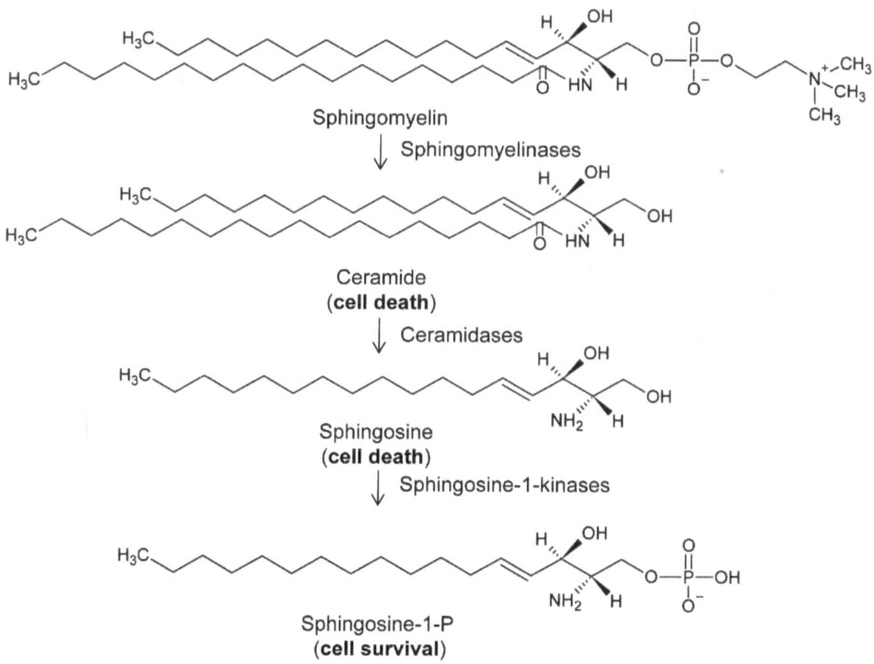

Figure 4.1 Skeletal formulas and biological effects of sphingolipids. Sphingomyelinases release the phosphorylcholine moiety from sphingomyelin, producing ceramide. Ceramide is a cytotoxic lipid that may be further hydrolyzed by ceramidases into sphingosine and free fatty acids, or used as a biosynthetic precursor to produce glycosphingolipids or phosphorylated ceramide derivatives, for example, ceramide-1-phosphate (not shown). The sphingosine produced by the activity of ceramidases also may be phosphorylated by sphingosine kinases to produce sphingosine-1-phosphate (S1P), a lipid that promotes cell survival. The main activity of rhASM for the treatment of cancer is the production of ceramide.

Based on the above, numerous strategies have been proposed and evaluated in cell culture and preclinical models to modulate the sphingolipid balance for cancer treatment. These include inhibiting the enzymes involved in generation of S1P (i.e., sphingosine kinases) (Zhang et al., 2012), blocking the enzymes that degrade ceramide (i.e., ceramidases) (Draper et al., 2011), or using synthetic ceramide analogues (Flowers et al., 2012).

By comparison, rhASM represents another viable approach to treatment, the safety of which has already been tested in Phase I studies in NPD patients (ClinicalTrials.gov Identifier: NCT00410566). An additional advantage comes from the knowledge that the primary target of ASM-generated ceramide in animal tumor models are endothelial cells of the tumor vasculature, likely leading to a more selective effect (see below). Furthermore, preclinical toxicity studies performed in normal rodents and large animals, in support of the enzyme replacement therapy program for NPD, have shown that rhASM can be used safely at very high doses (up to 30 mg/kg) in non-NPD animals.

2.3. The primary target of ASM-generated ceramide is tumor vasculature

The concept of tumor dependence on angiogenesis for continued growth was introduced by Judah Folkman four decades ago. Folkman postulated that preventing angiogenesis would require an antibody against the tumor angiogenesis factor which he and colleagues found responsible for inducing the proliferation of endothelial cells in animal models (Folkman, Merler, Abernathy, & Williams, 1971). An excellent review of Dr. Folkman's contributions to the development of current antiangiogenic therapies is available by Langer and colleagues (Cao et al., 2011). One of the first observations suggesting the potential role of ASM in radiation-induced ceramide generation and apoptosis of vascular endothelial cells came in 1996 (Santana et al., 1996). Radiation of wild-type 129/sv mice induced time- and radiation-dose-dependent ceramide generation and apoptosis in the lungs, primarily in the endothelium. No such apoptosis was observed in ASMKO mice, even at a 50% higher dose of radiation. The resistance of endothelial cells to cell death in animals lacking the active ASM was further observed in studies investigating endotoxic shock syndrome induced by lipopolysaccharide (LPS), a glycolipid from Gram-negative bacteria (Haimovitz-Friedman et al., 1997). C57BL/6 mice injected with 3.6 μg LPS/g of mouse weight had a measurable increase in ceramide and significant apoptosis in the vascular endothelial cells of the intestine, lungs, and thymus (~70% vs. only 5% apoptosis in sham-injected animals). Conversely, following the LPS injection, apoptosis in endothelial cells in organs from ASMKO mice was significantly lower (~10%) compared to wild-type controls (~70%). These studies provided further evidence that active ASM is needed for initiation of apoptosis in response to stress-induced elevation of ceramide, and that vascular endothelial cells are the primary targets of apoptotic cell death (Haimovitz-Friedman et al., 1997; Santana et al., 1996).

Importantly, another seminal paper by Kolesnick and colleagues established the role of ASM in tumor progression and response to irradiation (Garcia-Barros et al., 2003). The authors used ASMKO mice and littermates with normal expression of ASM (ASM +/+) to evaluate if the lack of ASM activity influenced the development of subcutaneous fibrosarcoma and melanoma tumors. With respect to tumor growth, two to four times greater tumor growth was observed in ASMKO mice compared to ASM +/+. To assess whether this difference was related to the ability of the host endothelium cells to undergo apoptosis, terminal deoxynucleotidyl transferase dUTP nick-end labeling (TUNEL) was performed on tumors grown in the ASMKO and wild-type animals. TUNEL staining revealed 2.4-fold greater baseline apoptosis of endothelial cells in ASM +/+ mice, suggesting that the enhanced tumor growth in ASMKO animals was due to the reduced apoptosis of the endothelial cells and increased vascularization of the tumors. To examine whether the presence of ASM in the host affected the response to radiation treatment, mice were treated with 10–15 Gy of radiation. This further showed the resistance of tumor endothelial cells to apoptosis in ASMKO host mice, but not in wild-type mice. To ensure that indeed the endothelial cells were the ones affording the resistance to irradiation, these cells were purified from the tumors (>95% purity) and irradiated. While tumors and their respective isolated endothelial cells from ASM +/+ mice were sensitive to radiation and underwent apoptosis, the endothelial cells from ASMKO mice remained largely resistant (Garcia-Barros et al., 2003). The above studies critically established the importance of active ASM in the vascular endothelium cells to achieve irradiation-induced antitumor effects and also established the concept of ASM-induced cell death of the endothelial cells as a potential adjunct for the treatment of cancer.

3. USE OF EXOGENOUS ASM AS AN ADJUNCT IN CANCER THERAPY

3.1. Mechanisms underlying cancer cell resistance to ceramide

As noted above, ceramide plays an important role in stress-induced cell death in cancer by increasing the levels of intracellular ceramide in response to (i) exogenous stress stimuli such as irradiation, heat shock, etc. (Andrieu-Abadie,

Gouaze, Salvayre, & Levade, 2001; Herr, Wilhelm, Bohler, Angel, & Debatin, 1997); (ii) chemotherapy (Devalapally, Duan, Seiden, & Amiji, 2008; Ogretmen & Hannun, 2001); (iii) exogenous administration of ceramide analogues (Crawford, Bittman, Chun, Byun, & Bowen, 2003; Macchia et al., 2001); or (iv) pharmacological modulation of ceramide breakdown (Morales et al., 2007; Proksch, Klein, & Arenz, 2011). In addition, the pro-apoptotic effects of ceramide can be maximized by inhibiting the production of antiapoptotic S1P (Pyne & Pyne, 2010). With regard to chemotherapy, a number of different chemotherapeutic agents, including topoisomerase inhibitors, DNA cross-linkers, nucleoside analogs, anthracyclines, and others, are known to increase ceramide content in cells, hence contributing to the anticancer activity of the drugs (Biswal, Datta, Acquaah-Mensah, & Kehrer, 2000; Kawase et al., 2002; Lacour et al., 2004; Litvak, Bilchik, & Cabot, 2003). However, resistance to drug treatment is a major hurdle to successful cancer therapy, and cells with aberrant ceramide metabolism have been shown to acquire such resistance (Sietsma, Veldman, & Kok, 2001).

For example, cancer cells can counter the actions of ceramide by increasing the expression of enzymes that eliminate or metabolize ceramide. Ceramide kinase (Sugiura et al., 2002) phosphorylates cytocidal ceramide into pro-survival ceramide-1-phosphate and is overexpressed, or more active, in a number of different tumors (Date, Mitsutake, & Igarashi, 2007; Mitra et al., 2007). Glucosylceramide synthase catalyzes the attachment of glucose to ceramide, creating the noncytotoxic glucosylceramide and leading to a reduction of cytocidal ceramide. Overexpression of glucosylceramide synthase is found in a variety of solid tumors and is particularly prominent in conferring resistance to chemotherapy (Liu, Han, Giuliano, & Cabot, 2001; Morjani et al., 2001; Ruckhaberle et al., 2009). Another notable strategy that cancer cells utilize to overcome the pro-apoptotic effects of ceramide is overexpression of the enzyme acid ceramidase, which hydrolyzes ceramide to sphingosine and free fatty acid. Sphingosine is the precursor of the proliferative S1P, and thus, upregulation of acid ceramidase has the dual effect of reducing cellular ceramide and potentially elevating S1P. Upregulation of acid ceramidase was first observed in prostate cancer cells and subsequently has been found in tumor cells from many different types of cancers (Liu et al., 2008; Park & Schuchman, 2006; Seelan et al., 2000). Cancer cells also can escape ceramide-induced apoptosis by upregulating the expression of antiapoptotic proteins, by loss-of-function mutations in pro-apoptotic genes or by downregulating the expression of death receptors (Igney & Krammer, 2002). For example, overexpression of Bcl-2 blocks the release of cytochrome *c* from mitochondria (Yang et al., 1997), which can inhibit ceramide-induced apoptosis (Geley, Hartmann, & Kofler, 1997).

Notwithstanding the above, addition of exogenous ASM, which allows generation of ceramide in the exoplasmic leaflet of the plasma membrane and initiation of signaling through cell death receptors (Bezombes et al., 2004; Grassme et al., 2001), may represent a viable approach to tumor treatment. Coupled with the fact that exogenously administered enzyme also is internalized by M6P receptors, the killing effect may be further maximized by the "trifecta" of cell surface activation of death receptors, increased generation of intracellular ceramide, which may induce mitochondrial and other abnormalities, and by combining this "enzyme therapy" approach with existing pharmacological treatments. The most likely target of ASM-induced cell death is the tumor microvasculature, although the direct effects on tumor cells themselves cannot be excluded and remain to be investigated in greater detail.

3.2. Proof of principle for ASM-mediated cancer treatment in mouse melanoma

Considering the above evidence documenting the involvement of ASM and ceramide in cancer, we used overexpression of the ASM cDNA and administration of exogenous rhASM to demonstrate the "proof of principle" of using ASM as an adjunct to radiation treatment of melanoma tumors in mice (Smith & Schuchman, 2008a). C57BL/6 mice were injected subcutaneously with melanoma B-16-F10-luc-G5, stably transduced with ASM or a sham retroviral vector, and irradiated (15 Gy) 4 days thereafter. Two weeks after the irradiation, significantly smaller tumors were observed in the group overexpressing ASM compared to tumors from parent cell lines (Fig. 4.2). In further proof of concept that ASM is important for the enhanced radiation tumor response, rhASM was administered locally by intratumoral injection 2–3 h before irradiation, which led to threefold reduction in tumors grown from cells with normal ASM expression (Smith & Schuchman, 2008a).

These seminal studies were the first demonstration of ASM-assisted sensitization of tumors to irradiation *in vivo*, including the administration of rhASM. Importantly, striking differences in tumor vascularization were observed as well. A significantly lower number of blood vessels were present 2 weeks after irradiation in mice overexpressing ASM than in control mice. The melanoma cells that overexpressed ASM also secreted the active enzyme at up to 10-fold greater levels than the parent cells. This suggested that

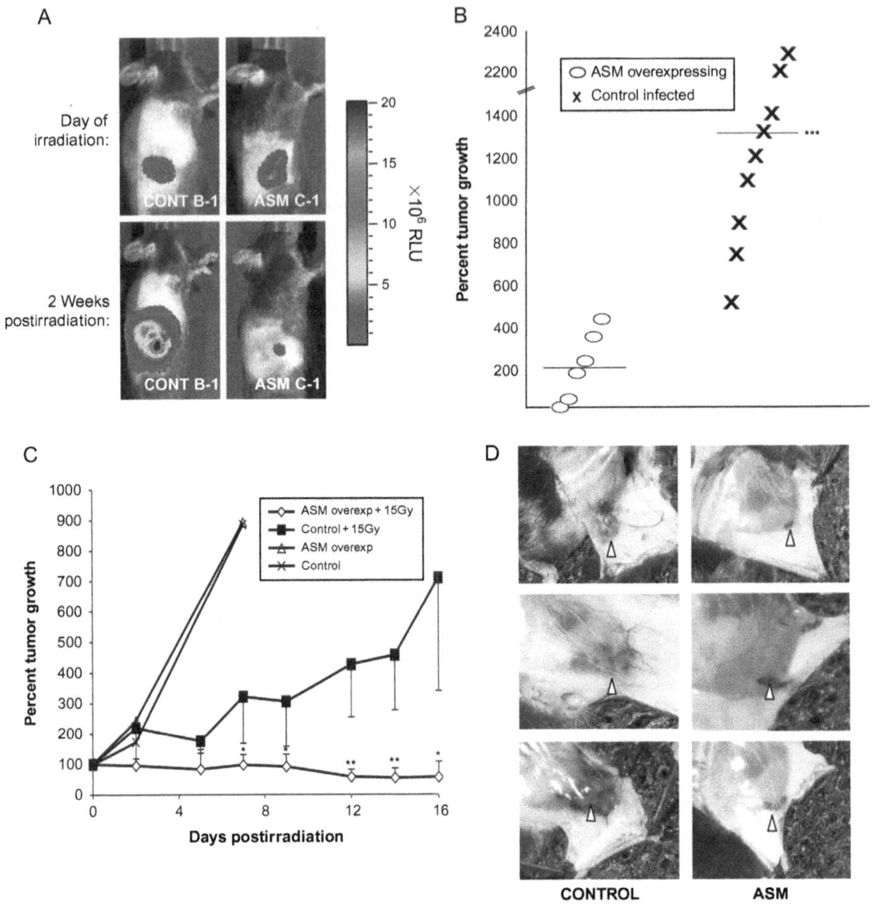

Figure 4.2 The effect of ASM overexpression in B16-F10 melanoma cells on subcutaneous tumor growth after linear irradiation *in vivo*. Mice were injected with 5×10^5 B16-F10 cells either infected with a retroviral vector expressing the ASM complementary DNA or vector alone. Tumors were allowed to grow for 4 days and then given a single dose of 15-Gy linear irradiation. (A) Bioluminescent images showing the tumor size from the same mice at the day of irradiation and 2 weeks after irradiation of ASM-overexpressing (ASM C-1) or control (CONT B-1) subcutaneous B16-F10 melanomas. (B) Scatter plot displaying percent tumor growth 2 weeks after irradiation as quantified by relative bioluminescent light units/second (***$P < 0.001$; each mark represents one tumor). (C) Mean percent change in tumor volume over time as quantified by caliper volume measurement. Sixteen days after irradiation, ASM-overexpressing tumors were 12.4-fold smaller than control tumors (*$P < 0.05$; **$P < 0.01$). (D) Representative gross images of ASM-overexpressing or control tumors displaying the size of melanomas and visible vasculature 2 weeks after linear irradiation. Note: *Reused from Smith and Schuchman (2008a) in accordance with the nonexclusive right to reuse figures in other works created by the author.* (See Page 3 in Color Section at the back of the book.)

the active ASM secreted from tumor cells could have induced apoptosis of nearby endothelial cells (bystander effect), thereby reducing the number of blood vessels as observed by histology.

To confirm this hypothesis, rhASM was used to treat the human microvascular endothelial cell line-1 *in vitro*, using acidification of media to mimic the microenvironment of the tumor and to create the most favorable pH for ASM activity (Smith & Schuchman, 2008a). Under acidified conditions (pH 6.8 and 6.3), rhASM treatment led to significantly lower cell viability (74% and 41%, respectively). It was also found that under nonacidic conditions, rhASM treatment (1 μM) had no reproducible effect on the viability of 60 cancer cell lines encompassing leukemia, nonsmall cell lung, colon, CNS, melanoma, ovarian, renal, prostate, and breast tumors, suggesting that the microenvironment of the tumor was important for the observed *in vivo* effects (NCI, 2011; Smith & Schuchman, 2008b).

In summary, considering the adjunctive role of ASM, and taking into the account the "hepato"tropic nature of rhASM following systemic administration (He et al., 1999), it is likely that an important target for this new therapy might be liver cancer, a highly heterogeneous and lethal malignancy with few effective therapies. The most prevalent form of liver cancer is hepatocellular carcinoma (HCC), which has numerous deregulated signaling pathways leading to elevated proliferation and angiogenesis, including RAF/MEK/ERK, PI3K/AKT/mTOR, WNT/β-catenin, IGF, and HGF/c-MET signaling pathways (Whittaker, Marais, & Zhu, 2010).

3.3. Potential use of rhASM in mouse HCC

Presently, HCC is the fastest growing tumor in the United States, with >20,000 newly diagnosed cases/year (El-Serag, Davila, Petersen, & McGlynn, 2003). It is estimated that the incidence of HCC will keep increasing in the coming years, and the plateau is expected between 2020 and 2025. Three of the main risk factors for HCC are on the rise in the United States: (1) chronic hepatitis C, which underlies almost half of the cases (Llovet, Burroughs, & Bruix, 2003; Moriya et al., 1998; Wurmbach et al., 2007); (2) obesity (Caldwell, Crespo, Kang, & Al-Osaimi, 2004); and (3) diabetes (El-Serag, Tran, & Everhart, 2004), which induces nonalcoholic steatohepatitis and metabolic syndrome. Sixty million Americans are obese, and over 20 million adults and children have been diagnosed with diabetes in the United States. The rest of the world, like the United States, is also experiencing a severe epidemic of "diabesity." The widespread increased incidence of these risk factors and predisposing conditions is likely to significantly impact the HCC rise in America and throughout the world.

The complex etiology, chemo-resistant properties, and poor survival rates necessitate the development of novel pharmacological interventions against HCC. The advent of sorafenib as an effective molecular-targeted agent in this cancer has completely changed the treatment approach in HCC (Llovet et al., 2008; Wilhelm et al., 2004). However, this is only the first, albeit a breakthrough, step toward effective treatment. The challenges that remain include improving the efficacy of sorafenib and establishing/maintaining acceptable adverse event profiles.

Recently, Kester and colleagues have demonstrated a superior response of SK-HEP-1 tumor xenografts to a C6-ceramide analogue delivered by nanosized liposomes (Tagaram et al., 2011). Nanoscience and nanotechnology have been at the forefront of research into the diagnosis and treatment of cancer over the past decade (Savic, Eisenberg, & Maysinger, 2006; Schroeder et al., 2012). The predominant use of nanosized particles is delivery of cytocidal molecules, although pro-survival treatment with nanodelivery systems is also achievable (Savic et al., 2009). Treatment of mice bearing subcutaneous HCC xenograft tumors with a dose of 36 mg/kg of nanoliposomal C6 ceramide every 48 h, over the course of 7 weeks, nearly prevented the tumor growth beyond $\sim$150 mm^3, while controls reached $\sim$600 mm^3. Histological examination of tumors revealed decreased staining of blood vessel markers and of VEGF, suggesting suppression of angiogenesis in the treated mice (Tagaram et al., 2011).

We have recently completed preliminary studies using Huh7 subcutaneous xenografts and rhASM treatment in combination with sorafenib and also observed a reduction in blood vessel density in mice treated with a combination of rhASM and sorafenib (unpublished observation, E.H.S). The distribution of rhASM heavily favored the liver, with modest activity detected in the subcutaneous tumors, suggesting that orthotopic liver HCC mouse models should be used in the future to evaluate the extent of rhASM benefit as an adjunct to sorafenib. Given the modest biodistribution to the subcutaneous tumors, it was encouraging that significant treatment end points were reached, including a reduction in tumor volume, providing another proof of principle for the use of rhASM as an adjuvant treatment in cancer therapy. The technological challenges in using the orthotopic models, as well as higher costs of such studies, will have to be considered (Richmond & Su, 2008).

4. CONCLUSIONS AND FUTURE DIRECTIONS

In the light of our recently enhanced understanding of the one-molecule/one-target approach to drug development in oncology (Kola & Landis, 2004), it has become clear that greater efficacy may be achieved with either single drugs that affect multiple targets, for example, multikinase inhibitors (Lorusso & Eder, 2008), and/or by combination therapies utilizing novel adjunctive treatments that target newly discovered pathways, e.g., such as modulation of sphingolipid signaling. This is particularly important given the low success rates in drug development in oncology (1/20 drugs approved), and high failure rates during clinical testing (Kola & Landis, 2004). In addition, even those drugs that do achieve regulatory approval often have very limited efficacy.

Our current understanding of the molecular pathogenesis of cancer has led to the identification of only a few disease-specific causes, which benefit from the blockade of specific proliferation pathways, for example, Her2/neu amplifications in breast cancer (blocked by trastuzumab) or BCR–ABL fusion protein in chronic myeloid leukemia (blocked by imatinib). Most of the recently approved molecular-targeted drugs (e.g., sorafenib, sunitinib) block multiple kinases, or signaling cascades common to all cancers (e.g., bevacizumab-blocked angiogenesis).

As presented in this review, and as evident from the extensive literature (only partially reviewed here), research on sphingolipids has moved from characterizing these molecules as simple structural components of cells and toward the identification of a vast array of lipid molecules and their metabolites that can influence many major aspects of cell fate. Ceramides, in particular, are the main cytocidal mediators and a focus of much investigation, although a more detailed understanding of their functions remains to be further elucidated (Hannun & Obeid, 2011). rhASM is being evaluated in patients with NPD and can be administered safely at high doses to non-NPD animals. The concept of using rhASM for cancer therapy is based on the idea that many cancer cells have evolved mechanisms to reduce ceramide and/or increase S1P. rhASM may be used to overcome these barriers and to achieve the pro-apoptotic threshold by increasing the ceramide levels beyond the coping capabilities of the cells. Tumors are likely to be more susceptible to this treatment approach than healthy tissues due to the high degree of microvascularization and/or the nature of the tumor microenvironment, including an acidified pH that facilitates higher rhASM activity. However, it is unlikely that rhASM therapy alone will suffice and we therefore envision the use of this enzyme in combination with other molecular-targeted therapies.

The choice of tumors appropriate for rhASM therapy should be based on the known biodistribution of the drug, which favors reticuloendothelial cells and organs, and in particular, the liver. Chemical modification(s) of rhASM might provide more efficient targeting of the enzyme to endothelial cells, e.g., by using the ICAM-1 adhesion protein (Garnacho et al., 2008), enhancing the efficacy of rhASM in cancer. We conclude this review by stressing the importance of using the current advances in, lipidomics and systems biology to further elucidate the mechanisms of sphingolipid-mediated tumor cell death and to identify the predominant ceramide species involved in ASM-mediated tumor sensitivity and resistance. Together, these approaches should facilitate development of more effective cancer therapeutics and biopharmaceuticals (Schadt, 2009; Sethi & Kang, 2011; Wenk, 2005). The reader is further directed to the expert reviews in the present thematic issue of *Advances in Cancer Treatment* for additional insights into the roles of sphingolipids in cancer pathology and treatment.

ACKNOWLEDGMENTS

Dr. E. H. S. acknowledges the contributions of the many laboratory members and colleagues who have contributed to his research on ASM and NPD. This research has been supported, in part, by the NIH grant 5 R01 HD 28607.

REFERENCES

Akgul, C., Moulding, D. A., & Edwards, S. W. (2004). Alternative splicing of Bcl-2-related genes: Functional consequences and potential therapeutic applications. *Cellular and Molecular Life Sciences*, *61*, 2189–2199.

Andrieu-Abadie, N., Gouaze, V., Salvayre, R., & Levade, T. (2001). Ceramide in apoptosis signaling: Relationship with oxidative stress. *Free Radical Biology & Medicine*, *31*, 717–728.

Avdeef, A. (2001). Physicochemical profiling (solubility, permeability and charge state). *Current Topics in Medicinal Chemistry*, *1*, 277–351.

Bernardo, K., Krut, O., Wiegmann, K., Kreder, D., Micheli, M., Schafer, R., et al. (2000). Purification and characterization of a magnesium-dependent neutral sphingomyelinase from bovine brain. *The Journal of Biological Chemistry*, *275*, 7641–7647.

Bertolaso, L., Gibellini, D., Secchiero, P., Previati, M., Falgione, D., Visani, G., et al. (1998). Accumulation of catalytically active PKC-zeta into the nucleus of HL-60 cell line plays a key role in the induction of granulocytic differentiation mediated by all-trans retinoic acid. *British Journal of Haematology*, *100*, 541–549.

Bezombes, C., Grazide, S., Garret, C., Fabre, C., Quillet-Mary, A., Muller, S., et al. (2004). Rituximab antiproliferative effect in B-lymphoma cells is associated with acid-sphingomyelinase activation in raft microdomains. *Blood*, *104*, 1166–1173.

Biswal, S. S., Datta, K., Acquaah-Mensah, G. K., & Kehrer, J. P. (2000). Changes in ceramide and sphingomyelin following fludarabine treatment of human chronic B-cell leukemia cells. *Toxicology*, *154*, 45–53.

Brady, R. O., Kanfer, J. N., Mock, M. B., & Fredrickson, D. S. (1966). The metabolism of sphingomyelin. II. Evidence of an enzymatic deficiency in Niemann-Pick disease. *Proceedings of the National Academy of Sciences of the United States of America*, *55*, 366–369.

Bretscher, M. S. (1973). Membrane structure: Some general principles. *Science*, *181*, 622–629.

Caldwell, S. H., Crespo, D. M., Kang, H. S., & Al-Osaimi, A. M. (2004). Obesity and hepatocellular carcinoma. *Gastroenterology*, *127*, S97–S103.

Cao, Y., Arbiser, J., D'Amato, R. J., D'Amore, P. A., Ingber, D. E., Kerbel, R., et al. (2011). Forty-year journey of angiogenesis translational research. *Science Translational Medicine*, *3*, 114rv3.

Carstea, E. D., Morris, J. A., Coleman, K. G., Loftus, S. K., Zhang, D., Cummings, C., et al. (1997). Niemann-Pick C1 disease gene: Homology to mediators of cholesterol homeostasis. *Science*, *277*, 228–231.

Chalfant, C. E., Ogretmen, B., Galadari, S., Kroesen, B. J., Pettus, B. J., & Hannun, Y. A. (2001). FAS activation induces dephosphorylation of SR proteins; dependence on the de novo generation of ceramide and activation of protein phosphatase 1. *The Journal of Biological Chemistry*, *276*, 44848–44855.

Chalfant, C. E., Szulc, Z., Roddy, P., Bielawska, A., & Hannun, Y. A. (2004). The structural requirements for ceramide activation of serine-threonine protein phosphatases. *Journal of Lipid Research*, *45*, 496–506.

Crawford, K. W., Bittman, R., Chun, J., Byun, H. S., & Bowen, W. D. (2003). Novel ceramide analogues display selective cytotoxicity in drug-resistant breast tumor cell lines compared to normal breast epithelial cells. *Cellular and Molecular Biology (Noisy-le-Grand, France)*, *49*, 1017–1023.

Crocker, A. C. (1961). The cerebral defect in Tay-Sachs disease and Niemann-Pick disease. *Journal of Neurochemistry*, *7*, 69–80.

Crocker, A. C., & Farber, S. (1958). Niemann-Pick disease: A review of eighteen patients. *Medicine (Baltimore)*, *37*, 1–95.

Cuvillier, O., Pirianov, G., Kleuser, B., Vanek, P. G., Coso, O. A., Gutkind, S., et al. (1996). Suppression of ceramide-mediated programmed cell death by sphingosine-1-phosphate. *Nature*, *381*, 800–803.

Date, T., Mitsutake, S., & Igarashi, Y. (2007). Ceramide kinase expression is altered during macrophage-like cell differentiation of the leukemia cell line HL-60. *In Vitro Cellular and Developmental Biology. Animal*, *43*, 321–323.

da Veiga Pereira, L., Desnick, R. J., Adler, D. A., Disteche, C. M., & Schuchman, E. H. (1991). Regional assignment of the human acid sphingomyelinase gene (SMPD1) by PCR analysis of somatic cell hybrids and in situ hybridization to 11p15.1–p15.4. *Genomics*, *9*, 229–234.

Devalapally, H., Duan, Z., Seiden, M. V., & Amiji, M. M. (2008). Modulation of drug resistance in ovarian adenocarcinoma by enhancing intracellular ceramide using tamoxifen-loaded biodegradable polymeric nanoparticles. *Clinical Cancer Research*, *14*, 3193–3203.

Draper, J. M., Xia, Z., Smith, R. A., Zhuang, Y., Wang, W., & Smith, C. D. (2011). Discovery and evaluation of inhibitors of human ceramidase. *Molecular Cancer Therapeutics*, *10*, 2052–2061.

Duan, R. D., Hertervig, E., Nyberg, L., Hauge, T., Sternby, B., Lillienau, J., et al. (1996). Distribution of alkaline sphingomyelinase activity in human beings and animals. Tissue and species differences. *Digestive Diseases and Sciences*, *41*, 1801–1806.

El-Serag, H. B., Davila, J. A., Petersen, N. J., & McGlynn, K. A. (2003). The continuing increase in the incidence of hepatocellular carcinoma in the United States: An update. *Annals of Internal Medicine*, *139*, 817–823.

El-Serag, H. B., Tran, T., & Everhart, J. E. (2004). Diabetes increases the risk of chronic liver disease and hepatocellular carcinoma. *Gastroenterology*, *126*, 460–468.
EMA. (2001). Orphan designation EU/3/01/056. http://www.ema.europa.eu/ema/index.jsp?curl=pages/medicines/human/orphans/2009/11/human_orphan_000157.jsp&mid=WC0b01ac058001d12b&murl=menus/medicines/medicines.jsp&jsenabled=true. Nov 12, 2012.
Fadok, V. A., Bratton, D. L., Frasch, S. C., Warner, M. L., & Henson, P. M. (1998). The role of phosphatidylserine in recognition of apoptotic cells by phagocytes. *Cell Death and Differentiation*, *5*, 551–562.
FDA. (2003). List of orphan products designated and approved as of December 31, 2003. http://www.fda.gov/ohrms/dockets/dailys/04/June04/060304/84n-0102-lst0101-01.pdf. Nov 12, 2012.
Ferlinz, K., Hurwitz, R., Moczall, H., Lansmann, S., Schuchman, E. H., & Sandhoff, K. (1997). Functional characterization of the N-glycosylation sites of human acid sphingomyelinase by site-directed mutagenesis. *European Journal of Biochemistry*, *243*, 511–517.
Flowers, M., Fabrias, G., Delgado, A., Casas, J., Abad, J. L., & Cabot, M. C. (2012). C6-Ceramide and targeted inhibition of acid ceramidase induce synergistic decreases in breast cancer cell growth. *Breast Cancer Research and Treatment*, *133*, 447–458.
Folkman, J., Merler, E., Abernathy, C., & Williams, G. (1971). Isolation of a tumor factor responsible for angiogenesis. *The Journal of Experimental Medicine*, *133*, 275–288.
Garcia-Barros, M., Paris, F., Cordon-Cardo, C., Lyden, D., Rafii, S., Haimovitz-Friedman, A., et al. (2003). Tumor response to radiotherapy regulated by endothelial cell apoptosis. *Science*, *300*, 1155–1159.
Garnacho, C., Dhami, R., Simone, E., Dziubla, T., Leferovich, J., Schuchman, E. H., et al. (2008). Delivery of acid sphingomyelinase in normal and Niemann-Pick disease mice using intercellular adhesion molecule-1-targeted polymer nanocarriers. *The Journal of Pharmacology and Experimental Therapeutics*, *325*, 400–408.
Geley, S., Hartmann, B. L., & Kofler, R. (1997). Ceramides induce a form of apoptosis in human acute lymphoblastic leukemia cells that is inhibited by Bcl-2, but not by CrmA. *FEBS Letters*, *400*, 15–18.
Goni, F. M., & Alonso, A. (2002). Sphingomyelinases: Enzymology and membrane activity. *FEBS Letters*, *531*, 38–46.
Grassme, H., Jekle, A., Riehle, A., Schwarz, H., Berger, J., Sandhoff, K., et al. (2001). CD95 signaling via ceramide-rich membrane rafts. *The Journal of Biological Chemistry*, *276*, 20589–20596.
Grassme, H., Riethmuller, J., & Gulbins, E. (2007). Biological aspects of ceramide-enriched membrane domains. *Progress in Lipid Research*, *46*, 161–170.
Haimovitz-Friedman, A., Cordon-Cardo, C., Bayoumy, S., Garzotto, M., McLoughlin, M., Gallily, R., et al. (1997). Lipopolysaccharide induces disseminated endothelial apoptosis requiring ceramide generation. *The Journal of Experimental Medicine*, *186*, 1831–1841.
Hakomori, S. (1973). Glycolipids of tumor cell membrane. *Advances in Cancer Research*, *18*, 265–315.
Hannun, Y. A., & Bell, R. M. (1989). Functions of sphingolipids and sphingolipid breakdown products in cellular regulation. *Science*, *243*, 500–507.
Hannun, Y. A., & Obeid, L. M. (2011). Many ceramides. *The Journal of Biological Chemistry*, *286*, 27855–27862.
He, X., Miranda, S. R., Xiong, X., Dagan, A., Gatt, S., & Schuchman, E. H. (1999). Characterization of human acid sphingomyelinase purified from the media of overexpressing Chinese hamster ovary cells. *Biochimica et Biophysica Acta*, *1432*, 251–264.
Heinrich, M., Neumeyer, J., Jakob, M., Hallas, C., Tchikov, V., Winoto-Morbach, S., et al. (2004). Cathepsin D links TNF-induced acid sphingomyelinase to Bid-mediated caspase-9 and -3 activation. *Cell Death and Differentiation*, *11*, 550–563.

Hemsley, K. M., & Hopwood, J. J. (2011). Emerging therapies for neurodegenerative lysosomal storage disorders—From concept to reality. *Journal of Inherited Metabolic Disease*, *34*, 1003–1012.

Herr, I., Wilhelm, D., Bohler, T., Angel, P., & Debatin, K. M. (1997). Activation of CD95 (APO-1/Fas) signaling by ceramide mediates cancer therapy-induced apoptosis. *The EMBO Journal*, *16*, 6200–6208.

Holopainen, J. M., Subramanian, M., & Kinnunen, P. K. (1998). Sphingomyelinase induces lipid microdomain formation in a fluid phosphatidylcholine/sphingomyelin membrane. *Biochemistry*, *37*, 17562–17570.

Horinouchi, K., Erlich, S., Perl, D. P., Ferlinz, K., Bisgaier, C. L., Sandhoff, K., et al. (1995). Acid sphingomyelinase deficient mice: A model of types A and B Niemann-Pick disease. *Nature Genetics*, *10*, 288–293.

Huang, C., Ma, W., Ding, M., Bowden, G. T., & Dong, Z. (1997). Direct evidence for an important role of sphingomyelinase in ultraviolet-induced activation of c-Jun N-terminal kinase. *The Journal of Biological Chemistry*, *272*, 27753–27757.

Hurwitz, R., Ferlinz, K., Vielhaber, G., Moczall, H., & Sandhoff, K. (1994). Processing of human acid sphingomyelinase in normal and I-cell fibroblasts. *The Journal of Biological Chemistry*, *269*, 5440–5445.

Huwiler, A., Brunner, J., Hummel, R., Vervoordeldonk, M., Stabel, S., van den Bosch, H., et al. (1996). Ceramide-binding and activation defines protein kinase c-Raf as a ceramide-activated protein kinase. *Proceedings of the National Academy of Sciences of the United States of America*, *93*, 6959–6963.

Igney, F. H., & Krammer, P. H. (2002). Death and anti-death: Tumour resistance to apoptosis. *Nature Reviews. Cancer*, *2*, 277–288.

Jacotot, E., Costantini, P., Laboureau, E., Zamzami, N., Susin, S. A., & Kroemer, G. (1999). Mitochondrial membrane permeabilization during the apoptotic process. *Annals of the New York Academy of Sciences*, *887*, 18–30.

Kanfer, J. N., Young, O. M., Shapiro, D., & Brady, R. O. (1966). The metabolism of sphingomyelin. I. Purification and properties of a sphingomyelin-cleaving enzyme from rat liver tissue. *The Journal of Biological Chemistry*, *241*, 1081–1084.

Kashkar, H., Wiegmann, K., Yazdanpanah, B., Haubert, D., & Kronke, M. (2005). Acid sphingomyelinase is indispensable for UV light-induced Bax conformational change at the mitochondrial membrane. *The Journal of Biological Chemistry*, *280*, 20804–20813.

Kawase, M., Watanabe, M., Kondo, T., Yabu, T., Taguchi, Y., Umehara, H., et al. (2002). Increase of ceramide in adriamycin-induced HL-60 cell apoptosis: Detection by a novel anti-ceramide antibody. *Biochimica et Biophysica Acta*, *1584*, 104–114.

Kerr, J. F., Wyllie, A. H., & Currie, A. R. (1972). Apoptosis: A basic biological phenomenon with wide-ranging implications in tissue kinetics. *British Journal of Cancer*, *26*, 239–257.

Kim, M. Y., Linardic, C., Obeid, L., & Hannun, Y. (1991). Identification of sphingomyelin turnover as an effector mechanism for the action of tumor necrosis factor alpha and gamma-interferon. Specific role in cell differentiation. *The Journal of Biological Chemistry*, *266*, 484–489.

Klenk, E. (1934). Uber die Natur der Phosphatide in der Milz in Niemann-Pickscher Krankheit. *Zeitschrift fur physikalische Chemie*, *229*, 151–156.

Kola, I., & Landis, J. (2004). Can the pharmaceutical industry reduce attrition rates? *Nature Reviews. Drug Discovery*, *3*, 711–715.

Komatsu, M., Takahashi, T., Abe, T., Takahashi, I., Ida, H., & Takada, G. (2001). Evidence for the association of ultraviolet-C and H(2)O(2)-induced apoptosis with acid sphingomyelinase activation. *Biochimica et Biophysica Acta*, *1533*, 47–54.

Koval, M., & Pagano, R. E. (1991). Intracellular transport and metabolism of sphingomyelin. *Biochimica et Biophysica Acta*, *1082*, 113–125.

Lachmann, R. H. (2011). Enzyme replacement therapy for lysosomal storage diseases. *Current Opinion in Pediatrics*, *23*, 588–593.

Lacour, S., Hammann, A., Grazide, S., Lagadic-Gossmann, D., Athias, A., Sergent, O., et al. (2004). Cisplatin-induced CD95 redistribution into membrane lipid rafts of HT29 human colon cancer cells. *Cancer Research*, *64*, 3593–3598.

Lee, O. H., Kim, Y. M., Lee, Y. M., Moon, E. J., Lee, D. J., Kim, J. H., et al. (1999). Sphingosine 1-phosphate induces angiogenesis: Its angiogenic action and signaling mechanism in human umbilical vein endothelial cells. *Biochemical and Biophysical Research Communications*, *264*, 743–750.

Licht, T., Tsirulnikov, L., Reuveni, H., Yarnitzky, T., & Ben-Sasson, S. A. (2003). Induction of pro-angiogenic signaling by a synthetic peptide derived from the second intracellular loop of S1P3 (EDG3). *Blood*, *102*, 2099–2107.

Lin, C. F., Chen, C. L., & Lin, Y. S. (2006). Ceramide in apoptotic signaling and anticancer therapy. *Current Medicinal Chemistry*, *13*, 1609–1616.

Litvak, D. A., Bilchik, A. J., & Cabot, M. C. (2003). Modulators of ceramide metabolism sensitize colorectal cancer cells to chemotherapy: A novel treatment strategy. *Journal of Gastrointestinal Surgery*, 7, 140–148 discussion 148.

Liu, X., Elojeimy, S., Turner, L. S., Mahdy, A. E., Zeidan, Y. H., Bielawska, A., et al. (2008). Acid ceramidase inhibition: A novel target for cancer therapy. *Frontiers in Bioscience*, *13*, 2293–2298.

Liu, Y. Y., Han, T. Y., Giuliano, A. E., & Cabot, M. C. (2001). Ceramide glycosylation potentiates cellular multidrug resistance. *The FASEB Journal*, *15*, 719–730.

Llovet, J. M., Burroughs, A., & Bruix, J. (2003). Hepatocellular carcinoma. *The Lancet*, *362*, 1907–1917.

Llovet, J. M., Ricci, S., Mazzaferro, V., Hilgard, P., Gane, E., Blanc, J. F., et al. (2008). Sorafenib in advanced hepatocellular carcinoma. *The New England Journal of Medicine*, *359*, 378–390.

Lorusso, P. M., & Eder, J. P. (2008). Therapeutic potential of novel selective-spectrum kinase inhibitors in oncology. *Expert Opinion on Investigational Drugs*, *17*, 1013–1028.

Macchia, M., Barontini, S., Bertini, S., Di Bussolo, V., Fogli, S., Giovannetti, E., et al. (2001). Design, synthesis, and characterization of the antitumor activity of novel ceramide analogues. *Journal of Medicinal Chemistry*, *44*, 3994–4000.

Mathias, S., Younes, A., Kan, C. C., Orlow, I., Joseph, C., & Kolesnick, R. N. (1993). Activation of the sphingomyelin signaling pathway in intact EL4 cells and in a cell-free system by IL-1 beta. *Science*, *259*, 519–522.

McGovern, M. M., Wasserstein, M. P., Kirmse, B., Duvall, L., Schiano, T., Thurberg, B. L., et al. (2009). A phase 1 trial of recombinant human acid sphingomyelinase (rhASM) enzyme replacement therapy in adults with ASM deficiency (ASMD). http://www.nnpdf.org/documents/niemannpickposterICIEM09FINAL_000.pdf. Nov 12, 2012.

Megha & London, E. (2004). Ceramide selectively displaces cholesterol from ordered lipid domains (rafts): Implications for lipid raft structure and function. *The Journal of Biological Chemistry*, *279*, 9997–10004.

Menten, M. L., & Welton, J. P. (1946). Lipid analysis in a case of Niemann-Pick disease. *American Journal of Diseases of Children (1911)*, *72*, 720–727.

Miranda, S. R., He, X., Simonaro, C. M., Gatt, S., Dagan, A., Desnick, R. J., et al. (2000). Infusion of recombinant human acid sphingomyelinase into Niemann-Pick disease mice leads to visceral, but not neurological, correction of the pathophysiology. *The FASEB Journal*, *14*, 1988–1995.

Mitra, P., Maceyka, M., Payne, S. G., Lamour, N., Milstien, S., Chalfant, C. E., et al. (2007). Ceramide kinase regulates growth and survival of A549 human lung adenocarcinoma cells. *FEBS Letters*, *581*, 735–740.

Morales, A., Paris, R., Villanueva, A., Llacuna, L., Garcia-Ruiz, C., & Fernandez-Checa, J. C. (2007). Pharmacological inhibition or small interfering RNA targeting acid ceramidase sensitizes hepatoma cells to chemotherapy and reduces tumor growth in vivo. *Oncogene, 26*, 905–916.

Moriya, K., Fujie, H., Shintani, Y., Yotsuyanagi, H., Tsutsumi, T., Ishibashi, K., et al. (1998). The core protein of hepatitis C virus induces hepatocellular carcinoma in transgenic mice. *Nature Medicine, 4*, 1065–1067.

Morjani, H., Aouali, N., Belhoussine, R., Veldman, R. J., Levade, T., & Manfait, M. (2001). Elevation of glucosylceramide in multidrug-resistant cancer cells and accumulation in cytoplasmic droplets. *International Journal of Cancer, 94*, 157–165.

Muller, G., Ayoub, M., Storz, P., Rennecke, J., Fabbro, D., & Pfizenmaier, K. (1995). PKC zeta is a molecular switch in signal transduction of TNF-alpha, bifunctionally regulated by ceramide and arachidonic acid. *The EMBO Journal, 14*, 1961–1969.

Naureckiene, S., Sleat, D. E., Lackland, H., Fensom, A., Vanier, M. T., Wattiaux, R., et al. (2000). Identification of HE1 as the second gene of Niemann-Pick C disease. *Science, 290*, 2298–2301.

Nava, V. E., Hobson, J. P., Murthy, S., Milstien, S., & Spiegel, S. (2002). Sphingosine kinase type 1 promotes estrogen-dependent tumorigenesis of breast cancer MCF-7 cells. *Experimental Cell Research, 281*, 115–127.

NCI. (2011). Developmental therapeutics program—Cell lines in the in vitro screen. http://dtp.nci.nih.gov/docs/misc/common_files/cell_list.html. Nov 12, 2012.

Niemann, A. (1914). Ein unbekanntes Krankheitsbild. *Jahrbuch fur Kindertheologie, 79*, 1–10.

Ogretmen, B., & Hannun, Y. A. (2001). Updates on functions of ceramide in chemotherapy-induced cell death and in multidrug resistance. *Drug Resistance Updates, 4*, 368–377.

Ogretmen, B., & Hannun, Y. A. (2004). Biologically active sphingolipids in cancer pathogenesis and treatment. *Nature Reviews. Cancer, 4*, 604–616.

Ogretmen, B., Kraveka, J. M., Schady, D., Usta, J., Hannun, Y. A., & Obeid, L. M. (2001). Molecular mechanisms of ceramide-mediated telomerase inhibition in the A549 human lung adenocarcinoma cell line. *The Journal of Biological Chemistry, 276*, 32506–32514.

Otterbach, B., & Stoffel, W. (1995). Acid sphingomyelinase-deficient mice mimic the neurovisceral form of human lysosomal storage disease (Niemann-Pick disease). *Cell, 81*, 1053–1061.

Park, J. H., & Schuchman, E. H. (2006). Acid ceramidase and human disease. *Biochimica et Biophysica Acta, 1758*, 2133–2138.

Pick, L. (1927). Über die lipoidzellige Splenohepatomegalie Typus Niemann-Pick als Stoffwechselerkrankung. *Medizinische Klinik, 23*, 1483–1488.

Proksch, D., Klein, J. J., & Arenz, C. (2011). Potent inhibition of acid ceramidase by novel B-13 analogues. *Journal of Lipids, 2011*, 971618.

PubChem. (2012). Chemical and physical properties of N-(octadecanoyl)-sphing-4-enine. http://pubchem.ncbi.nlm.nih.gov/summary/summarycgi?cid=5283565.

Pyne, N. J., & Pyne, S. (2010). Sphingosine 1-phosphate and cancer. *Nature Reviews. Cancer, 10*, 489–503.

Richmond, A., & Su, Y. (2008). Mouse xenograft models vs GEM models for human cancer therapeutics. *Disease Models & Mechanisms, 1*, 78–82.

Ruckhaberle, E., Karn, T., Hanker, L., Gatje, R., Metzler, D., Holtrich, U., et al. (2009). Prognostic relevance of glucosylceramide synthase (GCS) expression in breast cancer. *Journal of Cancer Research and Clinical Oncology, 135*, 81–90.

Ruvolo, P. P., Deng, X., Ito, T., Carr, B. K., & May, W. S. (1999). Ceramide induces Bcl2 dephosphorylation via a mechanism involving mitochondrial PP2A. *The Journal of Biological Chemistry, 274*, 20296–20300.

Santana, P., Pena, L. A., Haimovitz-Friedman, A., Martin, S., Green, D., McLoughlin, M., et al. (1996). Acid sphingomyelinase-deficient human lymphoblasts and mice are defective in radiation-induced apoptosis. *Cell, 86*, 189–199.

Savic, R., Azzam, T., Eisenberg, A., Nedev, H., Rosenberg, L., & Maysinger, D. (2009). Block-copolymer micelles as carriers of cell signaling modulators for the inhibition of JNK in human islets of Langerhans. *Biomaterials, 30*, 3597–3604.

Savic, R., Eisenberg, A., & Maysinger, D. (2006). Block copolymer micelles as delivery vehicles of hydrophobic drugs: Micelle-cell interactions. *Journal of Drug Targeting, 14*, 343–355.

Schadt, E. E. (2009). Molecular networks as sensors and drivers of common human diseases. *Nature, 461*, 218–223.

Schissel, S. L., Jiang, X., Tweedie-Hardman, J., Jeong, T., Camejo, E. H., Najib, J., et al. (1998). Secretory sphingomyelinase, a product of the acid sphingomyelinase gene, can hydrolyze atherogenic lipoproteins at neutral pH. Implications for atherosclerotic lesion development. *The Journal of Biological Chemistry, 273*, 2738–2746.

Schissel, S. L., Keesler, G. A., Schuchman, E. H., Williams, K. J., & Tabas, I. (1998). The cellular trafficking and zinc dependence of secretory and lysosomal sphingomyelinase, two products of the acid sphingomyelinase gene. *The Journal of Biological Chemistry, 273*, 18250–18259.

Schissel, S. L., Schuchman, E. H., Williams, K. J., & Tabas, I. (1996). Zn2+-stimulated sphingomyelinase is secreted by many cell types and is a product of the acid sphingomyelinase gene. *The Journal of Biological Chemistry, 271*, 18431–18436.

Schneider, P. B., & Kennedy, E. P. (1967). Sphingomyelinase in normal human spleens and in spleens from subjects with Niemann-Pick disease. *Journal of Lipid Research, 8*, 202–209.

Schroeder, A., Heller, D. A., Winslow, M. M., Dahlman, J. E., Pratt, G. W., Langer, R., et al. (2012). Treating metastatic cancer with nanotechnology. *Nature Reviews. Cancer, 12*, 39–50.

Schuchman, E. H. (2010). Acid sphingomyelinase, cell membranes and human disease: Lessons from Niemann-Pick disease. *FEBS Letters, 584*, 1895–1900.

Schuchman, E. H., Levran, O., Pereira, L. V., & Desnick, R. J. (1992). Structural organization and complete nucleotide sequence of the gene encoding human acid sphingomyelinase (SMPD1). *Genomics, 12*, 197–205.

Schuchman, E. H., Suchi, M., Takahashi, T., Sandhoff, K., & Desnick, R. J. (1991). Human acid sphingomyelinase. Isolation, nucleotide sequence and expression of the full-length and alternatively spliced cDNAs. *The Journal of Biological Chemistry, 266*, 8531–8539.

Seelan, R. S., Qian, C., Yokomizo, A., Bostwick, D. G., Smith, D. I., & Liu, W. (2000). Human acid ceramidase is overexpressed but not mutated in prostate cancer. *Genes, Chromosomes & Cancer, 29*, 137–146.

Sethi, N., & Kang, Y. (2011). Unravelling the complexity of metastasis—Molecular understanding and targeted therapies. *Nature Reviews. Cancer, 11*, 735–748.

Sietsma, H., Veldman, R. J., & Kok, J. W. (2001). The involvement of sphingolipids in multidrug resistance. *The Journal of Membrane Biology, 181*, 153–162.

Simons, K., & Ikonen, E. (1997). Functional rafts in cell membranes. *Nature, 387*, 569–572.

Siskind, L. J. (2005). Mitochondrial ceramide and the induction of apoptosis. *Journal of Bioenergetics and Biomembranes, 37*, 143–153.

Slotte, J. P., Hedstrom, G., Rannstrom, S., & Ekman, S. (1989). Effects of sphingomyelin degradation on cell cholesterol oxidizability and steady-state distribution between the cell surface and the cell interior. *Biochimica et Biophysica Acta, 985*, 90–96.

Smith, E. L., & Schuchman, E. H. (2008a). Acid sphingomyelinase overexpression enhances the antineoplastic effects of irradiation in vitro and in vivo. *Molecular Therapy, 16*, 1565–1571.

Smith, E. L., & Schuchman, E. H. (2008). Acid sphingomyelinase enhances the anti-neoplastic effects of irradiation in vitro and in vivo. Mount Sinai Graduate School of Biological Sciences, Vol. Ph.D. New York: Mount Sinai School of Medicine of NYU.

Spence, M. W., Byers, D. M., Palmer, F. B., & Cook, H. W. (1989). A new Zn2+-stimulated sphingomyelinase in fetal bovine serum. *The Journal of Biological Chemistry*, *264*, 5358–5363.

Sugiura, M., Kono, K., Liu, H., Shimizugawa, T., Minekura, H., Spiegel, S., et al. (2002). Ceramide kinase, a novel lipid kinase. Molecular cloning and functional characterization. *The Journal of Biological Chemistry*, *277*, 23294–23300.

Tagaram, H. R., Divittore, N. A., Barth, B. M., Kaiser, J. M., Avella, D., Kimchi, E. T., et al. (2011). Nanoliposomal ceramide prevents in vivo growth of hepatocellular carcinoma. *Gut*, *60*, 695–701.

Valcarcel, J., & Green, M. R. (1996). The SR protein family: Pleiotropic functions in pre-mRNA splicing. *Trends in Biochemical Sciences*, *21*, 296–301.

Wan, Q., & Schuchman, E. H. (1995). A novel polymorphism in the human acid sphingomyelinase gene due to size variation of the signal peptide region. *Biochimica et Biophysica Acta*, *1270*, 207–210.

Wenk, M. R. (2005). The emerging field of lipidomics. *Nature Reviews. Drug Discovery*, *4*, 594–610.

Whittaker, S., Marais, R., & Zhu, A. X. (2010). The role of signaling pathways in the development and treatment of hepatocellular carcinoma. *Oncogene*, *29*, 4989–5005.

Wilhelm, S. M., Carter, C., Tang, L., Wilkie, D., McNabola, A., Rong, H., et al. (2004). BAY 43-9006 exhibits broad spectrum oral antitumor activity and targets the RAF/MEK/ERK pathway and receptor tyrosine kinases involved in tumor progression and angiogenesis. *Cancer Research*, *64*, 7099–7109.

Wurmbach, E., Chen, Y. B., Khitrov, G., Zhang, W., Roayaie, S., Schwartz, M., et al. (2007). Genome-wide molecular profiles of HCV-induced dysplasia and hepatocellular carcinoma. *Hepatology*, *45*, 938–947.

Xu, X., Bittman, R., Duportail, G., Heissler, D., Vilcheze, C., & London, E. (2001). Effect of the structure of natural sterols and sphingolipids on the formation of ordered sphingolipid/sterol domains (rafts). Comparison of cholesterol to plant, fungal, and disease-associated sterols and comparison of sphingomyelin, cerebrosides, and ceramide. *The Journal of Biological Chemistry*, *276*, 33540–33546.

Yang, J., Liu, X., Bhalla, K., Kim, C. N., Ibrado, A. M., Cai, J., et al. (1997). Prevention of apoptosis by Bcl-2: Release of cytochrome c from mitochondria blocked. *Science*, *275*, 1129–1132.

Zhang, H., Li, W., Sun, S., Yu, S., Zhang, M., & Zou, F. (2012). Inhibition of sphingosine kinase 1 suppresses proliferation of glioma cells under hypoxia by attenuating activity of extracellular signal-regulated kinase. *Cell Proliferation*, *45*, 167–175.

CHAPTER FIVE

Hypoxia, Therapeutic Resistance, and Sphingosine 1-Phosphate

Olivier Cuvillier[*,†,1], **Isabelle Ader**[*,†], **Pierre Bouquerel**[*,†], **Leyre Brizuela**[*,†], **Cécile Gstalder**[*,†], **Bernard Malavaud**[*,†]
[*]CNRS, Institut de Pharmacologie et de Biologie Structurale, Toulouse, France
[†]Université de Toulouse, UPS, IPBS, Toulouse, France
[1]Corresponding author: e-mail address: olivier.cuvillier@inserm.fr

Contents

Abstract

Hypoxia, defined as a poor oxygenation, has been long recognized as a hallmark of solid tumors and a negative prognostic factor for response to therapeutics and survival of patients. Cancer cells have evolved biochemical mechanisms that allow them to react and adapt to hypoxia. At the cellular level, this adaptation is under the control of two related transcription factors, HIF-1 and HIF-2 (hypoxia-inducible factor), that respond rapidly to decreased oxygen levels to activate the expression of a broad range of genes promoting neoangiogenesis, glycolysis, metastasis, increased tumor growth, and resistance to treatments.

Recent studies have identified the sphingosine kinase 1/sphingosine 1-phosphate (SphK1/S1P) signaling pathway—which elicits various cellular processes including cell proliferation, cell survival, or angiogenesis—as a new regulator of HIF-1 or HIF-2 activity. In this review, we will focus on how the inhibition/neutralization of the SphK1/S1P signaling could be exploited for cancer therapy.

Advances in Cancer Research, Volume 117
ISSN 0065-230X
http://dx.doi.org/10.1016/B978-0-12-394274-6.00005-4

1. THE HYPOXIA-INDUCIBLE FACTORS FAMILY: REGULATION, ROLE IN CANCER AND THERAPEUTIC IMPLICATION

Hypoxia, or low oxygen tension, occurs in a variety of physiological conditions including embryonic development, as well as pathological conditions such as chronic inflammation, neurodegenerative disorders, ischemic cardiovascular disease, and cancer (Semenza, 2011). There is ample evidence that hypoxia is a feature of many solid human tumors as the aberrant cell proliferation rate is associated with a disequilibrium between oxygen supply and consumption (Harris, 2002). Because of their rapid proliferation, cancer cells become located too far from functional blood vessels for adequate diffusion of oxygen and nutrients. As a consequence, the establishment of a neovasculature is critical to the growth of the tumor (Hanahan & Folkman, 1996). However, this neoangiogenesis displays abnormalities—with irregular and leaky vessels, disordered network structure, and impaired transport characteristics—resulting in spatial and temporal inadequacies in delivery of oxygen leading to highly heterogeneous tissue oxygenation and areas of hypoxia (Jain, 2005; Liao & Johnson, 2007).

The many effects of hypoxia on cancer biology include not only promotion of progression and metastasis (Lu & Kang, 2010; Sullivan & Graham, 2007) but also resistance to radiotherapy or chemotherapy (Cosse & Michiels, 2008; Rohwer & Cramer, 2011; Tredan, Galmarini, Patel, & Tannock, 2007). The presence or not of oxygen influences the efficacy of ionizing radiation, a phenomenon known as the "oxygen effect," and hypoxic fraction of human cancers is resistant due to reduced generation of oxygen radicals and failure to induce DNA breaks (Bristow & Hill, 2008; Moeller, Richardson, & Dewhirst, 2007; Wilson & Hay, 2011).

For chemotherapeutics, because of the leakiness of intratumoral vasculature, the delivery of drugs via blood flow is inefficient (Jain, 2005). In addition, the contribution of hypoxia to drug resistance involves increased drug efflux through the upregulation of the multidrug resistance 1 (*MDR1*) gene encoding the P-gp membrane-resident glycoprotein (Comerford et al., 2002; Ding et al., 2010; Wartenberg et al., 2003), or the multidrug-associated protein 1 (MRP1) (Chen et al., 2009; Liu et al., 2008; Zhu et al., 2005), both belonging to a family of ATP-binding cassette (ABC) transporters. By acting as drug efflux pumps, MDR1 and MRP1 can lower the intracellular concentration of chemotherapeutic drugs (Gottesman, Fojo, & Bates, 2002).

The mechanisms of resistance of hypoxic cells to chemotherapeutics also include cell cycle arrest, inhibition of apoptosis, inhibition of DNA damage, and likely induction of autophagy (Daskalow et al., 2010; Rohwer & Cramer, 2011; Sullivan & Graham, 2009; Wilson & Hay, 2011; Wirthner, Wrann, Balamurugan, Wenger, & Stiehl, 2008).

Reflecting these major roles in tumor biology and therapy, there is compelling evidence that hypoxia can compromise clinical outcomes (Vaupel & Mayer, 2007). Numerous studies using the Eppendorf pO_2 probe have shown a connection between hypoxia and disease progression in a wide array of human tumors including carcinomas of the cervix, prostate, the head and neck, and soft tissue sarcomas (Milosevic, Fyles, Hedley, & Hill, 2004; Vaupel, 2004). The tumor oxygen status is currently considered as a prognostic marker that impacts on malignant progression and outcome of tumor therapy (Milosevic et al., 2012).

Hypoxia should result in the death of cancer and normal cells but cancer cells can adapt to this hostile environment by undergoing genetic changes, which allow them to survive and even proliferate. One way that cancer cells respond to reduced oxygen levels is through the hypoxia-inducible factors (HIFs), a family of transcription factors identified as important regulators of the cellular response to hypoxia (Majmundar, Wong, & Simon, 2010; Semenza, 2012). HIFs consist of a stable oxygen-insensitive HIF-1β subunit (also known as aryl hydrocarbon receptor nuclear translocator, ARNT) and an oxygen-labile HIF-α subunit. Three HIF-α homologues have been identified: HIF-1α (Wang, Jiang, Rue, & Semenza, 1995), HIF-2α (Ema et al., 1997; Flamme et al., 1997; Hogenesch et al., 1997; Tian, McKnight, & Russell, 1997), and HIF-3α (Makino, Kanopka, Wilson, Tanaka, & Poellinger, 2002). HIF-1α and HIF-2α are by far the best characterized α-subunits as little is known about the function of HIF-3α on tumor progression in hypoxic conditions (Heikkila, Pasanen, Kivirikko, & Myllyharju, 2011), although it may function as an inhibitor of HIF-1α, being involved in feedback regulation as its expression is transcriptionally regulated by HIF-1α (Makino et al., 2007).

Whereas HIF-1α is expressed ubiquitously in all cells, HIF-2α and HIF-3α have a limited tissue expression. For example, HIF-2α is particularly abundant in endothelial cells, from which it was first described (Ema et al., 1997; Tian et al., 1997). HIF-2α is also detected in highly vascularized organs or hypoxic tissues including kidney epithelial cells, cardiac myocytes, bone marrow macrophages, or pancreatic parenchymal cells (Talks et al., 2000; Wiesener et al., 2003).

Under normoxic conditions, the HIF-α subunit is hydroxylated on either one or two proline residues found in an oxygen-dependent domain (ODD), which controls the protein stability (Ivan et al., 2001; Jaakkola et al., 2001). This reaction is mediated by members of the oxygen-dependent specific prolyl-4-hydroxylase domain (PHD) family (Bruick & McKnight, 2001; Ivan et al., 2001; Jaakkola et al., 2001). The hydroxylation of HIF-α subunit is required for its recognition by the von Hippel–Lindau tumor suppressor gene product (pVHL) of the E3 ubiquitin ligase complex (Iwai et al., 1999; Lisztwan, Imbert, Wirbelauer, Gstaiger, & Krek, 1999), leading to its polyubiquitination and proteasomal degradation (Cockman et al., 2000; Ohh et al., 2000; Tanimoto, Makino, Pereira, & Poellinger, 2000).

Under hypoxia or in cells lacking functional pVHL (renal cell carcinomas and other tumors regrouped in the VHL syndrome), the HIF-α subunit remains unhydroxylated, and therefore accumulates and then translocates to the nucleus where it dimerizes with HIF-1β, and associates with coactivators such as CBP (cAMP Response Element-Binding Protein-binding protein)/p300 (Arany et al., 1996). The HIF-1 or HIF-2 heterodimer binds to hypoxia-response elements (HREs) located in the promoter region of its multiple target genes (Fig. 5.1): angiogenesis-promoting factors, glucose transporters, enzymes of glycolytic pathway, proteins involved in extracellular matrix remodeling, cell proliferation, and epithelial to mesenchymal transition (Hickey & Simon, 2006; Patiar & Harris, 2006; Semenza, 2010; Thiery, Acloque, Huang, & Nieto, 2009). It also includes those that are directly involved in therapeutic resistance such as the MDR1 (Comerford et al., 2002; Ding et al., 2010; Wartenberg et al., 2003) or MRP1 (Chen et al., 2009; Liu et al., 2008; Zhu et al., 2005).

A wealth of reports based on immunohistochemical studies of human tumor sections indicate that HIF-1α and/or HIF-2α are overexpressed in the majority of human cancers and these elevated levels correlate with cancer-related death (Semenza, 2010) (Keith, Johnson, & Simon, 2012). Despite their extensive sequence similarity and coexpression in many cell types, HIF-1α and HIF-2α play nonoverlapping roles in tumor progression and sometimes opposing activities (Keith et al., 2012). Differences exist in their targets, with HIF-1α activating glucose metabolism genes whereas HIF-2α preferentially promotes activation of genes regulating angiogenesis (VEGF) or proliferation (TGF-α, cyclin D1) (Hu, Wang, Chodosh, Keith, & Simon, 2003; Raval et al., 2005; Wang, Davis, Haque, Huang, & Yarchoan, 2005). Recent evidence also indicates that HIF-1α and HIF-2α can exert distinct, often opposite, direct or indirect interaction with oncoproteins and tumor suppressors such as c-myc, p53, or mTOR (An et al., 1998; Brugarolas et al., 2004; Gordan, Bertout, Hu, Diehl, & Simon, 2007). The distinct roles of HIF-1α and HIF-2α in promoting tumor growth

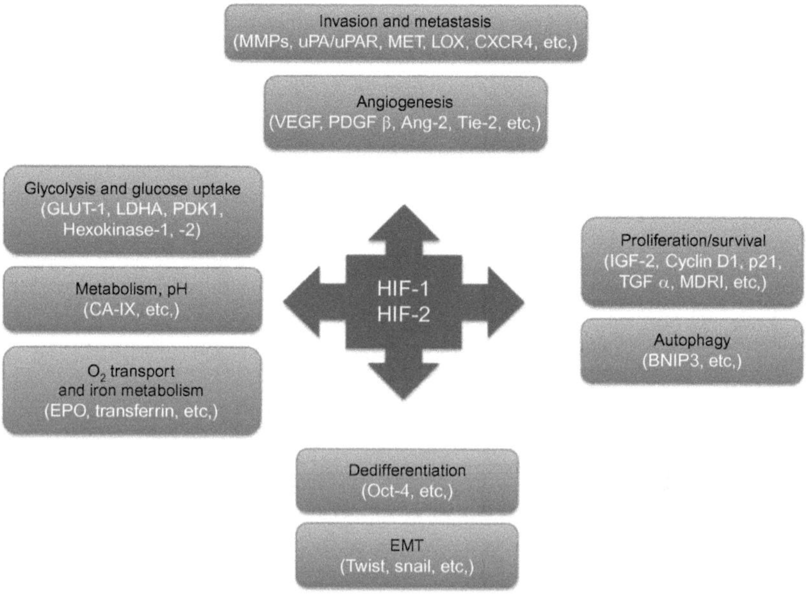

Figure 5.1 Genes activated by hypoxia-inducible factors (HIF-1 and HIF-2) that are involved in tumor progression or resistance to therapeutics. Genes encoding proteins involved in numerous aspects of tumor growth, extracellular matrix remodeling, metabolism, epithelial to mesenchymal transition, or drug resistance by either HIF-1 or HIF-2. Ang-2, Angiopoietin 2; BNIP, BCL2/adenovirus E1B 19 kDa protein-interacting protein 3; CA-IX, carbonic anhydrase IX; CXCR4, C-X-C chemokine receptor type 4; EMT, epithelial–mesenchymal transition; EPO, erythropoietin; GLUT-1, glucose transporter 1; IGF2, insulin-like growth factor 2; LDHA, lactate dehydrogenase A; LOX, lysyl oxidase; MDR1, multidrug resistance protein 1; MET, hepatocyte growth factor receptor; MMPs, Matrix metalloproteinases; Oct-4, octamer-binding transcription factor 4; PDGFβ, platelet-derived growth factor β; PDK1, pyruvate dehydrogenase kinase 1; TGFα, transforming growth factor α; Tie-2, tyrosine kinase with immunoglobulin and EGF homology domains; uPA/uPAR, urokinase-type plasminogen activator and its cellular receptor; VEGF, vascular endothelial growth factor. (See Page 3 in Color Section at the back of the book.)

have been mainly defined in von Hippel–Lindau disease-associated clear cell renal carcinoma (ccRCC) (Kaelin, 2008), which can produce either HIF-1α and HIF-2α or HIF-2α alone, and where the role for HIF-2α as a driver of a more aggressive disease has been established (Gordan et al., 2008).

2. WHAT DOES SPHINGOSINE 1-PHOSPHATE SIGNALING HAVE TO DO WITH HYPOXIA?

Sphingosine 1-phosphate (S1P) is a bioactive phospholipid regulating pleiotropic activities in cancer biology such as proliferation, survival, migration, inflammation, or angiogenesis (Pitson, 2011; Pyne & Pyne, 2011;

Yester, Tizazu, Harikumar, & Kordula, 2011). S1P is generated from sphingosine, the backbone component of all sphingolipids and a proapoptotic sphingolipid (Cuvillier, 2002; Cuvillier, Edsall, & Spiegel, 2000; Cuvillier et al., 2001), in a reaction catalyzed by two sphingosine kinase isoforms, SphK1 and SphK2 (Fig. 5.2). The balance between the levels of S1P and its metabolic precursors ceramide and sphingosine has been regarded as a switch that could determine whether a cell proliferates or dies (Cuvillier et al., 1996). S1P can be secreted even if the mechanism of extracellular transport is not clear and is present at high nanomolar concentration in the circulation (Pappu et al., 2007). Extracellularly, S1P signals as a ligand of five high-affinity G-coupled receptors, named $S1P_{1-5}$ (Rosen, Gonzalez-Cabrera, Sanna, & Brown, 2009). The receptors differ in their tissue distribution and the specific effect of S1P is driven by the predominance of the S1P receptor subtypes expressed (Rosen et al., 2009). Intracellular functions exist with recent studies suggesting that S1P could regulate histone

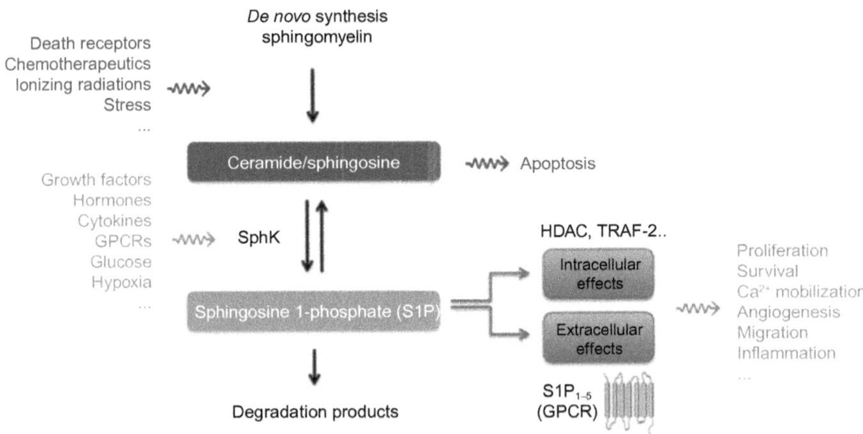

Figure 5.2 The ceramide/sphingosine 1-phosphate biostat. Ceramide, a proapoptotic sphingolipid, can be produced by *de novo* synthesis or degradation of membrane sphingomyelin in response to stress-associated stimuli (death receptors, chemotherapeutics, ionizing radiations, hormone deprivation, etc.). Sphingosine, the metabolite of ceramide, which can be easily converted into ceramide, has been associated with apoptosis. Conversely, sphingosine 1-phosphate, produced by the stimulation of sphingosine kinase activity by cell growth-associated stimuli, acts as an antiapoptotic, prosurvival, proinflammatory, proangiogenic sphingolipid. Sphingosine 1-phosphate exhibits both intracellular effects beginning to be elucidated, and extracellular effects (autocrine and paracrine) through the binding to five dedicated G protein-coupled receptors ($S1P_1$ to $S1P_5$). (For color version of this figure, the reader is referred to the online version of this chapter.)

acetylation in the nucleus, linking S1P to epigenetic regulation of gene expression (Hait et al., 2009). S1P was also reported to act as a cofactor for the ubiquitin ligase activity of tumor necrosis factor receptor-associated factor 2 necessary for the activation of NF-κB (nuclear factor-kappa B) (Alvarez et al., 2010). Although SphK1 and SphK2 differ in size, they share a high degree of sequence similarity (Liu et al., 2000). SphK1 and SphK2 have different tissue distribution and subcellular localization, suggesting distinct physiological roles (Pitson, 2011; Spiegel & Milstien, 2007). SphK1 is activated by various stimuli, among which growth and survival factors are prominent, thus primarily promoting cell survival and proliferation (Strub, Maceyka, Hait, Milstien, & Spiegel, 2010). On the contrary, the role of SphK2 appears by far more complex generally displaying a proapoptotic role but antiapoptotic effects have also been reported (Strub et al., 2010). The subcellular localization of both SphK1 and SphK2 is likely to be critical in determining their function (Strub et al., 2010).

Importantly, the agonist-induced S1P production as well as its downstream effects can be impeded by inhibition of the SphK1 gene expression or enzymatic activity illustrating that SphK1 plays a crucial role in the observed effects played by S1P. A wide array of studies support the convincing role of SphK1 in the promotion of oncogenesis in addition to being a cellular target for many anticancer treatments (Cuvillier, 2007). SphK1 expression has been found upregulated in a number of solid tumors, and high SphK1 expression has been correlated with a significant decrease in survival rate in patients with several forms of cancer (Cuvillier et al., 2010; Pyne & Pyne, 2010). On the other hand, anticancer regimens (chemotherapeutics, radiation therapy, hormonotherapy, chemopreventive agents) have been shown to downregulate SphK1 activity and decrease S1P levels in various cancer cell and animal models (Bonhoure et al., 2008, 2006; Brizuela et al., 2010; Dayon et al., 2009; Li et al., 2007; Nava et al., 2000; Pchejetski et al., 2005; Taha et al., 2004), suggesting that SphK1 could act as a "sensor" to anticancer therapies (Pchejetski et al., 2005). A number of preclinical studies have shown that pharmacological inhibition of SphK1 could be efficacious in decreasing tumor size (French et al., 2003, 2006; Kapitonov et al., 2009; Paugh et al., 2008) or sensitize to chemotherapy or radiotherapy (Pchejetski et al., 2010, 2008), whereas its enforced expression can accelerate tumor proliferation (Nava, Hobson, Murthy, Milstien, & Spiegel, 2002; Pchejetski et al., 2005).

Accumulating evidence has linked SphK1/S1P signaling with adaptation of noncancer cells to hypoxia notably in cardiomyocytes (Karliner, Honbo,

Summers, Gray, & Goetzl, 2001; Tao, Zhang, Vessey, Honbo, & Karliner, 2007; Zhang et al., 2007). Hypoxia also stimulates the proliferation of smooth muscle cells in a S1P-dependent manner (Ahmad, Long, Pyne, & Pyne, 2006; Yun & Kester, 2002). Of note, S1P could act as a nonhypoxic activator of HIF-1α independently of the level of oxygen, by a mechanism requiring a $S1P_2$ receptor-dependent signaling in endothelial cells (Michaud, Robitaille, Gratton, & Richard, 2009). In line with data showing that $S1P_2$ activation in endothelial cells increases vascular permeability (Sanchez et al., 2007), the $S1P_2$ receptor was also shown to be induced in hypoxia-triggered pathological angiogenesis of the mouse retina characterized by evident abnormal vascular growth (vessels were tortuous and dilated) (Skoura et al., 2007). Interestingly, neovascularization was suppressed in the $S1p_2{}^{-/-}$ animals subjected to ischemia-driven retinopathy, suggesting that $S1P_2$ is crucial for endothelial cell responses to hypoxia (Skoura et al., 2007).

Despite the crucial role of hypoxia in solid tumors, the link between SphK1/S1P signaling and adaptation to hypoxic conditions has recently emerged in cancer cells. In a study conducted with five distinct tumor cell models (glioblastoma, prostate, breast, lung, renal cell carcinoma), SphK1 activity was shown to be rapidly stimulated under hypoxic conditions suggesting a posttranslational effect (Ader, Brizuela, Bouquerel, Malavaud, & Cuvillier, 2008). Hypoxia-induced SphK1 activation was dependent on reactive oxygen species (ROS) production as the ROS scavenger *N*-acetyl cysteine could prevent both SphK1 stimulation and HIF-1α accumulation (Fig. 5.3). Numerous studies have reported that exposing cells or tissues to hypoxia increases oxidative stress, although the mechanism(s) by which reduced oxygen concentration augments the production of ROS is not known (Poyton, Ball, & Castello, 2009). Moreover, the addition of exogenous pro-oxidants such as hydrogen peroxide has been shown to be sufficient to trigger HIF-1α accumulation under normoxia (Bell et al., 2007; Simon, 2006). Mitochondria, as the primary oxygen-consuming organelles in cells, have been proposed to be the source of ROS under hypoxia (Chandel et al., 1998, 2000; Guzy et al., 2005; Klimova & Chandel, 2008). The neutralization of ROS by antioxidants, the inhibition of the mitochondrial electron transport chain, or the use of cells depleted in mitochondria were able to abolish the hypoxic induction of HIF-1α (Bell et al., 2007; Chandel et al., 1998).

A large body of evidence suggests that stability of the HIF-α protein subunit is regulated by hydroxylation of two proline residues located within the oxygen-dependent degradation domain (ODD) via the PHD enzymes

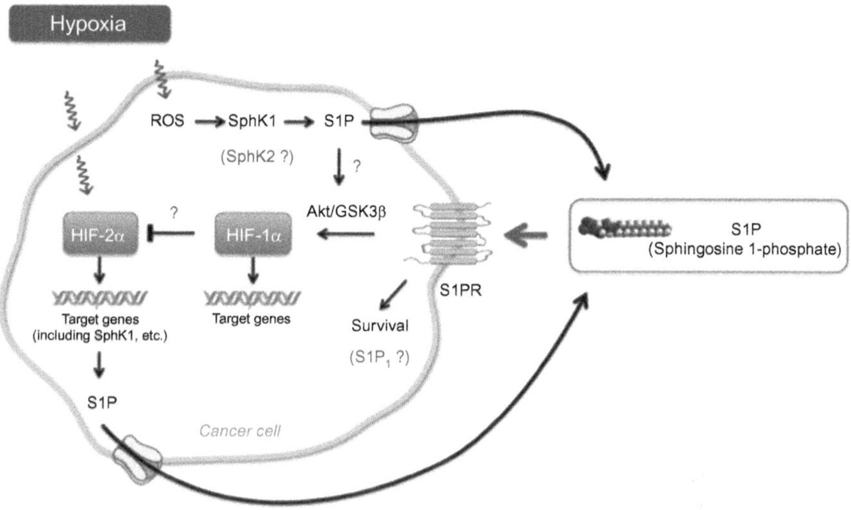

Figure 5.3 Schematic depicting the relationship between HIF-1α, HIF-2α and the SphK1/S1P signaling pathway in cancer cells subjected to hypoxia. Under hypoxia, SphK1 activity can be quickly but transiently stimulated in a ROS-dependent manner. Once formed, S1P through the binding to one of its G-coupled receptor (S1PRs) can trigger the activation the Akt/GSK3β signaling that regulate HIF-1α level and activity. Alternatively, the accumulation of HIF-2α under hypoxia can also lead to the delayed activation of the SphK1/S1P signaling with secretion of S1P. The S1P released from hypoxic tumors could possibly display an autocrine activity on tumor cells leading to a potential amplification of the HIF-1/2 signaling and survival to therapeutics (see text for details). (See Page 4 in Color Section at the back of the book.)

(Bruick & McKnight, 2001; Epstein et al., 2001; Ivan et al., 2001; Jaakkola et al., 2001; Masson, Willam, Maxwell, Pugh, & Ratcliffe, 2001; Yu, White, Zhao, & Lee, 2001). ROS can not only directly modulate HIF-α level through direct inhibition of prolyl hydroxylases or their cofactors (Kaelin & Ratcliffe, 2008) but also indirectly by activating signaling cascades upstream of HIF-α such as mitogen-activated protein kinases (Du et al., 2011; Mylonis et al., 2006; Richard, Berra, Gothie, Roux, & Pouyssegur, 1999) or the phosphatidylinositol 3-kinase (PI3K)/Akt (Du et al., 2011; Flugel, Gorlach, Michiels, & Kietzmann, 2007; Gao et al., 2002; Koshikawa, Hayashi, Nakagawara, & Takenaga, 2009; Liu et al., 2006; Mylonis et al., 2006; Zhou et al., 2007). PI3K/Akt has several downstream targets (Galanis et al., 2008) including the glycogen synthase kinase 3-β (GSK3β), which has been involved in the destabilization of HIF-1α (Chen, Mazure, Cooper, & Giaccia, 2001; Flugel et al., 2007; Mottet et al., 2003; Skuli et al., 2006; Sodhi, Montaner, Miyazaki, & Gutkind, 2001). Akt inactivates the phosphorylation of GSK3β in the ODD and therefore promotes HIF-1α accumulation (Sodhi et al., 2001). Interestingly, the SphK1-mediated accumulation of HIF-1α levels under hypoxia was found to be Akt/GSK3β pathway dependent (Ader et al., 2008).

How does S1P produced upon SphK1 stimulation activate the Akt/GSK3β signaling? The main hypothesis is that S1P could act through the S1P receptors (Fig. 5.3), as the PI3K/Akt signaling can be activated by G_i-coupling of all subtypes of S1P receptors (Lappano & Maggiolini, 2011). Activation of the PI3K/Akt pathway leading to GSK3β phosphorylation by addition of exogenous S1P has been reported in endothelial cells (Gonzalez, Kou, & Michel, 2006). Work conducted in our lab establishes that neutralizing exogenous S1P by an antibody-based approach (Visentin et al., 2006; Wojciak et al., 2009) controls HIF-1α expression under hypoxia in various cancer cell lineages (I. Ader & O. Cuvillier, unpublished data).

The inhibition (pharmacological inhibition or RNA interference) of SphK1 activity prevents activation of the Akt/GSK3β signaling, accumulation of HIF-1α, and its transcriptional activity in all human cancer cell lineages (Ader et al., 2008). As the seminal paper by Wang et al., the main regulatory mechanism of HIF-1α accumulation under hypoxia is its pVHL-mediated proteasomal degradation (Wang et al., 1995) although alternative mechanisms have been described such as the HAF (hypoxia-associated factor) pathway (Koh, Darnay, & Powis, 2008). The degradation of HIF-1α induced by inhibition of SphK1 is mediated by the proteasome through a pVHL-dependent mechanism as shown by inhibition of the proteasome by the MG132 compound or using pVHL-deficient and reconstituted pVHL cell models (Ader et al., 2008). The ROS/SphK1/Akt/GSK3β signaling cascade activated by hypoxic conditions was recently confirmed by Kim and coworkers in prostate cancer cells (Cho et al., 2011).

Obeid and coworkers suggested a relationship between the SphK1/S1P pathway and HIF-2α in U87 glioblastoma cells (Anelli, Gault, Cheng, & Obeid, 2008). Chemically induced hypoxia by cobalt chloride ($CoCl_2$) or 0.5% oxygen conditions triggered the accumulation of HIF-1α and HIF-2α, the increase in SphK1 mRNA (but not SphK2) and protein, and secretion of S1P (Fig. 5.3). Under $CoCl_2$, the silencing of HIF-2α by RNA interference increased HIF-1α content, abolished the induction of SphK1, and decreased extracellular S1P content. Conversely, HIF-1α downregulation resulted in slight increased HIF-2α and SphK1 expression. With regard to the regulatory mechanisms, it was suggested that both HIF-1 and HIF-2 transcription factors could compete for binding to putative HREs located in the promoter region SphK1 under normoxia, and HIF-2 association increasing and HIF-1 decreasing under hypoxia (Anelli et al., 2008). Although in apparent contradiction with our data where SphK1 activity was found to be an upstream regulator of HIF-1α (Ader et al., 2008; Cho et al., 2011), it should be noted that the influence of SphK1 activity on HIF-1α protein levels was not evaluated in this study. Hypoxia mimetics such as $CoCl_2$ used for the most part in this study should not be considered equivalent to *bona fide* hypoxia, and results should be interpreted with caution. Nevertheless, it cannot be ruled out that SphK1 activity might first regulate HIF-1α (and potentially HIF-2α) activity, which in turn could transcriptionally regulate(s) the proangiogenic and prosurvival SphK1/S1P pathway (Fig. 5.3). More work is needed to elucidate whether (i) SphK1 can be a target gene of HIF-2α as proposed by Obeid and coworkers (Anelli et al., 2008) and (ii) SphK1 activity can regulate HIF-2α, as it does for HIF-1α (Ader et al., 2008), particularly in a relevant model such as the von Hippel–Lindau-associated ccRCC. Recent studies conducted in our lab indicate that SphK1 signaling controls HIF-2α expression under hypoxia (P. Bouquerel & O. Cuvillier, unpublished data).

With regard to the SphK2 isoform, contradictory data have been reported for its potential role in mediating adaptation to hypoxia. In cerebral microvascular endothelial cells (in which SphK2 is the most prominent isoform) of Swiss-Webster ND4 mice, SphK2 was found activated (but not SphK1) during hypoxic preconditioning (Wacker, Park, & Gidday, 2009), whereas SphK2 activity and mRNA expression were unchanged in hypoxic EA.hy 926 human endothelial cells (Schwalm et al., 2008). In cancer cells, a single study has reported the upregulation of SphK2 under hypoxia in A549 lung cells, associated with a release of S1P able to protect against chemotherapeutics by a mechanism relying potentially on $S1P_{1/3}$ signaling (inhibitory effect of the VPC23019 pharmacological antagonist), yet the relationship between SphK2 and HIF-1α or HIF-2α was not examined (Schnitzer, Weigert, Zhou, & Brune, 2009). In hypoxic U87 glioblastoma cells, whereas SphK1 expression was augmented, SphK2 expression was unaltered (Anelli et al., 2008) or downregulated (Zhang et al., 2012). Further studies are obviously required to elucidate whether SphK2 represents or not a mediator of hypoxia.

It is important to point out that (when examined) the stimulation of SphK1 or SphK2 was associated with the release of S1P from the hypoxic tumor cells (Anelli et al., 2008; Schnitzer et al., 2009). These findings suggest that S1P released from hypoxic tumors could exhibit an autocrine activity (Fig. 5.3) on cancer cells including (i) a potential feed-forward amplification loop of the HIF1/2 signaling (yet to be proven) or (ii) the induction of tumor cells survival in response to anticancer therapies as shown in hypoxic A549 cells where the conditioned medium could protect these cells to etoposide by a mechanism suggesting the involvement of $S1P_{1/3}$ signaling (Schnitzer et al., 2009). A paracrine proangiogenic effect of S1P (Fig. 5.4) released from hypoxic tumor cells can effect endothelial cells. HUVECs have been shown to lose their ability to form tubes and migrate in the presence of hypoxic U87-conditioned medium if pretreated with VPC23019, the antagonist of $S1P_1$ and $S1P_3$, or silenced by siRNA (Anelli et al., 2008). Another potential paracrine effect of S1P can interfere with cells from innate immunity such as macrophages (Fig. 5.4). Hypoxic microenvironment is well known to favorize the infiltration of macrophages (Yotnda, Wu, & Swanson, 2010), which once present in the tumor are identified as

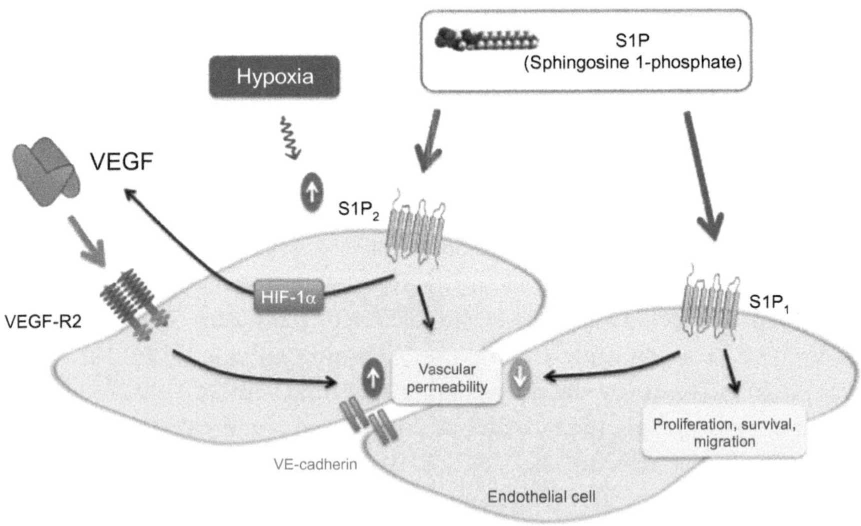

Figure 5.4 The distinctive biological effects of S1P on endothelial cells. This schematic depicts the contrasting effects of S1P on endothelial cells depending on the S1P receptor isoform activated. Like the VEGF, S1P can trigger vascular permeability or weakening of VE-cadherin junctions through $S1P_2$ signaling, which results in loss of association of endothelial cells. Hypoxia can also upregulate the expression of $S1P_2$ receptor subtype, which is associated with HIF-1α activation and release of the vascular permeability factor VEGF. Conversely, S1P can promote endothelial cell integrity via $S1P_1$ signaling (see text for details). (For color version of this figure, the reader is referred to the online version of this chapter.)

tumor-associated macrophages (TAMs). The high number of macrophages correlates with tumor progression and unfavorable clinical prognosis in many cancer types (Lewis & Pollard, 2006). Hypoxic TAMs switch from immune to protumor cells, being polarized toward the M2 phenotype (Yotnda et al., 2010). In conjunction with inhibition of antitumor response, they play a critical role in promoting tumor progression through stimulation of angiogenesis and extracellular matrix changes (Condeelis & Pollard, 2006; Mantovani, Schioppa, Porta, Allavena, & Sica, 2006). It is now well established that S1P can trigger the polarization of macrophages toward the M2 phenotype when added to the cells (Hughes et al., 2008) or when being present in the supernatant of apoptotic tumor cells (Weigert et al., 2007; Weis, Weigert, von Knethen, & Brune, 2009). A recent work from Brüne and coworkers has shown that S1P by binding to $S1P_1$ receptors expressed on macrophages provokes their activation by increasing HIF-1α expression and activity, which could amplify tumor angiogenesis (Herr et al., 2009).

3. SPHINGOSINE 1-PHOSPHATE SIGNALING AS A TARGET FOR ANTIHYPOXIC STRATEGY?

Given the critical role of HIF-1 and HIF-2 in the regulation of hypoxia, it is clear that decreasing their activity should represent a valid strategy to control tumor hypoxia and its molecular consequences: increased potential for invasion, angiogenesis, metastasis, and patient mortality. The search for specific HIF inhibitors is not easy as transcription factors are conventionally considered complex if not impractical targets for the discovery of small molecule inhibitors (Melillo, 2007). A growing number of chemical compounds have been shown to inhibit HIF activity through various mechanisms including decreased HIF-1α synthesis, increased degradation, decreased interaction with the HIF-1β subunit, decreased DNA-binding capabilities or decreased transcriptional activity (Rodriguez-Jimenez & Moreno-Manzano, 2012; Semenza, 2012). Signal transduction pathways involved in HIF-1 or HIF-2 stabilization occurring during hypoxic stress can also be targeted to inhibit HIF-1 activity. This has been nicely illustrated with the inhibition of the PI3K/Akt signaling, which downregulates HIF-1α (and VEGF expression) and sensitizes to ionizing radiations (Pore et al., 2006).

Because the SphK1/S1P pathway appears to act as a key regulator of HIF-1 activity, we hypothesize that its inhibition should represent an unconventional and valid strategy to control tumor hypoxia and its consequences, notably the formation of an aberrant tumor vasculature due to unleashed angiogenesis. For instance, *in vitro* studies have shown that SphK1-mediated inhibition of HIF-1 blocks the release of the vascular permeability factor VEGF and the glucose transporter GLUT-1 from hypoxic tumor cells (Ader et al., 2008; Cho et al., 2011). Anti-SphK1 strategies also decrease S1P secretion from hypoxic tumor cells hence impeding the potential prosurvival effect of S1P on hypoxic tumor cells (Schnitzer et al., 2009), or interfering within the tumor microenvironment by inhibiting the effects of S1P on tube formation and migration of endothelial cells (Anelli et al., 2008) or the activation of TAMs (Herr et al., 2009). Similarly, strategies aimed at (i) neutralizing circulating S1P by anti-S1P antibody to prevent its binding to its receptors (Visentin et al., 2006) or (ii) targeting a specific S1P receptor such as $S1P_1$ or $S1P_2$, the roles of which have been suggested in hypoxic conditions, could be employed. For instance, $S1P_1$ could mediate the effect of S1P on tumor cell survival (Schnitzer et al., 2009) and on tube formation and migration of endothelial cells (Anelli et al., 2008), and its therapeutic targeting seems to be beneficial to prevent tumor growth and angiogenesis in animal models (Chae, Paik, Furneaux, & Hla, 2004; LaMontagne et al., 2006).

Similarly to anti-PI3K strategies that downregulate VEGF production (Fokas et al., 2012; Pore et al., 2006) or to direct anti-VEGF directed strategies (Winkler et al., 2004), one can envision that inhibiting SphK1/S1P signaling might also increase tumor oxygenation according to the concept of "vascular normalization" (Jain, 2005), and increase the efficacy of irradiation and chemotherapy. As aforementioned, the overproduction of VEGF and other proangiogenic factors (such as S1P) leads to the formation of an abnormal tumor vasculature, continuously remodeled and characterized by a marked increase in vessel leakiness, consequence of a decrease in interendothelial junctions, and loss of coverage by perivascular stromal cells, namely, pericytes, known to stabilize blood vessels (Goel et al., 2011). By improving connections between adjacent endothelial cells and increasing pericyte coverage, strategies designed to "normalize" vessels can improve delivery and efficacy of cytotoxic therapies in preclinical and clinical settings (Goel et al., 2011).

Mature endothelial cells are connected by adherens junctions including vascular endothelial (VE)–cadherin (Dejana, Tournier-Lasserve, & Weinstein, 2009), that foster vessel stabilization (Fig. 5.4). Downstream signaling from VEGF–VEGFR2 coupling promotes the weakening of VE–cadherin junctions, and thus a loosening of association of endothelial cells (Dejana et al.,

2009). As a consequence, endothelial cells within hypoxic tumors are poorly connected, with less VE–cadherin junctions (Hashizume et al., 2000). Interestingly, despite the fact that S1P has originally been shown to promote endothelial cell integrity (Garcia et al., 2001; Lee et al., 1999) through $S1P_1$ signaling, it has been reported that S1P could increase vascular permeability (Sanchez et al., 2007) similar to VEGF, the canonical vascular permeability factor (Fig. 5.4). The mechanism of action involved the activation of the $S1P_2$ subtype S1P receptor and required the signal transduction of the Rho-ROCK-PTEN signaling. This observation was extended to the *in vivo* model of vascular permeability in the rat lung, in which the $S1P_2$ antagonist JTE013 significantly inhibited H_2O_2-induced permeability in the rat lung perfused model (Sanchez et al., 2007). Although $S1P_1$ receptor is regarded as the principal S1P receptor isoform expressed in VE cells (Michel, Mulders, Jongsma, Alewijnse, & Peters, 2007), *in vivo* studies have shown that $S1P_2$ was strongly upregulated under hypoxic stress in pathological angiogenesis of the mouse retina, establishing its essential role in pathological neovascularization (Skoura et al., 2007). One might hypothesize that the balance between $S1P_1$ and $S1P_2$ receptors in the endothelium could be altered under hypoxia, with a shift toward higher $S1P_2$ expression that could increase vascular permeability. With respect to $S1P_2$, the addition of exogenous S1P to normoxic endothelial cells also induces the activation of HIF-1 and subsequent rise in VEGF release, suggesting a potential amplification of VEGF signaling (Michaud et al., 2009). Last, $S1P_2$-deficient ($S1P2^{-/-}$) mice implanted with Lewis lung carcinoma or B16 melanoma cells displayed increased number of maturated and functional tumor vessels, showing increased pericyte coverage (Du et al., 2010). Although it remains to demonstrate whether (or not) anti-S1P strategies might directly impact vascular permeability, it has been recently established that neutralizing exogenous S1P by using anti-S1P antibody could reduce plasma levels of VEGF in xenograft experimental models (Visentin et al., 2006).

4. CONCLUDING REMARKS

Since the discovery that HIF-1α was overexpressed in common human cancers, a large body of evidence has since decrypted many of the mechanisms and consequences of hypoxia-inducible factor overexpression during cancer progression and resistance to conventional therapeutics. Accordingly, the discovery and development of novel agents targeting hypoxia signaling pathways is an attractive area of cancer therapeutics. More studies will be needed to better understand the role of the SphK1/S1P metabolism in regulating hypoxia-inducible factors-dependent pathways and provide opportunities for therapeutic intervention. Based on the currently available data, the inhibition/neutralization of the S1P signaling in hypoxic tumors is anticipated to interfere directly or indirectly with tumor cells, immune cells such as TAMs and the tumor vasculature. Of particular interest, interrupting the S1P signaling by reducing VEGF levels could be associated with a decrease vascular permeability resulting in normalization of tumor vessels, a prerequisite for improving delivery and efficacy of cytotoxic therapies.

ACKNOWLEDGMENTS

O. C. is supported by grants from the Ligue Contre le Cancer (Equipe Labellisée), the Institut National du Cancer (INCa), the Association pour la Recherche sur la Cancer (ARC), the Fondation de France, the Fondation pour la Recherche Médicale (FRM), the Association pour la Recherche sur les Tumeurs de la Prostate (ARTP), and the Bourses Roche.

REFERENCES

Ader, I., Brizuela, L., Bouquerel, P., Malavaud, B., & Cuvillier, O. (2008). Sphingosine kinase 1: A new modulator of hypoxia inducible factor 1alpha during hypoxia in human cancer cells. *Cancer Research*, *68*, 8635–8642.

Ahmad, M., Long, J. S., Pyne, N. J., & Pyne, S. (2006). The effect of hypoxia on lipid phosphate receptor and sphingosine kinase expression and mitogen-activated protein kinase signaling in human pulmonary smooth muscle cells. *Prostaglandins & Other Lipid Mediators*, *79*, 278–286.

Alvarez, S. E., Harikumar, K. B., Hait, N. C., Allegood, J., Strub, G. M., Kim, E. Y., et al. (2010). Sphingosine-1-phosphate is a missing cofactor for the E3 ubiquitin ligase TRAF2. *Nature*, *465*, 1084–1088.

An, W. G., Kanekal, M., Simon, M. C., Maltepe, E., Blagosklonny, M. V., & Neckers, L. M. (1998). Stabilization of wild-type p53 by hypoxia-inducible factor 1alpha. *Nature*, *392*, 405–408.

Anelli, V., Gault, C. R., Cheng, A. B., & Obeid, L. M. (2008). Sphingosine Kinase 1 Is Up-regulated during Hypoxia in U87MG glioma cells: Role of hypoxia-inducible factors 1 and 2. *The Journal of Biological Chemistry*, *283*, 3365–3375.

Arany, Z., Huang, L. E., Eckner, R., Bhattacharya, S., Jiang, C., Goldberg, M. A., et al. (1996). An essential role for p300/CBP in the cellular response to hypoxia. *Proceedings of the National Academy of Sciences of the United States of America*, *93*, 12969–12973.

Bell, E. L., Klimova, T. A., Eisenbart, J., Moraes, C. T., Murphy, M. P., Budinger, G. R., et al. (2007). The Qo site of the mitochondrial complex III is required for the transduction of hypoxic signaling via reactive oxygen species production. *The Journal of Cell Biology*, *177*, 1029–1036.

Bonhoure, E., Lauret, A., Barnes, D. J., Martin, C., Malavaud, B., Kohama, T., et al. (2008). Sphingosine kinase-1 is a downstream regulator of imatinib-induced apoptosis in chronic myeloid leukemia cells. *Leukemia*, *22*, 971–979.

Bonhoure, E., Pchejetski, D., Aouali, N., Morjani, H., Levade, T., Kohama, T., et al. (2006). Overcoming MDR-associated chemoresistance in HL-60 acute myeloid leukemia cells by targeting sphingosine kinase-1. *Leukemia*, *20*, 95–102.

Bristow, R. G., & Hill, R. P. (2008). Hypoxia and metabolism. Hypoxia, DNA repair and genetic instability. *Nature Reviews. Cancer*, *8*, 180–192.

Brizuela, L., Dayon, A., Doumerc, N., Ader, I., Golzio, M., Izard, J. C., et al. (2010). The sphingosine kinase-1 survival pathway is a molecular target for the tumor-suppressive tea and wine polyphenols in prostate cancer. *The FASEB Journal*, *24*, 3882–3894.

Brugarolas, J., Lei, K., Hurley, R. L., Manning, B. D., Reiling, J. H., Hafen, E., et al. (2004). Regulation of mTOR function in response to hypoxia by REDD1 and the TSC1/TSC2 tumor suppressor complex. *Genes & Development*, *18*, 2893–2904.

Bruick, R. K., & McKnight, S. L. (2001). A conserved family of prolyl-4-hydroxylases that modify HIF. *Science*, *294*, 1337–1340.

Chae, S. S., Paik, J. H., Furneaux, H., & Hla, T. (2004). Requirement for sphingosine 1-phosphate receptor-1 in tumor angiogenesis demonstrated by in vivo RNA interference. *The Journal of Clinical Investigation*, *114*, 1082–1089.

Chandel, N. S., Maltepe, E., Goldwasser, E., Mathieu, C. E., Simon, M. C., & Schumacker, P. T. (1998). Mitochondrial reactive oxygen species trigger hypoxia-induced transcription. *Proceedings of the National Academy of Sciences of the United States of America*, *95*, 11715–11720.

Chandel, N. S., McClintock, D. S., Feliciano, C. E., Wood, T. M., Melendez, J. A., Rodriguez, A. M., et al. (2000). Reactive oxygen species generated at mitochondrial complex III stabilize hypoxia-inducible factor-1alpha during hypoxia: A mechanism of O2 sensing. *The Journal of Biological Chemistry*, *275*, 25130–25138.

Chen, L., Feng, P., Li, S., Long, D., Cheng, J., Lu, Y., et al. (2009). Effect of hypoxia-inducible factor-1alpha silencing on the sensitivity of human brain glioma cells to doxorubicin and etoposide. *Neurochemical Research*, *34*, 984–990.

Chen, E. Y., Mazure, N. M., Cooper, J. A., & Giaccia, A. J. (2001). Hypoxia activates a platelet-derived growth factor receptor/phosphatidylinositol 3-kinase/Akt pathway that results in glycogen synthase kinase-3 inactivation. *Cancer Research*, *61*, 2429–2433.

Cho, S. Y., Lee, H. J., Jeong, S. J., Kim, H. S., Chen, C. Y., Lee, E. O., et al. (2011). Sphingosine kinase 1 pathway is involved in melatonin-induced HIF-1alpha inactivation in hypoxic PC-3 prostate cancer cells. *Journal of Pineal Research*, *51*, 87–93.

Cockman, M. E., Masson, N., Mole, D. R., Jaakkola, P., Chang, G. W., Clifford, S. C., et al. (2000). Hypoxia inducible factor-alpha binding and ubiquitylation by the von Hippel-Lindau tumor suppressor protein. *The Journal of Biological Chemistry*, *275*, 25733–25741.

Comerford, K. M., Wallace, T. J., Karhausen, J., Louis, N. A., Montalto, M. C., & Colgan, S. P. (2002). Hypoxia-inducible factor-1-dependent regulation of the multidrug resistance (MDR1) gene. *Cancer Research*, *62*, 3387–3394.

Condeelis, J., & Pollard, J. W. (2006). Macrophages: Obligate partners for tumor cell migration, invasion, and metastasis. *Cell*, *124*, 263–266.

Cosse, J. P., & Michiels, C. (2008). Tumour hypoxia affects the responsiveness of cancer cells to chemotherapy and promotes cancer progression. *Anti-Cancer Agents in Medicinal Chemistry*, *8*, 790–797.

Cuvillier, O. (2002). Sphingosine in apoptosis signaling. *Biochimica et Biophysica Acta*, *1585*, 153–162.

Cuvillier, O. (2007). Sphingosine kinase-1—A potential therapeutic target in cancer. *Anti-cancer Drugs*, *18*, 105–110.

Cuvillier, O., Ader, I., Bouquerel, P., Brizuela, L., Malavaud, B., Mazerolles, C., et al. (2010). Activation of Sphingosine Kinase-1 in cancer: Implications for therapeutic targeting. *Current Molecular Pharmacology*, *3*, 53–65.

Cuvillier, O., Edsall, L., & Spiegel, S. (2000). Involvement of sphingosine in mitochondria-dependent Fas-induced apoptosis of type II Jurkat T cells. *The Journal of Biological Chemistry*, *275*, 15691–15700.

Cuvillier, O., Nava, V. E., Murthy, S. K., Edsall, L. C., Levade, T., Milstien, S., et al. (2001). Sphingosine generation, cytochrome c release, and activation of caspase-7 in doxorubicin-induced apoptosis of MCF7 breast adenocarcinoma cells. *Cell Death and Differentiation, 8*, 162–171.

Cuvillier, O., Pirianov, G., Kleuser, B., Vanek, P. G., Coso, O. A., Gutkind, S., et al. (1996). Suppression of ceramide-mediated programmed cell death by sphingosine-1-phosphate. *Nature, 381*, 800–803.

Daskalow, K., Rohwer, N., Raskopf, E., Dupuy, E., Kuhl, A., Loddenkemper, C., et al. (2010). Role of hypoxia-inducible transcription factor 1alpha for progression and chemosensitivity of murine hepatocellular carcinoma. *Journal of molecular medicine (Berlin), 88*, 817–827.

Dayon, A., Brizuela, L., Martin, C., Mazerolles, C., Pirot, N., Doumerc, N., et al. (2009). Sphingosine kinase-1 is central to androgen-regulated prostate cancer growth and survival. *PloS One, 4*, e8048.

Dejana, E., Tournier-Lasserve, E., & Weinstein, B. M. (2009). The control of vascular integrity by endothelial cell junctions: Molecular basis and pathological implications. *Developmental Cell, 16*, 209–221.

Ding, Z., Yang, L., Xie, X., Xie, F., Pan, F., Li, J., et al. (2010). Expression and significance of hypoxia-inducible factor-1 alpha and MDR1/P-glycoprotein in human colon carcinoma tissue and cells. *Journal of Cancer Research and Clinical Oncology, 136*, 1697–1707.

Du, W., Takuwa, N., Yoshioka, K., Okamoto, Y., Gonda, K., Sugihara, K., et al. (2010). S1P (2), the G protein-coupled receptor for sphingosine-1-phosphate, negatively regulates tumor angiogenesis and tumor growth in vivo in mice. *Cancer Research, 70*, 772–781.

Du, J., Xu, R., Hu, Z., Tian, Y., Zhu, Y., Gu, L., et al. (2011). PI3K and ERK-induced Rac1 activation mediates hypoxia-induced HIF-1alpha expression in MCF-7 breast cancer cells. *PloS One, 6*, e25213.

Ema, M., Taya, S., Yokotani, N., Sogawa, K., Matsuda, Y., & Fujii-Kuriyama, Y. (1997). A novel bHLH-PAS factor with close sequence similarity to hypoxia-inducible factor 1alpha regulates the VEGF expression and is potentially involved in lung and vascular development. *Proceedings of the National Academy of Sciences of the United States of America, 94*, 4273–4278.

Epstein, A. C., Gleadle, J. M., McNeill, L. A., Hewitson, K. S., O'Rourke, J., Mole, D. R., et al. (2001). C. elegans EGL-9 and mammalian homologs define a family of dioxygenases that regulate HIF by prolyl hydroxylation. *Cell, 107*, 43–54.

Flamme, I., Frohlich, T., von Reutern, M., Kappel, A., Damert, A., & Risau, W. (1997). HRF, a putative basic helix-loop-helix-PAS-domain transcription factor is closely related to hypoxia-inducible factor-1 alpha and developmentally expressed in blood vessels. *Mechanisms of Development, 63*, 51–60.

Flugel, D., Gorlach, A., Michiels, C., & Kietzmann, T. (2007). Glycogen synthase kinase 3 phosphorylates hypoxia-inducible factor 1alpha and mediates its destabilization in a VHL-independent manner. *Molecular and Cellular Biology, 27*, 3253–3265.

Fokas, E., Im, J. H., Hill, S., Yameen, S., Stratford, M., Beech, J., et al. (2012). Dual inhibition of the PI3K/mTOR pathway increases tumor radiosensitivity by normalizing tumor vasculature. *Cancer Research, 72*, 239–248.

French, K. J., Schrecengost, R. S., Lee, B. D., Zhuang, Y., Smith, S. N., Eberly, J. L., et al. (2003). Discovery and evaluation of inhibitors of human sphingosine kinase. *Cancer Research, 63*, 5962–5969.

French, K. J., Upson, J. J., Keller, S. N., Zhuang, Y., Yun, J. K., & Smith, C. D. (2006). Antitumor activity of sphingosine kinase inhibitors. *The Journal of Pharmacology and Experimental Therapeutics, 318*, 596–603.

Galanis, A., Pappa, A., Giannakakis, A., Lanitis, E., Dangaj, D., & Sandaltzopoulos, R. (2008). Reactive oxygen species and HIF-1 signalling in cancer. *Cancer Letters, 266*, 12–20.

Gao, N., Ding, M., Zheng, J. Z., Zhang, Z., Leonard, S. S., Liu, K. J., et al. (2002). Vanadate-induced expression of hypoxia-inducible factor 1 alpha and vascular endothelial growth factor through phosphatidylinositol 3-kinase/Akt pathway and reactive oxygen species. *The Journal of Biological Chemistry*, *277*, 31963–31971.

Garcia, J. G., Liu, F., Verin, A. D., Birukova, A., Dechert, M. A., Gerthoffer, W. T., et al. (2001). Sphingosine 1-phosphate promotes endothelial cell barrier integrity by Edg-dependent cytoskeletal rearrangement. *The Journal of Clinical Investigation*, *108*, 689–701.

Goel, S., Duda, D. G., Xu, L., Munn, L. L., Boucher, Y., Fukumura, D., et al. (2011). Normalization of the vasculature for treatment of cancer and other diseases. *Physiological Reviews*, *91*, 1071–1121.

Gonzalez, E., Kou, R., & Michel, T. (2006). Rac1 modulates sphingosine 1-phosphate-mediated activation of phosphoinositide 3-kinase/Akt signaling pathways in vascular endothelial cells. *The Journal of Biological Chemistry*, *281*, 3210–3216.

Gordan, J. D., Bertout, J. A., Hu, C. J., Diehl, J. A., & Simon, M. C. (2007). HIF-2alpha promotes hypoxic cell proliferation by enhancing c-myc transcriptional activity. *Cancer Cell*, *11*, 335–347.

Gordan, J. D., Lal, P., Dondeti, V. R., Letrero, R., Parekh, K. N., Oquendo, C. E., et al. (2008). HIF-alpha effects on c-Myc distinguish two subtypes of sporadic VHL-deficient clear cell renal carcinoma. *Cancer Cell*, *14*, 435–446.

Gottesman, M. M., Fojo, T., & Bates, S. E. (2002). Multidrug resistance in cancer: Role of ATP-dependent transporters. *Nature Reviews. Cancer*, *2*, 48–58.

Guzy, R. D., Hoyos, B., Robin, E., Chen, H., Liu, L., Mansfield, K. D., et al. (2005). Mitochondrial complex III is required for hypoxia-induced ROS production and cellular oxygen sensing. *Cell Metabolism*, *1*, 401–408.

Hait, N. C., Allegood, J., Maceyka, M., Strub, G. M., Harikumar, K. B., Singh, S. K., et al. (2009). Regulation of histone acetylation in the nucleus by sphingosine-1-phosphate. *Science*, *325*, 1254–1257.

Hanahan, D., & Folkman, J. (1996). Patterns and emerging mechanisms of the angiogenic switch during tumorigenesis. *Cell*, *86*, 353–364.

Harris, A. L. (2002). Hypoxia—A key regulatory factor in tumour growth. *Nature Reviews. Cancer*, *2*, 38–47.

Hashizume, H., Baluk, P., Morikawa, S., McLean, J. W., Thurston, G., Roberge, S., et al. (2000). Openings between defective endothelial cells explain tumor vessel leakiness. *The American Journal of Pathology*, *156*, 1363–1380.

Heikkila, M., Pasanen, A., Kivirikko, K. I., & Myllyharju, J. (2011). Roles of the human hypoxia-inducible factor (HIF)-3alpha variants in the hypoxia response. *Cellular and Molecular Life Sciences*, *68*, 3885–3901.

Herr, B., Zhou, J., Werno, C., Menrad, H., Namgaladze, D., Weigert, A., et al. (2009). The supernatant of apoptotic cells causes transcriptional activation of hypoxia-inducible factor-1alpha in macrophages via sphingosine-1-phosphate and transforming growth factor-beta. *Blood*, *114*, 2140–2148.

Hickey, M. M., & Simon, M. C. (2006). Regulation of angiogenesis by hypoxia and hypoxia-inducible factors. *Current Topics in Developmental Biology*, *76*, 217–257.

Hogenesch, J. B., Chan, W. K., Jackiw, V. H., Brown, R. C., Gu, Y. Z., Pray-Grant, M., et al. (1997). Characterization of a subset of the basic-helix-loop-helix-PAS superfamily that interacts with components of the dioxin signaling pathway. *The Journal of Biological Chemistry*, *272*, 8581–8593.

Hu, C. J., Wang, L. Y., Chodosh, L. A., Keith, B., & Simon, M. C. (2003). Differential roles of hypoxia-inducible factor 1alpha (HIF-1alpha) and HIF-2alpha in hypoxic gene regulation. *Molecular and Cellular Biology*, *23*, 9361–9374.

Hughes, J. E., Srinivasan, S., Lynch, K. R., Proia, R. L., Ferdek, P., & Hedrick, C. C. (2008). Sphingosine-1-phosphate induces an antiinflammatory phenotype in macrophages. *Circulation Research, 102*, 950–958.

Ivan, M., Kondo, K., Yang, H., Kim, W., Valiando, J., Ohh, M., et al. (2001). HIFalpha targeted for VHL-mediated destruction by proline hydroxylation: Implications for O2 sensing. *Science, 292*, 464–468.

Iwai, K., Yamanaka, K., Kamura, T., Minato, N., Conaway, R. C., Conaway, J. W., et al. (1999). Identification of the von Hippel-Lindau tumor-suppressor protein as part of an active E3 ubiquitin ligase complex. *Proceedings of the National Academy of Sciences of the United States of America, 96*, 12436–12441.

Jaakkola, P., Mole, D. R., Tian, Y. M., Wilson, M. I., Gielbert, J., Gaskell, S. J., et al. (2001). Targeting of HIF-alpha to the von Hippel-Lindau ubiquitylation complex by O2-regulated prolyl hydroxylation. *Science, 292*, 468–472.

Jain, R. K. (2005). Normalization of tumor vasculature: An emerging concept in anti-angiogenic therapy. *Science, 307*, 58–62.

Kaelin, W. G., Jr. (2008). Kidney cancer: Now available in a new flavor. *Cancer Cell, 14*, 423–424.

Kaelin, W. G., Jr., & Ratcliffe, P. J. (2008). Oxygen sensing by metazoans: The central role of the HIF hydroxylase pathway. *Molecular Cell, 30*, 393–402.

Kapitonov, D., Allegood, J. C., Mitchell, C., Hait, N. C., Almenara, J. A., Adams, J. K., et al. (2009). Targeting sphingosine kinase 1 inhibits Akt signaling, induces apoptosis, and suppresses growth of human glioblastoma cells and xenografts. *Cancer Research, 69*, 6915–6923.

Karliner, J. S., Honbo, N., Summers, K., Gray, M. O., & Goetzl, E. J. (2001). The lysophospholipids sphingosine-1-phosphate and lysophosphatidic acid enhance survival during hypoxia in neonatal rat cardiac myocytes. *Journal of Molecular and Cellular Cardiology, 33*, 1713–1717.

Keith, B., Johnson, R. S., & Simon, M. C. (2012). HIF1alpha and HIF2alpha: Sibling rivalry in hypoxic tumour growth and progression. *Nature Reviews. Cancer, 12*, 9–22.

Klimova, T., & Chandel, N. S. (2008). Mitochondrial complex III regulates hypoxic activation of HIF. *Cell Death and Differentiation, 15*, 660–666.

Koh, M. Y., Darnay, B. G., & Powis, G. (2008). Hypoxia-associated factor, a novel E3-ubiquitin ligase, binds and ubiquitinates hypoxia-inducible factor 1alpha, leading to its oxygen-independent degradation. *Molecular and Cellular Biology, 28*, 7081–7095.

Koshikawa, N., Hayashi, J., Nakagawara, A., & Takenaga, K. (2009). Reactive oxygen species-generating mitochondrial DNA mutation up-regulates hypoxia-inducible factor-1alpha gene transcription via phosphatidylinositol 3-kinase-Akt/protein kinase C/histone deacetylase pathway. *The Journal of Biological Chemistry, 284*, 33185–33194.

LaMontagne, K., Littlewood-Evans, A., Schnell, C., O'Reilly, T., Wyder, L., Sanchez, T., et al. (2006). Antagonism of sphingosine-1-phosphate receptors by FTY720 inhibits angiogenesis and tumor vascularization. *Cancer Research, 66*, 221–231.

Lappano, R., & Maggiolini, M. (2011). G protein-coupled receptors: Novel targets for drug discovery in cancer. *Nature Reviews. Drug Discovery, 10*, 47–60.

Lee, M. J., Thangada, S., Claffey, K. P., Ancellin, N., Liu, C. H., Kluk, M., et al. (1999). Vascular endothelial cell adherens junction assembly and morphogenesis induced by sphingosine-1-phosphate. *Cell, 99*, 301–312.

Lewis, C. E., & Pollard, J. W. (2006). Distinct role of macrophages in different tumor microenvironments. *Cancer Research, 66*, 605–612.

Li, Q. F., Huang, W. R., Duan, H. F., Wang, H., Wu, C. T., & Wang, L. S. (2007). Sphingosine kinase-1 mediates BCR/ABL-induced upregulation of Mcl-1 in chronic myeloid leukemia cells. *Oncogene, 26*, 7904–7908.

Liao, D., & Johnson, R. S. (2007). Hypoxia: A key regulator of angiogenesis in cancer. *Cancer Metastasis Reviews*, *26*, 281–290.

Lisztwan, J., Imbert, G., Wirbelauer, C., Gstaiger, M., & Krek, W. (1999). The von Hippel-Lindau tumor suppressor protein is a component of an E3 ubiquitin-protein ligase activity. *Genes & Development*, *13*, 1822–1833.

Liu, L. Z., Hu, X. W., Xia, C., He, J., Zhou, Q., Shi, X., et al. (2006). Reactive oxygen species regulate epidermal growth factor-induced vascular endothelial growth factor and hypoxia-inducible factor-1alpha expression through activation of AKT and P70S6K1 in human ovarian cancer cells. *Free Radical Biology & Medicine*, *41*, 1521–1533.

Liu, L., Ning, X., Sun, L., Zhang, H., Shi, Y., Guo, C., et al. (2008). Hypoxia-inducible factor-1 alpha contributes to hypoxia-induced chemoresistance in gastric cancer. *Cancer Science*, *99*, 121–128.

Liu, H., Sugiura, M., Nava, V. E., Edsall, L. C., Kono, K., Poulton, S., et al. (2000). Molecular cloning and functional characterization of a novel mammalian sphingosine kinase type 2 isoform. *The Journal of Biological Chemistry*, *275*, 19513–19520.

Lu, X., & Kang, Y. (2010). Hypoxia and hypoxia-inducible factors: Master regulators of metastasis. *Clinical Cancer Research*, *16*, 5928–5935.

Majmundar, A. J., Wong, W. J., & Simon, M. C. (2010). Hypoxia-inducible factors and the response to hypoxic stress. *Molecular Cell*, *40*, 294–309.

Makino, Y., Kanopka, A., Wilson, W. J., Tanaka, H., & Poellinger, L. (2002). Inhibitory PAS domain protein (IPAS) is a hypoxia-inducible splicing variant of the hypoxia-inducible factor-3alpha locus. *The Journal of Biological Chemistry*, *277*, 32405–32408.

Makino, Y., Uenishi, R., Okamoto, K., Isoe, T., Hosono, O., Tanaka, H., et al. (2007). Transcriptional up-regulation of inhibitory PAS domain protein gene expression by hypoxia-inducible factor 1 (HIF-1): A negative feedback regulatory circuit in HIF-1-mediated signaling in hypoxic cells. *The Journal of Biological Chemistry*, *282*, 14073–14082.

Mantovani, A., Schioppa, T., Porta, C., Allavena, P., & Sica, A. (2006). Role of tumor-associated macrophages in tumor progression and invasion. *Cancer Metastasis Reviews*, *25*, 315–322.

Masson, N., Willam, C., Maxwell, P. H., Pugh, C. W., & Ratcliffe, P. J. (2001). Independent function of two destruction domains in hypoxia-inducible factor-alpha chains activated by prolyl hydroxylation. *The EMBO Journal*, *20*, 5197–5206.

Melillo, G. (2007). Hypoxia-inducible factor 1 inhibitors. *Methods in Enzymology*, *435*, 385–402.

Michaud, M. D., Robitaille, G. A., Gratton, J. P., & Richard, D. E. (2009). Sphingosine-1-phosphate: A novel nonhypoxic activator of hypoxia-inducible factor-1 in vascular cells. *Arteriosclerosis, Thrombosis, and Vascular Biology*, *29*, 902–908.

Michel, M. C., Mulders, A. C., Jongsma, M., Alewijnse, A. E., & Peters, S. L. (2007). Vascular effects of sphingolipids. *Acta Paediatrica. Supplement*, *96*, 44–48.

Milosevic, M., Fyles, A., Hedley, D., & Hill, R. (2004). The human tumor microenvironment: Invasive (needle) measurement of oxygen and interstitial fluid pressure. *Seminars in Radiation Oncology*, *14*, 249–258.

Milosevic, M., Warde, P., Menard, C., Chung, P., Toi, A., Ishkanian, A., et al. (2012). Tumor hypoxia predicts biochemical failure following radiotherapy for clinically localized prostate cancer. *Clinical Cancer Research*, *18*, 2108–2114.

Moeller, B. J., Richardson, R. A., & Dewhirst, M. W. (2007). Hypoxia and radiotherapy: Opportunities for improved outcomes in cancer treatment. *Cancer Metastasis Reviews*, *26*, 241–248.

Mottet, D., Dumont, V., Deccache, Y., Demazy, C., Ninane, N., Raes, M., et al. (2003). Regulation of hypoxia-inducible factor-1alpha protein level during hypoxic conditions

by the phosphatidylinositol 3-kinase/Akt/glycogen synthase kinase 3beta pathway in HepG2 cells. *The Journal of Biological Chemistry, 278*, 31277–31285.
Mylonis, I., Chachami, G., Samiotaki, M., Panayotou, G., Paraskeva, E., Kalousi, A., et al. (2006). Identification of MAPK phosphorylation sites and their role in the localization and activity of hypoxia-inducible factor-1alpha. *The Journal of Biological Chemistry, 281*, 33095–33106.
Nava, V. E., Cuvillier, O., Edsall, L. C., Kimura, K., Milstien, S., Gelmann, E. P., et al. (2000). Sphingosine enhances apoptosis of radiation-resistant prostate cancer cells. *Cancer Research, 60*, 4468–4474.
Nava, V. E., Hobson, J. P., Murthy, S., Milstien, S., & Spiegel, S. (2002). Sphingosine kinase type 1 promotes estrogen-dependent tumorigenesis of breast cancer MCF-7 cells. *Experimental Cell Research, 281*, 115–127.
Ohh, M., Park, C. W., Ivan, M., Hoffman, M. A., Kim, T. Y., Huang, L. E., et al. (2000). Ubiquitination of hypoxia-inducible factor requires direct binding to the beta-domain of the von Hippel-Lindau protein. *Nature Cell Biology, 2*, 423–427.
Pappu, R., Schwab, S. R., Cornelissen, I., Pereira, J. P., Regard, J. B., Xu, Y., et al. (2007). Promotion of lymphocyte egress into blood and lymph by distinct sources of sphingosine-1-phosphate. *Science, 316*, 295–298.
Patiar, S., & Harris, A. L. (2006). Role of hypoxia-inducible factor-1alpha as a cancer therapy target. *Endocrine-Related Cancer, 13*(Suppl. 1), S61–S75.
Paugh, S. W., Paugh, B. S., Rahmani, M., Kapitonov, D., Almenara, J. A., Kordula, T., et al. (2008). A selective sphingosine kinase 1 inhibitor integrates multiple molecular therapeutic targets in human leukemia. *Blood, 112*, 1382–1391.
Pchejetski, D., Boehler, T., Brizuela, L., Sauer, L., Doumerc, N., Golzio, M., et al. (2010). FTY720 (Fingolimod) sensitizes prostate cancer cells to radiotherapy by inhibition of sphingosine kinase-1. *Cancer Research, 70*, 8651–8661.
Pchejetski, D., Doumerc, N., Golzio, M., Naymark, M., Teissie, J., Kohama, T., et al. (2008). Chemosensitizing effects of sphingosine kinase-1 inhibition in prostate cancer cell and animal models. *Molecular Cancer Therapeutics, 7*, 1836–1845.
Pchejetski, D., Golzio, M., Bonhoure, E., Calvet, C., Doumerc, N., Garcia, V., et al. (2005). Sphingosine kinase-1 as a chemotherapy sensor in prostate adenocarcinoma cell and mouse models. *Cancer Research, 65*, 11667–11675.
Pitson, S. M. (2011). Regulation of sphingosine kinase and sphingolipid signaling. *Trends in Biochemical Sciences, 36*, 97–107.
Pore, N., Gupta, A. K., Cerniglia, G. J., Jiang, Z., Bernhard, E. J., Evans, S. M., et al. (2006). Nelfinavir down-regulates hypoxia-inducible factor 1alpha and VEGF expression and increases tumor oxygenation: Implications for radiotherapy. *Cancer Research, 66*, 9252–9259.
Poyton, R. O., Ball, K. A., & Castello, P. R. (2009). Mitochondrial generation of free radicals and hypoxic signaling. *Trends in Endocrinology and Metabolism, 20*, 332–340.
Pyne, N. J., & Pyne, S. (2010). Sphingosine 1-phosphate and cancer. *Nature Reviews. Cancer, 10*, 489–503.
Pyne, S., & Pyne, N. J. (2011). Translational aspects of sphingosine 1-phosphate biology. *Trends in Molecular Medicine, 17*, 463–472.
Raval, R. R., Lau, K. W., Tran, M. G., Sowter, H. M., Mandriota, S. J., Li, J. L., et al. (2005). Contrasting properties of hypoxia-inducible factor 1 (HIF-1) and HIF-2 in von Hippel-Lindau-associated renal cell carcinoma. *Molecular and Cellular Biology, 25*, 5675–5686.
Richard, D. E., Berra, E., Gothie, E., Roux, D., & Pouyssegur, J. (1999). p42/p44 mitogen-activated protein kinases phosphorylate hypoxia-inducible factor 1alpha (HIF-1alpha) and enhance the transcriptional activity of HIF-1. *The Journal of Biological Chemistry, 274*, 32631–32637.

Rodriguez-Jimenez, F. J., & Moreno-Manzano, V. (2012). Modulation of hypoxia-inducible factors (HIF) from an integrative pharmacological perspective. *Cellular and Molecular Life Sciences*, *69*, 519–534.
Rohwer, N., & Cramer, T. (2011). Hypoxia-mediated drug resistance: Novel insights on the functional interaction of HIFs and cell death pathways. *Drug Resistance Updates*, *14*, 191–201.
Rosen, H., Gonzalez-Cabrera, P. J., Sanna, M. G., & Brown, S. (2009). Sphingosine 1-phosphate receptor signaling. *Annual Review of Biochemistry*, *78*, 743–768.
Sanchez, T., Skoura, A., Wu, M. T., Casserly, B., Harrington, E. O., & Hla, T. (2007). Induction of vascular permeability by the sphingosine-1-phosphate receptor-2 (S1P2R) and its downstream effectors ROCK and PTEN. *Arteriosclerosis, Thrombosis, and Vascular Biology*, *27*, 1312–1318.
Schnitzer, S. E., Weigert, A., Zhou, J., & Brune, B. (2009). Hypoxia enhances sphingosine kinase 2 activity and provokes sphingosine-1-phosphate-mediated chemoresistance in A549 lung cancer cells. *Molecular Cancer Research*, 7, 393–401.
Schwalm, S., Doll, F., Romer, I., Bubnova, S., Pfeilschifter, J., & Huwiler, A. (2008). Sphingosine kinase-1 is a hypoxia-regulated gene that stimulates migration of human endothelial cells. *Biochemical and Biophysical Research Communications*, *368*, 1020–1025.
Semenza, G. L. (2010). Defining the role of hypoxia-inducible factor 1 in cancer biology and therapeutics. *Oncogene*, *29*, 625–634.
Semenza, G. L. (2011). Oxygen sensing, homeostasis, and disease. *The New England Journal of Medicine*, *365*, 537–547.
Semenza, G. L. (2012). Hypoxia-inducible factors: Mediators of cancer progression and targets for cancer therapy. *Trends in Pharmacological Sciences*, *33*, 207–214.
Simon, M. C. (2006). Mitochondrial reactive oxygen species are required for hypoxic HIF alpha stabilization. *Advances in Experimental Medicine and Biology*, *588*, 165–170.
Skoura, A., Sanchez, T., Claffey, K., Mandala, S. M., Proia, R. L., & Hla, T. (2007). Essential role of sphingosine 1-phosphate receptor 2 in pathological angiogenesis of the mouse retina. *The Journal of Clinical Investigation*, *117*, 2506–2516.
Skuli, N., Monferran, S., Delmas, C., Lajoie-Mazenc, I., Favre, G., Toulas, C., et al. (2006). Activation of RhoB by hypoxia controls hypoxia-inducible factor-1alpha stabilization through glycogen synthase kinase-3 in U87 glioblastoma cells. *Cancer Research*, *66*, 482–489.
Sodhi, A., Montaner, S., Miyazaki, H., & Gutkind, J. S. (2001). MAPK and Akt act cooperatively but independently on hypoxia inducible factor-1alpha in rasV12 upregulation of VEGF. *Biochemical and Biophysical Research Communications*, *287*, 292–300.
Spiegel, S., & Milstien, S. (2007). Functions of the multifaceted family of sphingosine kinases and some close relatives. *The Journal of Biological Chemistry*, *282*, 2125–2129.
Strub, G. M., Maceyka, M., Hait, N. C., Milstien, S., & Spiegel, S. (2010). Extracellular and intracellular actions of sphingosine-1-phosphate. *Advances in Experimental Medicine and Biology*, *688*, 141–155.
Sullivan, R., & Graham, C. H. (2007). Hypoxia-driven selection of the metastatic phenotype. *Cancer Metastasis Reviews*, *26*, 319–331.
Sullivan, R., & Graham, C. H. (2009). Hypoxia prevents etoposide-induced DNA damage in cancer cells through a mechanism involving hypoxia-inducible factor 1. *Molecular Cancer Therapeutics*, *8*, 1702–1713.
Taha, T. A., Osta, W., Kozhaya, L., Bielawski, J., Johnson, K. R., Gillanders, W. E., et al. (2004). Down-regulation of sphingosine kinase-1 by DNA damage: Dependence on proteases and p53. *The Journal of Biological Chemistry*, *279*, 20546–20554.
Talks, K. L., Turley, H., Gatter, K. C., Maxwell, P. H., Pugh, C. W., Ratcliffe, P. J., et al. (2000). The expression and distribution of the hypoxia-inducible factors HIF-1alpha and HIF-2alpha in normal human tissues, cancers, and tumor-associated macrophages. *The American Journal of Pathology*, *157*, 411–421.

Tanimoto, K., Makino, Y., Pereira, T., & Poellinger, L. (2000). Mechanism of regulation of the hypoxia-inducible factor-1 alpha by the von Hippel-Lindau tumor suppressor protein. *The EMBO Journal, 19*, 4298–4309.
Tao, R., Zhang, J., Vessey, D. A., Honbo, N., & Karliner, J. S. (2007). Deletion of the sphingosine kinase-1 gene influences cell fate during hypoxia and glucose deprivation in adult mouse cardiomyocytes. *Cardiovascular Research, 74*, 56–63.
Thiery, J. P., Acloque, H., Huang, R. Y., & Nieto, M. A. (2009). Epithelial-mesenchymal transitions in development and disease. *Cell, 139*, 871–890.
Tian, H., McKnight, S. L., & Russell, D. W. (1997). Endothelial PAS domain protein 1 (EPAS1), a transcription factor selectively expressed in endothelial cells. *Genes & Development, 11*, 72–82.
Tredan, O., Galmarini, C. M., Patel, K., & Tannock, I. F. (2007). Drug resistance and the solid tumor microenvironment. *Journal of the National Cancer Institute, 99*, 1441–1454.
Vaupel, P. (2004). The role of hypoxia-induced factors in tumor progression. *The Oncologist, 9*(Suppl. 5), 10–17.
Vaupel, P., & Mayer, A. (2007). Hypoxia in cancer: Significance and impact on clinical outcome. *Cancer Metastasis Reviews, 26*, 225–239.
Visentin, B., Vekich, J. A., Sibbald, B. J., Cavalli, A. L., Moreno, K. M., Matteo, R. G., et al. (2006). Validation of an anti-sphingosine-1-phosphate antibody as a potential therapeutic in reducing growth, invasion, and angiogenesis in multiple tumor lineages. *Cancer Cell, 9*, 225–238.
Wacker, B. K., Park, T. S., & Gidday, J. M. (2009). Hypoxic preconditioning-induced cerebral ischemic tolerance: Role of microvascular sphingosine kinase 2. *Stroke, 40*, 3342–3348.
Wang, V., Davis, D. A., Haque, M., Huang, L. E., & Yarchoan, R. (2005). Differential gene up-regulation by hypoxia-inducible factor-1alpha and hypoxia-inducible factor-2alpha in HEK293T cells. *Cancer Research, 65*, 3299–3306.
Wang, G. L., Jiang, B. H., Rue, E. A., & Semenza, G. L. (1995). Hypoxia-inducible factor 1 is a basic-helix-loop-helix-PAS heterodimer regulated by cellular O2 tension. *Proceedings of the National Academy of Sciences of the United States of America, 92*, 5510–5514.
Wartenberg, M., Ling, F. C., Muschen, M., Klein, F., Acker, H., Gassmann, M., et al. (2003). Regulation of the multidrug resistance transporter P-glycoprotein in multicellular tumor spheroids by hypoxia-inducible factor (HIF-1) and reactive oxygen species. *The FASEB Journal, 17*, 503–505.
Weigert, A., Tzieply, N., von Knethen, A., Johann, A. M., Schmidt, H., Geisslinger, G., et al. (2007). Tumor cell apoptosis polarizes macrophages role of sphingosine-1-phosphate. *Molecular Biology of the Cell, 18*, 3810–3819.
Weis, N., Weigert, A., von Knethen, A., & Brune, B. (2009). Heme oxygenase-1 contributes to an alternative macrophage activation profile induced by apoptotic cell supernatants. *Molecular Biology of the Cell, 20*, 1280–1288.
Wiesener, M. S., Jurgensen, J. S., Rosenberger, C., Scholze, C. K., Horstrup, J. H., Warnecke, C., et al. (2003). Widespread hypoxia-inducible expression of HIF-2alpha in distinct cell populations of different organs. *The FASEB Journal, 17*, 271–273.
Wilson, W. R., & Hay, M. P. (2011). Targeting hypoxia in cancer therapy. *Nature Reviews. Cancer, 11*, 393–410.
Winkler, F., Kozin, S. V., Tong, R. T., Chae, S. S., Booth, M. F., Garkavtsev, I., et al. (2004). Kinetics of vascular normalization by VEGFR2 blockade governs brain tumor response to radiation: Role of oxygenation, angiopoietin-1, and matrix metalloproteinases. *Cancer Cell, 6*, 553–563.
Wirthner, R., Wrann, S., Balamurugan, K., Wenger, R. H., & Stiehl, D. P. (2008). Impaired DNA double-strand break repair contributes to chemoresistance in HIF-1 alpha-deficient mouse embryonic fibroblasts. *Carcinogenesis, 29*, 2306–2316.

Wojciak, J. M., Zhu, N., Schuerenberg, K. T., Moreno, K., Shestowsky, W. S., Hiraiwa, M., et al. (2009). The crystal structure of sphingosine-1-phosphate in complex with a Fab fragment reveals metal bridging of an antibody and its antigen. *Proceedings of the National Academy of Sciences of the United States of America, 106*, 17717–17722.

Yester, J. W., Tizazu, E., Harikumar, K. B., & Kordula, T. (2011). Extracellular and intracellular sphingosine-1-phosphate in cancer. *Cancer Metastasis Reviews, 30*, 577–597.

Yotnda, P., Wu, D., & Swanson, A. M. (2010). Hypoxic tumors and their effect on immune cells and cancer therapy. *Methods in Molecular Biology, 651*, 1–29.

Yu, F., White, S. B., Zhao, Q., & Lee, F. S. (2001). HIF-1alpha binding to VHL is regulated by stimulus-sensitive proline hydroxylation. *Proceedings of the National Academy of Sciences of the United States of America, 98*, 9630–9635.

Yun, J. K., & Kester, M. (2002). Regulatory role of sphingomyelin metabolites in hypoxia-induced vascular smooth muscle cell proliferation. *Archives of Biochemistry and Biophysics, 408*, 78–86.

Zhang, J., Honbo, N., Goetzl, E. J., Chatterjee, K., Karliner, J. S., & Gray, M. O. (2007). Signals from type 1 sphingosine 1-phosphate receptors enhance adult mouse cardiac myocyte survival during hypoxia. *American Journal of Physiology. Heart and Circulatory Physiology, 293*, H3150–H3158.

Zhang, H., Li, W., Sun, S., Yu, S., Zhang, M., & Zou, F. (2012). Inhibition of sphingosine kinase 1 suppresses proliferation of glioma cells under hypoxia by attenuating activity of extracellular signal-regulated kinase. *Cell Proliferation, 45*, 167–175.

Zhou, Q., Liu, L. Z., Fu, B., Hu, X., Shi, X., Fang, J., et al. (2007). Reactive oxygen species regulate insulin-induced VEGF and HIF-1alpha expression through the activation of p70S6K1 in human prostate cancer cells. *Carcinogenesis, 28*, 28–37.

Zhu, H., Chen, X. P., Luo, S. F., Guan, J., Zhang, W. G., & Zhang, B. X. (2005). Involvement of hypoxia-inducible factor-1-alpha in multidrug resistance induced by hypoxia in HepG2 cells. *Journal of Experimental & Clinical Cancer Research, 24*, 565–574.

CHAPTER SIX

Therapeutic Potential of Targeting SK1 in Human Cancers

Heba Alshaker*, Lysann Sauer*, Danielle Monteil*, Silvia Ottaviani*, Shyam Srivats*, Torsten Böhler†, Dmitri Pchejetski*,[1]

*Department of Surgery and Cancer, Faculty of Medicine, Imperial College London, Hammersmith Hospital, London, United Kingdom

†Research and Development, Gambro, Hechingen, Germany

[1]Corresponding author: e-mail address: d.pshezhetskiy@imperial.ac.uk

Contents

Advances in Cancer Research, Volume 117
ISSN 0065-230X
http://dx.doi.org/10.1016/B978-0-12-394274-6.00006-6

Abstract

Sphingosine kinase 1 (SK1) is a lipid enzyme with oncogenic properties that converts the proapoptotic lipids ceramide and sphingosine into the antiapoptotic lipid sphingosine-1-phosphate and activates the signal transduction pathways that lead to cell proliferation, migration, the activation of the inflammatory response, and the impairment of apoptosis.

There is compelling evidence that SK1 activation contributes to cancer progression leading to increased oncogenic transformation, tumor growth, resistance to therapies, tumor neovascularization, and metastatic spread. High levels of SK1 expression or activity have been associated with a poor prognosis in several human cancers.

Recent studies using cancer cell and mouse models demonstrate a significant potential for SK1-targeting therapies to synergize with the effects of chemotherapy and radiotherapy; however, until recently the absence of clinically applicable SK1 inhibitors has limited the translation of these findings into patients. With the recent discovery of SK1 inhibiting properties of a clinically approved drug FTY720 (Fingolimod), SK1 has gained significant attention from both clinicians and the pharmaceutical industry and it is hoped that trials of newly developed SK1 inhibitors may follow soon.

This review provides an overview of the SK1 signaling, its relevance to cancer progression, and the potential clinical significance of targeting SK1 for improved local or systemic control of human cancers.

1. INTRODUCTION TO SPHINGOLIPID METABOLISM

1.1. Overview of sphingolipid metabolism

Sphingolipids are structural components of the plasma membrane in all eukaryotic cells. Aside from being an integral part of the cell membrane they play a significant role in regulation of intracellular signaling. Two sphingolipid metabolites, ceramide and sphingosine-1-phosphate (S1P), have drawn a particular attention because of their pleiotropic effects on cellular processes such as growth, survival, angiogenesis, lymphocyte trafficking, and cancer progression (recently reviewed in Maceyka, Harikumar, Milstien, & Spiegel, 2012; Mullen, Hannun, & Obeid, 2012). Sphingolipid biosynthesis and catabolism are a highly complex network of metabolic steps where ceramide occupies a central position. *De novo* synthesis of ceramide starts with the condensation of palmitate and serine, and is a unique entry point of sphingolipid metabolism (Ogretmen & Hannun, 2004). Ceramide can also be formed from hydrolysis of sphingomyelin or cerebrosides, which is catalyzed by sphingomyelinases (SMases). The cleavage of ceramide by ceramidases yields sphingosine. Sphingosine is phosphorylated by sphingosine kinases 1 and 2 (SK1 and SK2) to generate S1P. S1P can be either dephosphorylated by S1P phosphatases back to sphingosine or degraded by S1P lyase to nonsphingolipids ethanolamide phosphate and hexadecenal, which can be reused for the biosynthesis of phosphatidylethanolamine and is the only exit point from the sphingosine pathway (Hannun & Obeid, 2008). Ceramide can be further converted to glycosphingolipids through glucosylceramidase. The simplest glycoshingolipids are glucosylceramide and galactosylceramide with a glucose or galactose group, respectively, from which more complex glycosphingolipids can be synthesized by adding glycose subunits (Fig. 6.1).

The balance between S1P production and degradation is tightly regulated because S1P is a potent bioactive lysophospholipid (reviewed in

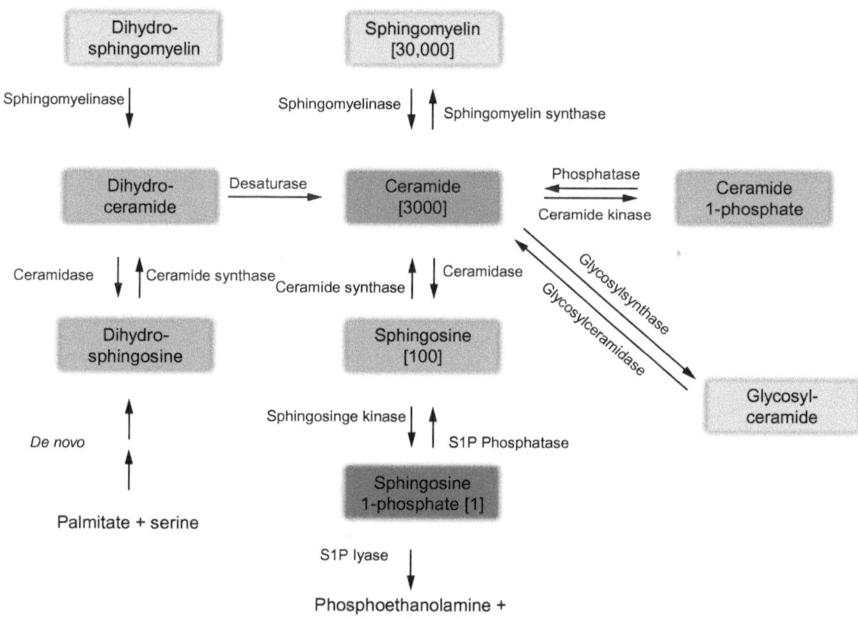

Figure 6.1 Sphingolipid metabolism. Ceramide is a central molecule of sphingolipid metabolism and can be converted to other interconnected bioactive lipid species. The cleavage of ceramide by ceramidases yields sphingosine, which can be phosphorylated by sphingosine kinases to generate S1P. S1P is degraded by S1P lyase into nonsphingolipid phosphoethanolamide phosphate and hexadecenal which can be reused for the biosynthesis of lipids. Degradation of S1P is the only exit pathway from the sphingolipid pathway. The numbers in brackets indicate the relative levels of these sphingolipids. *Reviewed in Hannun and Obeid (2008).* (See Page 4 in Color Section at the back of the book.)

Takabe, Paugh, Milstien, & Spiegel, 2008). S1P has a dual signaling action. On the one hand, S1P may act as an intracellular second messenger enhancing proliferation and suppressing apoptosis (reviewed in Spiegel & Kolesnick, 2002). On the other hand, S1P is secreted to activate specific cell surface receptors, initiating autocrine, and paracrine signaling cascades that induce cell migration, angiogenesis, and differentiation (reviewed in Pyne & Pyne, 2002). In contrast, ceramide is predominantly involved in apoptosis, cell cycle arrest, and senescence (reviewed in Hannun & Obeid, 2008).

1.2. Receptor-mediated S1P signaling

S1P functions intracellularly as a second messenger and extracellularly in an autocrine and paracrine fashion by activating S1P receptors (S1PRs). Intracellular S1P is unable to migrate across the plasma membrane due to the charge of its polar head group. ATP-binding cassette (ABC) transporter family members, ABCC1 (Mitra et al., 2006) and ABCG2 (Takabe et al., 2010), or a sphingolipid transporter spinster homolog 2 (SPNS2) (Hisano, Kobayashi, Kawahara, Yamaguchi, & Nishi, 2011) has been suggested to play a role in the transport of S1P across the membrane. A number of studies indicate a cellular specificity of S1P release depending on the efficiency and type of these transporters. Endothelial cells synthesize and release endogenous S1P more efficiently than fibroblasts and colon cancer cells (Lee, Venkataraman, Hwang, Han, & Hla, 2007). In mast and breast cancer cells, S1P release is mediated via the ABCC1 (MRP1) transporter (Mitra et al., 2006; Takabe et al., 2010) and in astrocytes and platelets via the ABCA1 transporter (Kobayashi et al., 2006; Sato et al., 2007), while in endothelial cells both of these are important (Lee et al., 2007). Recently, both ABCC1 and ABCG2 transporters have been shown to mediate the export of S1P in breast cancer cells (Takabe et al., 2010). SPNS2 has been shown to export S1P in zebrafish myocardial precursors (Kawahara et al., 2009) and Chinese hamster ovary (CHO) cells (Hisano et al., 2011). Moreover it was shown to mediate the release of phosphorylated FTY720 after its intracellular phosphorylation, suggesting an important intermediary mechanism for this key antagonist of S1P1 receptor (Hisano et al., 2011). On the other hand, cystic fibrosis transmembrane regulator, a member of the ABC transporter family, was suggested to be implicated in the uptake of S1P in the cell (Boujaoude et al., 2001).

When secreted outside of the cell, S1P can bind with high affinity (except S1P4; Candelore et al. (2002)) to five transmembrane G-protein-coupled receptors, namely, S1P1, S1P2, S1P3, S1P4, and S1P5, previously referred to as endothelial differentiation gene-1, -5, -3, -6, and -8 and regulate diverse signaling pathways, depending on the cell type, their expression pattern, and the interacting G-proteins (reviewed in Sanchez & Hla, 2004). S1P1, S1P2, and S1P3 are widely expressed in mammalian tissues, while the expression of S1P4 is mostly confined to lymphoid and hematopoietic tissue and S1P5 to the central nervous system (Sanchez & Hla, 2004; Young & Van Brocklyn, 2006). These S1PRs couple to a variety of G-proteins that regulate numerous downstream signaling pathways (Fig. 6.2). This endows S1P with the ability to stimulate diverse signal transduction pathways in different cell types as well as within the same cell, resulting in the possibility of diverse biological outcomes, depending on the cell type, and the relative expression of G-proteins and S1PRs (Payne, Milstien, & Spiegel, 2002). The coupling of S1P1, S1P2, and S1P3 to G-proteins has been studied by determining the

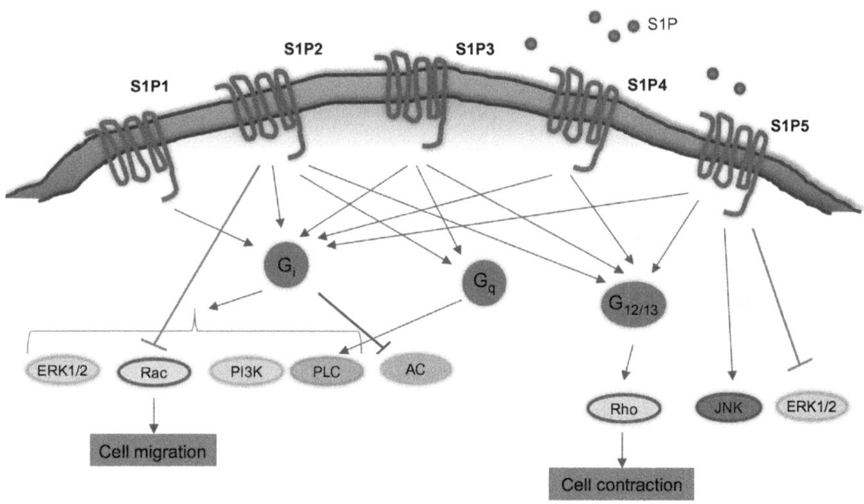

Figure 6.2 S1P receptor-mediated signaling. S1P in the serum is mainly secreted by platelets, erythrocytes. S1P binds to five specific G-protein coupled receptors termed S1P1–5. Couple with a variety of G-proteins leads to the activation or inhibition of various downstream signaling cascades resulting in diverse biological outcomes). S1P1 stimulation leads to activation of phospholipase C (PLC), extracellular-regulated kinase (ERK), and inhibition of adenylate cyclase (AC) via G_i. S1P3 is coupled via G_i to ERK, PLC activation and inhibition of AC, via $G_{q/11}$ to PLC and via $G_{12/13}$ to Rho GTPases (Rho). S1P2 activates PLC and ERK via G_i, it also activates PLC via $G_{q/11}$ and Rho via $G_{12/13}$. Both S1P1 and S1P3 induce Rac activation via G_i, leading to cell migration, whereas S1P2 mediates inhibition of Rac activity through an unknown mechanism, leading to inhibition of cell migration. Both S1P3 and S1P2 mediate Rho activation via $G_{12/13}$. *Reviewed in Spiegel and Milstien (2003) and Takuwa (2002).* (See Page 5 in Color Section at the back of the book.)

binding of ^{35}S-GTPγS to the four major subclasses of heterotrimeric G-protein, G_s, G_i, G_q, and $G_{12/13}$ in cell membranes using a heterologous expression system of insect Sf9 cells (Windh et al., 1999). The study revealed that S1P1 coupled exclusively to G_i, whereas both S1P2 and S1P3 coupled not only to G_i but also to G_q and G_{13}. Subsequently, S1P5 was shown to couple to G_i and G_{12} (Malek et al., 2001) and S1P4 to G_i and $G_{12/13}$ (Graler et al., 2003). The downstream signaling pathways activated by S1P binding to its receptors have been extensively studied in various types of mammalian cells including human embryonic kidney (HEK) cells, CHO cells, human erythroleukemia cells, Jurkat T cells, and HTC4 hepatoma cells (Kluk & Hla, 2002).

In CHO cells expressing each of the S1PRs, it was shown that S1P1, S1P2, and S1P3 coupled to overlapping, but distinct sets of signaling pathways (Fig. 6.2). S1P1 induces the stimulation of extracellular signal-regulated kinase (ERK) and phospholipase C (PLC), as well as inhibition of cyclic adenosine monophosphate (cAMP) accumulation (Okamoto et al., 1998). S1P2 and S1P3 also mediate activation of ERK in a G_i-dependent manner. Both these receptors can also mediate PLC activation through G_q and Rho activation through $G_{12/13}$. Lastly, the signaling through S1P4 and S1P5 has yet to be studied thoroughly. However, it is believed that S1P4 stimulates PLC and ERK via G_i (Van Brocklyn et al., 2000) and S1P5 mediates inhibition of cyclic AMP accumulation and inhibition of serum-induced ERK activation through G_i (Malek et al., 2001).

S1PRs are internalized upon ligand binding and can either be recycled back to the membrane (resensitization) or transported to the lysosome for degradation (downregulation) (Fig. 6.3;Verzijl et al., 2010). The underlying mechanism of this cycling pattern is not clear; however, the receptor fate is likely to be reflected by the ubiquitination status. S1P1 receptor is the most studied. S1P1 activation upon S1P binding generally results in monoubiquitination, while other agonists, for example, FTY720, induce polyubiquitination which leads to degradation (Oo et al., 2007). The difference in ubiquitination was suggested to be due to conformational or functional selectivity or ligand-directed signaling (Rosen, Gonzalez-Cabrera, Sanna, & Brown, 2009). Compelling evidence suggests that internalized S1PRs are also involved in the activation of intracellular signaling pathways. S1P1 receptor internalized after treatment with FTY720 retained a persistent signaling activity (Osinde, Mullershausen, & Dev, 2007). In endothelial cells, S1P1 receptor is rapidly internalized upon S1P stimuli, translocates to the nucleus to induce the expression of proangiogenic factors (Estrada et al., 2009).

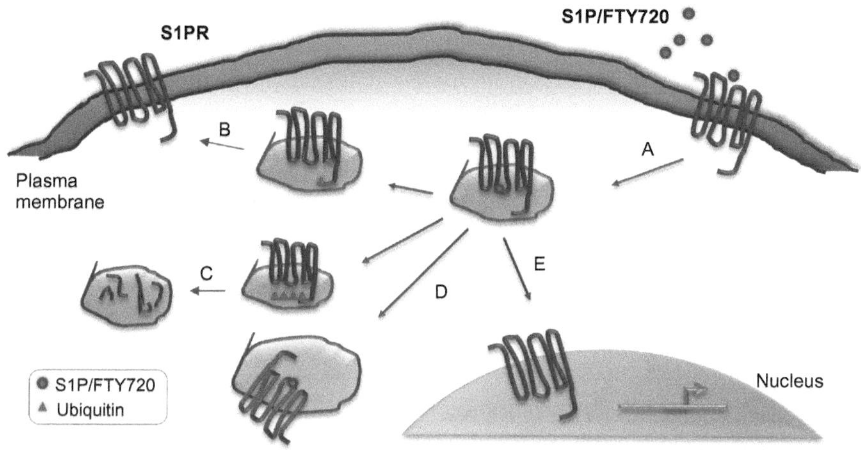

Figure 6.3 Regulation of S1P receptors. Upon binding of S1P, the S1P receptors are internalized (A). Depending on the ligand, S1P receptors are either mono- or polyuboquitinylated. The monoubiquitinylated receptors get recylced to the plasma membrane (B) while the polyubiquinitylated receptors are degraded (C). Internalized S1P receptors may act in intracellular compartments (D) or in the nucleus (E). *Reviewed in Verzijl, Peters, and Alewijnse (2010).* (See Page 5 in Color Section at the back of the book.)

In addition, S1P5 receptor translocates to the centrosomes where it might be involved in the regulation of microtubule and spindle dynamics (Gillies et al., 2009). Therefore, although, the majority of the studies have focused on the extracellular-induced S1P signaling via membrane bound S1PRs, there is growing evidence that internalization of S1PRs is not the end point but can be the beginning of activation of intracellular signaling pathways.

The effect of S1P on cells is cell type specific and largely depends on S1P receptor expression (reviewed in Takabe et al., 2008). S1P1 modulates cellular processes such as vascular maturation (Kono et al., 2004; Lee, Kim, et al., 1999) and lymphocyte trafficking (Pappu et al., 2007). S1P1 activates ERK and PLC and inhibits the accumulation of cyclic AMP (Okamoto et al., 1998). Activation of S1P1 and S1P3 can also lead to cytoskeletal rearrangement and increased cell motility, invasion, and angiogenesis (Lee et al., 2001; Rosenfeldt et al., 2003), notably through activation of integrin alpha v beta 3 in the lamellipodia (Wang, Lee, Lin, & Lee, 2008). In contrast, S1P2 inhibits cell motility (Goparaju et al., 2005; Sugimoto, Takuwa, Okamoto, Sakurada, & Takuwa, 2003). Similar to S1P1, S1P2 is involved in cell migration and regulation of cytoskeletal rearrangements and has an antiapoptotic role (Young & Van Brocklyn, 2007). S1P2 and S1P3 mediate activation of ERK in a G_i-dependent manner. Both of these receptors can mediate PLC activation through G_q and Rho family GTPases activation through $G_{12/13}$ (Sanchez & Hla, 2004). Similar to S1P1 and S1P3, S1P4 mediates cytoskeletal rearrangement (stress fibers formation and cell rounding), cell motility, and calcium mobilization (Graler et al., 2003; Yamazaki et al., 2000). Depending on the cell type, S1P5 may mediate inhibition of mitogen-activated protein kinase (MAPK) and cell proliferation (Malek et al., 2001; Niedernberg, Scherer, Busch, & Kostenis, 2002), or induce an activation of PLC and ERK via G_i (Van Brocklyn et al., 2000). Other important pathways upregulated by S1P include cortactin–Arp2/3 complex (Lee et al., 2006), the small GTPase of the Rho family Rac and cell division control protein 42 (CDC42) (Mehta, Konstantoulaki, Ahmmed, & Malik, 2005), Ras/Raf/MAPK (Usui et al., 2004), phosphatidylinositol 3-kinase (PI3K)/protein kinase B (Akt) (Kang et al., 2004), nuclear factor kappa B (NF-kappaB) (Lin, Chen, Chen, & Lee, 2006), and cyclooxygenase 2 (COX2)/prostaglandin E2 (PGE2) (Serrano-Sanchez, Tanfin, & Leiber, 2008).

1.3. Dual effect of S1P on cell migration

Among the diversity of biological activities of S1P, its bimodal action on cell motility is of particular interest (Table 6.1). S1P either stimulates or inhibits cell migration, depending on cell types (Takuwa, 2002). It was shown that S1P inhibited cell migration of vascular smooth muscle cells (Bornfeldt et al., 1995) and neutrophils (Kawa et al., 1997) that were directed toward PDGF and IL-8, respectively. In addition, S1P was observed to inhibit *in vitro* invasion of certain tumor cell lines, including mouse and human melanoma cells (Sadahira et al., 1992; Yamamura et al., 1997), osteosarcoma cells (Sadahira et al., 1992), and breast carcinoma cells (Wang, Van Brocklyn, Edsall, Nava, & Spiegel, 1999). On the other hand, it was shown that S1P by itself was able to induce migration of vascular endothelial cells (Lee, Thangada, et al., 1999; Wang, Van Brocklyn, Hobson, et al., 1999) and mouse embryonic fibroblasts (Liu, Wada, et al., 2000).

Molecular mechanisms underlying the bimodal regulation of cell migration by S1P remained elusive until recently. It is now revealed that the distinctive regulation of the Rho family GTPases, particularly Rho and Rac, by the S1PRs is responsible for the positive or negative effect of S1P on cell migration (Fig. 6.2). The Rho family of GTPases is important regulators of the actin cytoskeleton and cell motility (Hall, 1998). Rho GTPases mediate stress fiber formation and focal adhesion, while Rac and CDC42 direct peripheral actin polymerization that results in formation of lamellipodia and

Table 6.1 Effects of S1P on cell migration

Responses	Cell type	References
Formation of stress fibers and focal adhesions	HUVEC	Lee, Thangada, et al. (1999)
	CHO cells overexpressing S1P2	Gonda et al. (1999)
Pseudopodium formation	T-lymphoma cells	Stam, Michiels, van der Kammen, Moolenaar, and Collard (1998)
Membrane ruffle formation	CHO cells overexpressing S1P1 and S1P3	Okamoto et al. (2000)
Stimulation of migration	HUVEC, BAEC	Lee, Thangada, et al. (1999)
	HEK293 cells overexpressing S1P1	Wang, Van Brocklyn, Hobson, et al. (1999)
	CHO cells overexpressing S1P1 and S1P3	Okamoto et al. (2000)
Stimulation of invasion	T-lymphoma cells	Stam et al. (1998)
Inhibition of migration	Neutrophils	Kawa, Kimura, Hakomori, and Igarashi (1997)
	B16 melanoma cells	Sadahira, Ruan, Hakomori, and Igarashi (1992)
	MDA-MB-231 breast cancer cells	Wang, Van Brocklyn, Edsall, et al. (1999)
	Vascular smooth muscle cells	Bornfeldt et al. (1995)
	CHO cells overexpressing S1P2	Okamoto et al. (2000)
Inhibition of invasion	B16 melanoma cell	Sadahira et al. (1992)
	MDA-MB-231 breast cancer cells	Wang, Van Brocklyn, Edsall, et al. (1999)

HUVEC, human umbilical vein endothelial cells; BAEC, bovine aortic endothelial cells.

filopodia, respectively. Widely expressed S1P1 and S1P3 receptors typically promote cell migration by activating the small GTPase Rac, leading to actin polymerization and lamellipodia formation (Liu, Wada, et al., 2000; Matloubian et al., 2004; Paik, Chae, Lee, Thangada, & Hla, 2001). For this reason they are called "chemoattractant" receptors (Fig. 6.2). Conversely,

S1P2 has been called a "repellent" receptor as it stimulates Rho and antagonizes Rac, leading to stress fiber formation and cessation of migration (Goparaju et al., 2005; Sanchez et al., 2005; Sugimoto et al., 2003). Recently, it has been shown that S1P4, which is highly expressed in the lymphoid system, stimulates cell migration of CHO and Jurkat T cells (Graler et al., 2003), whereas S1P5, which is expressed in the central nervous system, inhibits cell migration of oligodendrocyte precursor cells (Novgorodov, El-Alwani, Bielawski, Obeid, & Gudz, 2007).

S1P signaling has been shown to promote other pathways involved in cell migration. In glioblastoma cells, upregulation of SK1/S1P/S1P1 axis induces urokinase-type plasminogen activator (uPA) and urokinase-type plasminogen activator receptor (uPAR) expression thereby enhancing invasion (Young, Pearl, & Van Brocklyn, 2009). S1P stimulation was also shown to induce plasminogen activator inhibitor 1 (PAI-1) expression promoting cancer cell invasion (Bryan et al., 2008).

1.4. Nonreceptor-mediated S1P signaling

Several studies indicate a possibility for a nonreceptor, intracellular action of S1P that leads to an activation of the same signaling pathways as by S1PRs, although the mechanism of this activation has not yet been identified (Strub, Maceyka, Hait, Milstien, & Spiegel, 2010). In a recent study, nuclear S1P was demonstrated to bind to the histone deacetylases (HDAC1 and HDAC2) and regulate gene expression epigenetically (Hait et al., 2009). S1P was shown to form complexes with tumor necrosis factor (TNF) receptor-associated factor 2 (TRAF2) and induce its activation leading to increased NF-kappaB signaling (Alvarez et al., 2010). Intracellular S1P production was shown to induce release of calcium from intracellular stores and its mobilization (Blom, Slotte, Pitson, & Tornquist, 2005; Meyer zu Heringdorf et al., 2001). Recently, a centrosomal localization of S1P5 was identified and suggested to be linked to mitosis regulation by intracellular S1P (Gillies et al., 2009). An S1PRs-independent mechanism of SK1-mediated cell migration has been proposed (Duan et al., 2004) suggesting a possible role for the intracellular S1P.

2. SPHINGOSINE KINASE: A REGULATOR OF THE SPHINGOLIPID SIGNALING

The contrasting effects of ceramide and S1P and the fact that both sphingolipid metabolites are interconvertible led to the concept of the so-called ceramide/S1P rheostat postulating that the ratio of these lipids determines cell fate (Cuvillier et al., 1996). SK1 is a lipid enzyme with oncogenic properties that converts proapoptotic lipid second messenger sphingosine into an antiapoptotic lipid second messenger S1P (Figs. 6.4 and 6.5). SK1 is tightly regulated by growth factors, cytokines, receptor tyrosine kinases, and prosurvival signaling pathways and plays a key role in several fundamental biological processes including: cell proliferation, regulation of apoptosis, cell migration, fibrosis, angiogenesis, nociception, and inflammatory responses (Table 6.2, Figs. 6.4 and 6.5), making it a key pathway in the search for targeted therapies.

2.1. Isoforms of sphingosine kinase

Two SK genes are expressed in humans: *SphK1* and *SphK2*. *SphK1* and *SphK2* are located in different genes (17q25.2 and 19q13.2, respectively), differ in size (43 kDa and 65kDA, respectively), and have a number of splice variants (Liu, Sugiura, et al., 2000). Both enzymes have five highly

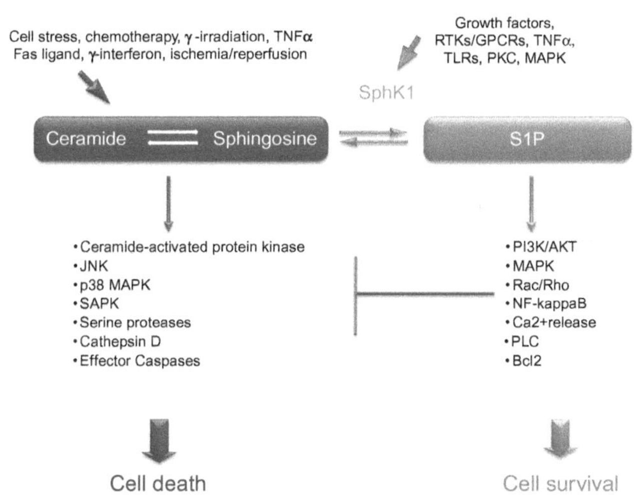

Figure 6.4 "Sphingolipid Rheostat." In response to stress stimuli there is an accumulation of intracellular lipid second messengers ceramide and sphingosine, which induce activation of apoptotic pathways. Prosurvival stimuli (GPCRs/RTKs, LPA, Toll-like receptors) induce activation of SK1 and generation of S1P, a lipid second messenger that activates antiapoptotic pathways and antagonizes the effects of ceramide and sphingosine. The intracellular balance between ceramide, sphingosine and S1P determines the cell fate. TNF, tumor necrosis factor; RTKs, receptor tyrosine kinases; GPCRs, G-protein coupled receptors; TLRs, toll-like receptors; PKC, protein kinase C; MAPK, mitogen-activated protein kinases; JNK, c-Jun N-terminal kinases; SAPK, stress activated protein kinase; NF-kappaB, Nuclear factor kappa B; PLC, phospholipase C; Bcl2, B cell lymphoma 2. *Reviewed in Pchejetski, Bohler, Stebbing, and Waxman (2011).* (See Page 6 in Color Section at the back of the book.)

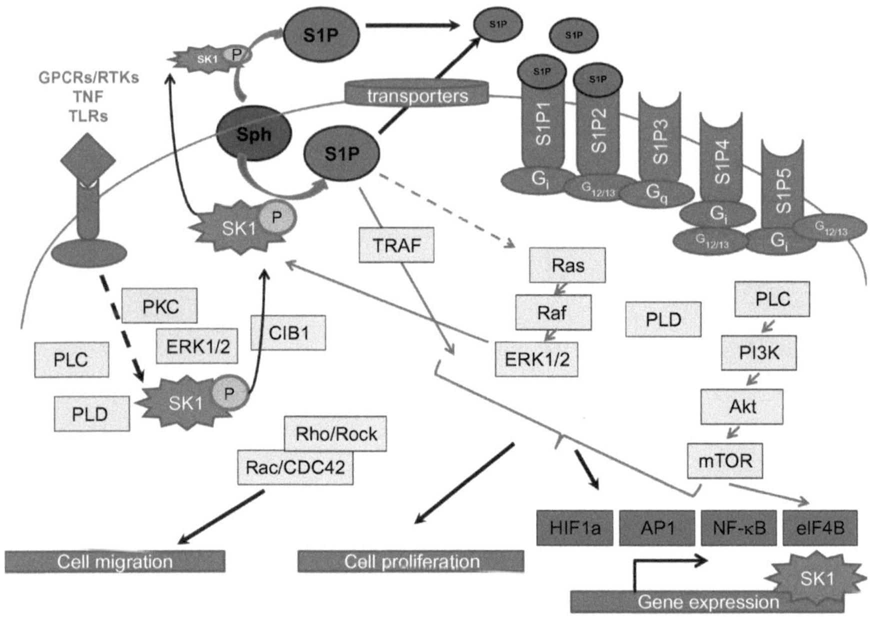

Figure 6.5 SK1/S1P signaling pathway. Tumor necrosis factor alpha, Receptor tyrosine kinases, G-protein coupled receptors, Toll-like receptors and other prosurvival stimuli induce phosphorylation of sphingosine kinase 1 (SK1), its translocation to plasma membrane, and generation of sphingosine-1-phosphate (S1P) from sphingosine (reviewed in Pitson, 2011). SK1 is then translocated to the plasma membrane (via CIB1) or outside the cell, where it converts sphingosine into S1P. S1P can then be exported outside the cell where it acts in a paracrine or autocrine manner and activates five specific S1P receptors (S1P1–5). Upon coupling with S1P, these receptors can activate downstream signaling pathways leading to cell proliferation, migration, and gene expression. Intracellular S1P was also reported to induce activation of prosurvival signaling pathways (e.g., TRAF2). ERK1/2 and PKC can directly phosphorylate SK1 making a signal amplification loop. TNF, tumor necrosis factor; RTKs, receptor tyrosine kinases; GPCRs, G-protein coupled receptors; TLRs, toll-like receptors; PKC, protein kinase C; MAPK, mitogen-activated protein kinases; PLC, phospholipase C; PLD, phospholipase D; Ras, Rat Sarcoma; NF-kappaB, nuclear factor kappa B; HIF1α, hypoxia-induced factor alpha; EIF4b, eukaryotic initiation factor 4B; CDC42, cell division control protein 42 homolog (reviewed in Pchejetski, Bohler, et al., 2011). (See Page 6 in Color Section at the back of the book.)

conserved regions (C1–C5), whereby the unique catalytic domain is within C1–C3 and the ATP-binding site (SGDGX(1721)K(R)) is within the C2 domain (Alemany, van Koppen, Danneberg, Ter Braak, & Meyer Zu Heringdorf, 2007) and their sequence similarity is 47% and 43% for the N- and C-terminal regions, respectively. Additionally, *SphK2* has four predicted transmembrane domains, while *SphK1* has none (Spiegel & Milstien, 2003).

Table 6.2 Major physiological functions of SK1/S1P pathway

1. Tissue homeostasis	
(a) Cardiac tissue	
Survival of cardiomyocytes	Jin, Goetzl, and Karliner (2004) and Pchejetski et al. (2007)
Protect from hypoxia	Zhang et al. (2007)
CM hypertrophy	Robert et al. (2001)
Regulation of heart rate	Himmel et al. (2000)
Mediation of TGF1β-induced cardiac fibrosis, collagen type I and TIMP1 expression in cardiac fibroblasts.	Gellings Lowe, Swaney, Moreno, and Sabbadini (2009) and Yamanaka et al. (2004)
(b) Neural tissue	
Survival of neuronal cells	Gomez-Brouchet et al. (2007)
(c) Liver tissue	
Survival of neuronal cells	Osawa et al. (2005)
(d) Bone remodeling	
Osteoclast chemoattraction	Ryu et al. (2006)
Coupling of osteoclast and osteoblast activity	Ryu et al. (2006)
Osteoblast survival, MAPK and PKC activation and IL-6 synthesis	Carpio, Stephan, Kamer, and Dziak (1999), Grey et al. (2002), Grey et al. (2004), Kozawa, Kawamura, Matsuno, and Uematsu (2000), Kozawa, Tokuda, Matsuno, and Uematsu (1997), Lampasso, Kamer, Margarone, and Dziak (2001), and Osamu, Haruhiko, Hiroyuki, and Toshihiko (1998)
Mediating the proliferative effects of androgen in osteoblasts	Martin, Lafosse, Malavaud, and Cuvillier (2010)
2. Inflammatory response	
Inflammatory response to TNFα and IL-1	Xia et al. (1998)
Interaction with TRAF2, activation of Akt and NF-kappaB	Radeff-Huang, Seasholtz, Chang, Smith, and Walsh (2006) and Xia et al. (2002)

Continued

Table 6.2 Major physiological functions of SK1/S1P pathway—cont'd

Ubiquitination of receptor interacting protein 1 (RIP1) and stimulation of IkappaB kinase	Alvarez et al. (2010)
Induction of COX2/PGE2 production	Pettus et al. (2003)
Release of proinflammatory cytokines	Melendez and Ibrahim (2004) and Pchejetski, Nunes, Coughlan, et al. (2010)
3. Fibrotic response	
Mediating TGFβ1-induced proliferation of dermal fibroblasts	Yamanaka et al. (2004)
Differentiation of myoblasts into myofibroblasts	Cencetti, Bernacchioni, Nincheri, Donati, and Bruni (2010)
Heart and liver fibrosis	Gellings Lowe et al. (2009) and Serriere-Lanneau et al. (2007)
4. Lymphocyte trafficking	
Chemoattraction	Reviewed in Hla and Brinkmann (2011)
FTY720 is a potent immunosuppressive drug through inhibition of lymphocyte egress from lymphoid organs	Chiba (2005)
5. Neuronal signaling and nociception	
Mediation of (NGF)-stimulated neurite outgrowth and survival of neurons and astrocytes in the central nervous system	Okada, Kajimoto, Jahangeer, and Nakamura (2009)
Increasing neuronal functions (e.g., glutamate secretion)	Kajimoto et al. (2007)
Mediating neurotransmitter release at neuromuscular junctions	Brailoiu, Cooper, and Dun (2002)

Three splice isoforms exist for *SphK1* (SK1a, -b, and -c), which differ only at the N-terminus. SK1a (RefSeq NM 001142601) is the shortest isoform, whereas SK1b (RefSeq NM 021972) has additional 14 amino acids and includes a putative palmitoylation site, which might explain its enhanced constitutive localization at the plasma membrane (Pitson, 2011). SK1c (RefSeq NM 182965) is the longest isoform and is 86-amino acids longer than SK1a.

SphK2 has two additional regions: at the N-terminus and in the central part. *SphK2* has two splice isoforms, which were reported to arise from alternative start codon usage: SK2a (SK2s, GenbankTM accession number AF245447) and SK2b (SK2l, RefSeq NM 020126), which differ in 36 amino acids. SK2b is expressed more abundantly in a range of human tissues and cultured cells (reviewed in Pitson, 2011).

Despite a high homology and the mutual product S1P, SKs display different catalytic properties, temporal expression, subcellular localization, and tissue expression patterns (Taha, Hannun, & Obeid, 2006), suggesting distinct physiological roles of these enzymes (reviewed in Pitson, 2010; Pyne, Lee, Long, & Pyne, 2009).

SphK1 is a constitutively active cytozolic enzyme, and its activity can be further increased by phosphorylation at Ser225 via ERK1/2 (Pitson et al., 2003) and translocation to plasma membrane, where it phosphorylates its substrate sphingosine and produces prosurvival S1P. *SphK2* is localized mainly to nucleus (Igarashi et al., 2003), where it was reported to inhibit DNA synthesis and induce G1/S cell cycle arrest (Igarashi et al., 2003), suppress cell growth and induce apoptosis via regulation of the cyclin-dependent kinase inhibitor p21 (Sankala et al., 2007) and through enhanced histone H3 acetylation (Hait et al., 2009). However, several studies showed a critical role of *SphK2* for EGF-stimulated migration of breast cancer cells (Hait et al., 2005), growth of tumor xenografts (Weigert et al., 2009) and lung cancer chemoresistance (Schnitzer, Weigert, Zhou, & Brune, 2009). Further, recent studies have indicated an antitumorigenic role of a specific *SphK2* inhibitor ABC294640 (Antoon et al., 2011; Beljanski, Lewis, & Smith, 2011). Interestingly, both *SphK1* and *SphK2* were recently shown to colocalize with S1P5 receptor at the centrosome (Gillies et al., 2009). In mice, neither *SphK1* nor *SphK2* single knockout has any effect on development and reproduction, suggesting that despite distinct cellular roles, deficiency of one SK isoform seems to be compensated by the other. However, the simultaneous disruption of both kinases results in S1P deficiency leading to severe malfunction of angiogenesis and neurogenesis causing embryonic lethality (Mizugishi et al., 2005), indicating a critical role for S1P during development.

2.2. Activation of sphingosine kinases

Sphingosine kinases possess an intrinsic catalytic activity that is independent on posttranslational modifications thereby generating basal levels of S1P in the cell (Pitson et al., 2000). This basal activity is rapidly accelerated upon phosphorylation which is stimulated by a variety of growth factors.

The activation of SK1 is triggered by ERK1/2-mediated phosphorylation at Ser225; thus, increasing the catalytic activity and inducing the translocation to the plasma membrane where sphingosine is concentrated (Pitson et al., 2000). This phosphorylation is essential for enhancing plasma membrane binding affinity and selectivity (Stahelin et al., 2005).

SK1 activation induces cytosolic calcium concentration, which enables calmodulin-related protein, calcium and integrin-binding protein 1 (CIB1) to interact with SK1 at its calmodulin-binding site (Jarman, Moretti, Zebol, & Pitson, 2010; Sutherland et al., 2006). CIB1 is a calcium-myristoyl switch protein, which translocates to the plasma membrane in a process dependent on its myristoylation and calcium binding following calcium fluxes (Jarman et al., 2010). Further, SK1 is targeted to membrane microdomains (lipid rafts), and phosphatidylserine or phosphatidic acid (PA) within facilitates the retention of SK1 at the plasma membrane (Delon et al., 2004). The localization of SK1 in the vicinity to its substrate enhances the S1P generation thereby increasing its intracellular levels and extracellular release (Johnson, Becker, Facchinetti, Hannun, & Obeid, 2002; Pitson et al., 2003; Pitson et al., 2005). Interestingly, splice variant SK1b possesses an additional 14 amino acids at the N-terminus, which harbor a potential palmitoylation site. It has been suggested that palmitoylation (a modification known to target proteins to lipid rafts) leads to the constitutive localization of SK1b at the plasma membrane (Pitson, 2011; Venkataraman et al., 2006). SK1 activity can be rapidly and transiently stimulated by a wide variety of growth factors such as EGF (Shu, Wu, Mosteller, & Broek, 2002), PDGF (Hobson et al., 2001), EGF (Sarkar et al., 2005), and many others (Alvarez, Milstien, & Spiegel, 2007).

SK2 regulation and function are less studied. SK2 has a roughly 10-fold lower specific activity, but phosphorylates a wider range of sphingoid base substrates (Liu, Sugiura, et al., 2000; Roberts et al., 2004). A recent report showed that ERK1/2 binds to SK2 and phosphorylates it on Ser351 and Thr578, enhancing its enzymatic activity (Hait, Bellamy, Milstien, Kordula, & Spiegel, 2007). EGF and phorbol ester, a protein kinase C activator, have been reported to activate SK2 (Hait et al., 2007, 2005).

2.3. Subcellular localization of sphingosine kinases

It appears that the localization of SKs is important for their physiological role. SK1 is located predominately in the cytosol and translocates upon stimulation to the plasma membrane, where it enhances S1P secretion from the cell and thereby proliferative signaling (Pitson et al., 2005). Interestingly, artificial localization of nonphosphorylatable SK1 mutant to the plasma membrane also induces cell proliferation (Pitson et al., 2005). The localization of SK2 is more complex and seems to depend on the cell type. Different reports suggest that SK2 is predominately localized in the nucleus and cytosol (Leclercq & Pitson, 2006; Maceyka et al., 2005). The phosphorylation of SK2 (at either Ser383 or Ser385) by protein kinase D causes its nuclear export (Ding et al., 2007; Igarashi et al., 2003). Moreover, SK2 can also be found in the endoplasmatic reticulum (ER) where it appears to have proapoptotic functions (Maceyka et al., 2005). Interestingly, artificial targeting of an antiapoptotic SK1 to the ER induces the same proapoptotic effect as SK2 (Maceyka et al., 2005). There is growing evidence that the SKs can be released into the extracellular environment. It has been suggested that this constitutive release is mediated via a nonclassical secretory pathway (reviewed in Tani, Ito, & Igarashi, 2007). Fibroblasts were shown to release SK1 in a complex with FGF1 in response to cellular stress (Soldi et al., 2007). Interestingly, SK2b can be secreted after cleavage of its N-terminus by caspase-1 in apoptotic cells (Wei, Wolin, & Colditz, 2010). The mechanism of its extracellular action is unclear, but it is believed that sphingomyelin degradation occurs at the outer leaflet of the plasma membrane and ATP is released by cells for direct extracellular S1P generation (reviewed in Tani et al., 2007). Recently, SK1 has been shown to be shed in vesicles from human tumor cells (Rigogliuso et al., 2010).

2.4. Potential S1P-independent signaling of SK1

Several studies indicate that SK1 interacts or forms complexes with various proteins (Table 6.3). While the mechanisms of SK1 interaction with these proteins and their physiological significance are not fully investigated, these findings indicate the potential of SK1 to mediate intracellular signaling independently of S1P production (e.g., modulation of cytoskeletal dynamics, cell migration) (Kohno et al., 2006). This assumption is partially supported by a critical finding that SK1 phosphorylation and translocation to plasma membrane, is critical for its prosurvival and oncogenic signaling irrespective of its catalytic activity and S1P production (Pitson et al., 2005). In this report, authors did not fully discard the role of plasma S1P production in these events; they have, however, clearly demonstrated that S1P generation in other cell compartments had no effect on SK1-mediated cell proliferation.

Table 6.3 SK1 interacting proteins

	Protein	Physiological effect or effect on SK1	References
1	ERK1/2	Direct phosphorylation	Pitson et al. (2000)
2	PKC	Direct phosphorylation	Shu et al. (2002)
3	Delta-catenin/NPRAP (neural plakophilin-related armadillo repeat protein)	SK1 activation	Fujita et al. (2004)
4	TRAF2	SK1 activation	Alvarez et al. (2010) and Xia et al. (2002)
5	Phosphatidylserine	Plasma membrane localization and activation of SK1	Stahelin et al. (2005)
6	Shp-2 tyrosine phosphatase	SK1 activation in response to HGF	Duan et al. (2006)
7	Actin	Maintenance of SK1 activity, modulation of cytoskeletal dynamics	Iyer and Kusner (2009) and Kusner et al. (2007)
8	Filamin A	Cell migration	Maceyka, Alvarez, Milstien, and Spiegel (2008)
9	RPK118	Not fully elucidated	Hayashi et al. (2002)
10	PECAM-1	Reduces SK1 activity	Fukuda, Aoyama, Wada, and Igarashi (2004)
11	Aminoacylase 1	Reduces SK1 activity	Maceyka, Nava, Milstien, and Spiegel (2004)
12	PP2a (protein phosphatase 2A)	Directly reduces SK1 activity through dephosphorylation	Barr, Lynn, Moretti, Khew-Goodall, and Pitson (2008)
13	SKIP/AKAP (sphingosine kinase type 1-interacting protein/protein kinase A anchoring protein)	Reduces SK1 activity	Kovanich et al. (2010) and Lacana, Maceyka, Milstien, and Spiegel (2002)

3. ROLE OF SK1 IN CANCER PROGRESSION

3.1. Oncogenicity, cancer cell growth, and resistance to stress

There is compelling evidence that SK1 activation contributes to cancer progression and leads to oncogenic transformation (Xia et al., 2000), increased tumor growth (Olivera et al., 1999), and impairment of apoptosis (Xia, Wang, Gamble, & Vadas, 1999). The SK1/S1P pathway contributes to cancer pathogenesis being induced by various mutations (e.g., Spi-1/PU.1, Le Scolan et al., 2005; MDR, Bonhoure et al., 2006) and subsequently activates various signal transduction pathways like PI3K/Akt (Song et al., 2011), MAPK (Sukocheva et al., 2003), NF-kappaB (Billich et al., 2005), or hypoxia inducible factor 1α (Ader, Brizuela, Bouquerel, Malavaud, & Cuvillier, 2008), leading to uncontrolled cell proliferation. SK1 has been found to be a tumor-associated enzyme: high levels of SK1 expression have been shown in various human tumors such as brain, breast, colon, lung, ovary, stomach, uterus, kidney, rectum, and small intestine (Bayerl et al., 2008; French et al., 2003; Johnson et al., 2005; Van Brocklyn et al., 2005), where they enhance tumor neovascularization (Licht, Tsirulnikov, Reuveni, Yarnitzky, & Ben-Sasson, 2003) and metastatic potential by promoting motility and invasion of cancer cells (Visentin et al., 2006).

High levels of SK1 expression or activity are associated with a poor prognosis in several human cancers (Table 6.4). Increased SK1 expression has been associated with tumor progression and has significantly decreased survival rate in glioblastoma multiforme patients (Van Brocklyn et al., 2005). Upregulation of SK1 has been linked to a poor survival outcome in estrogen-dependent and -independent breast cancer patients (Ruckhaberle et al., 2008). Upregulation of SK1 has been shown to significantly correlate with the histologic grade of astrocytoma and patients with high SK1 level exhibited shorter survival time (Li et al., 2008). Increased tumor stage-dependent expression of SK1 was shown in azoxymethane-induced colon carcinomas (Kawamori et al., 2006). High expression of S1P1 and S1P3 receptors, SK1, and ERK1/2 was associated with development of tamoxifen resistance in estrogen receptor-positive breast cancer patients (Watson et al., 2010). On the contrary, in estrogen receptor-negative breast cancer patients it is rather S1P4 and SK1 expression, which determine the clinical outcome (Ohotski et al., 2012). A preclinical study of prostate samples from 30 patients who underwent radical prostatectomy reported

Table 6.4 Cancers where high SK1 expression/activity was linked with cancer progression and patients survival

Cancer	Clinical correlates	References
1 Prostate	PSA, Gleason sum, positive resection margins, surgical failure	Malavaud et al. (2010)
2 Breast	Poor prognosis, shorter recurrence time, tamoxifen resistance	Ruckhaberle et al. (2007) and Watson et al. (2010)
3 Glioma	Histologic grade of astrocytoma, shorter patient survival	Li et al. (2008) and Van Brocklyn et al. (2005)
4 Gastric	Shorter overall survival time	Li et al. (2009)
5 Head and neck	Advanced tumor stage, nodal involvement, recurrence, shorter patient survival time and loss of p21 expression	Facchinetti et al. (2010), and Sinha et al. (2010)
6 Salivary gland	TNM status, tumor differentiation, shorter overall survival time	Liu, Zheng, et al. (2010)
7 Non-small cell lung	Tumor progression, poor survival	Song et al. (2011)

that tumor SK1 activity is associated with prostate cancer progression (PSA, tumor volumes, and Gleason score) and disease recurrence (positive margins and surgical failure) (Malavaud et al., 2010). Another study has shown that basal SK1 activity is $\sim$10-fold higher in androgen-independent, metastatic PC-3 cells than in androgen-dependent, nonmetastatic LNCaP cells (Pchejetski et al., 2005). In 2009, Dayon et al. have shown that SK1 is central to androgen-regulated prostate cancer growth and survival (Dayon et al., 2009). This study demonstrated that while SK1 is upregulated in response to androgen signaling through PI3K/Akt pathway, long-term androgen deprivation raises basal SK1 levels, proposing it as a potential compensatory mechanism leading to androgen-independence. Interestingly, a potential mechanism of SK1 overexpression in prostate cancer may come from a recent study showing that SK1 is regulated by eukaryotic elongation factor 1A and particularly by its truncated isoform termed prostate tumor inducer-1 (PTI-1), which was previously shown to be overexpressed in human prostate cancer (Sun, Lin, Katz, & Fisher, 1997). PTI-1 that lacks the G-protein domain of eEF1A1 was previously shown to induce neoplastic cell transformation through undefined mechanisms. A recent study identified that PTI-1 can directly activate SK1 and that this activation is essential for oncogenic transformation induced by PTI-1 (Leclercq, Moretti, & Pitson, 2011).

SK1 expression and activity are upregulated in various cancers, but so far there is no evidence of mutations in SK1. This led to the debate whether SK1 is an oncogene or a nononcogene addiction. NIH-3T3 fibroblasts stably transfected with SK1 showed accelerated cell proliferation in a serum- and anchorage-independent manner while expressing the dominant-negative SK1 mutant had no such effect. SK1-transfected cells have the ability to form tumors in NOD mice (Xia et al., 2000). In addition, an active Ras mutant activated SK1 while the inhibition of SK1 inhibited Ras transforming activity (Xia et al., 2000). Orthotopic prostate tumors transfected with SK1 showed enhanced tumor growth and chemoresistance (Pchejetski et al., 2005).

Besides genomic regulation, SK1 can also be upregulated in response to several anticancer treatments (Akao et al., 2006; Pchejetski, Bohler, et al., 2010; Pchejetski et al., 2005), which leads to a partial or complete resistance of cancer cells to these therapies. It was further shown that apoptosis-induced SK1 expression and subsequent release of S1P serve as a "come-and-get-me" signal for tumor-associated macrophages (TAMs) and may therefore promote inflammatory tumor microenvironment (Gude et al., 2008). SK1 expression can protect the cells against apoptosis induced by various factors such as tumor necrosis factor alpha (TNF-α) and Fas ligand (Bektas et al., 2005; Nava, Hobson, Murthy, Milstien, & Spiegel, 2002). Moreover, in cancer cells, SK1 mediates survival under stress conditions such as starvation (Sukocheva et al., 2003; Xia et al., 2000). The down-regulation of SK1 by small interfering RNAs (siRNAs) in melanoma cell lines has been demonstrated to decrease S1P levels and cancer cell resistance to ceramide-mediated apoptosis (Bektas et al., 2005).

3.2. Cancer cell migration

Several publications indicate that activation of SK1 induces cancer cell migration (Doll, Pfeilschifter, & Huwiler, 2007; Kaneider, Djanani, Fischer-Colbrie, & Wiedermann, 2002; Maceyka et al., 2008) and that SK1/S1P axis enhances metastatic potential of cancer cells (Takuwa, 2002; Visentin et al., 2006).

SK1 was shown to be activated downstream of several chemotactic receptors (e.g., LPA1, EGF, PDGF) and subsequently mediate cell motility and chemotaxis (Hobson et al., 2001; Shida et al., 2008). The upregulation of SK1 activity has been shown to increase cell migration, whereas the selective depletion of SK1 by siRNAs can decrease cell migration (Bergelin et al., 2009; Doll et al., 2007; Kaneider et al., 2002; Schwalm et al., 2008). However, a recent paper has shown that SK1 prevents formation of a migratory phenotype in response to S1P in estrogen receptor-positive breast cancer cells (Long et al., 2010).

The major proposed mechanism of SK1-induced cell migration is through S1P secretion and coupling to S1PRs (Hla, 2003) leading to activation of Rac signaling, leading to actin polymerization and lamellipodia formation (as described above). S1P1, S1P3, and S1P4 receptors have been shown to mediate mainly promigratory responses (Berdyshev et al., 2011; Graler et al., 2003; Maeda et al., 2007; Van Brocklyn, Young, & Roof, 2003), while S1P2 and S1P5 stimulate Rho and antagonize Rac, leading to stress fiber formation and the cessation of migration (Arikawa et al., 2003; Novgorodov et al., 2007). Due to the differential modulation of the Rac and Rho pathways by different S1PRs, the effect of S1P on cell motility depends on the differential expression of S1PRs in a specific cell type. For example, although antimigratory in other cell types, in fibroblasts, S1P2 has been reported to promote cell migration (Hashimoto et al., 2008). SK1/S1P-induced cancer cell migration has also been linked with expression of PAI-1 and uPAR, leading to enhanced adhesion and invasion of U373 glioblastoma cells (Bryan et al., 2008).

The nonreceptor intracellular effect of SK1/S1P axis has been considered as an alternate mechanism of action. In support of this proposed mechanism, hepatocyte growth factor (HGF)-induced migration of endothelial cells involves intracellular S1P and/or SK1 but not S1PRs (Duan et al., 2004). Importantly, a recent study describes the formation of a signaling complex between SK1, S1P1, and the cytoskeletal protein Filamin A that localizes to membrane ruffles of migrating cells to promote cell movement (Maceyka et al., 2008).

3.3. Tumor angiogenesis

Several reports suggest that SK1/S1P signaling is involved in tumor angiogenesis. S1P promotes endothelial-cell growth and interacts with vascular endothelial growth factor (VEGF) signaling, which is crucial for angiogenesis (Liu et al., 2001). VEGF was shown to stimulate S1P production in the T24 bladder tumor cell line, which in turn mediated VEGF-induced activation of RAS and MAPKs in these cells (Wu, Shu, Hovsepyan,

Mosteller, & Broek, 2003). It has been demonstrated that S1P can be secreted from the tumor tissue to the neighboring sites where it can have access to peripheral blood or lymph flow and act as a chemoattractant and draw various cells (including vascular endothelial cells) to the site of the tumor growth (Licht et al., 2003). The expression of S1P1 is strongly induced in tumor vessels and the specific knockdown of S1P1 significantly abrogates angiogenesis resulting in suppressed tumor growth in murine models (Chae, Paik, Furneaux, & Hla, 2004). It was then shown that secreted S1P initiated endothelial-cell sprouting in three-dimensional collagen matrices, which is a representative of a multistep angiogenic process (Anelli, Gault, Snider, & Obeid, 2010). Antibodies to S1P have been recently demonstrated to have antitumor potential (Visentin et al., 2006), notably by inhibition of cell proliferation, the release of proangiogenic cytokines such as VEGF, IL-8, and IL-6, and blocking S1P-related angiogenesis (Visentin et al., 2006).

3.4. Tumor inflammation

TAMs have been shown to mediate tumor growth, progression, and metastasis (Balkwill, 2004) and strongly associated with a poor prognostic outcome in cancer patients. TAMs are able to induce TNFα-dependent activation of JNK and NF-kappaB in adjacent tumor cells to promote their growth, motility, and invasion (Condeelis & Pollard, 2006; Karin & Greten, 2005). The molecular mechanisms underlying TAM-mediated tumor cell migration are not very well defined, but are likely to involve the secretion of promigratory cytokines or chemokines, including those released in response to activation of the SK1/S1P pathway (Johann et al., 2008; Pettus et al., 2003).

The SK1/S1P pathway is involved in inflammatory responses to cytokines such as TNFα and IL-1 (Xia et al., 1998). TNFα activates SK1 via a TRAF2-dependent mechanism and leads to activation of the prosurvival and proinflammatory pathways mediated by Akt (Radeff-Huang et al., 2006; Xia et al., 2002) and NF-kappaB (Xia et al., 2002). This may form part of a positive feedback loop since SK1 itself can promote proinflammatory cytokine release (Melendez & Ibrahim, 2004), notably through activation of NF-kappaB, itself a proinflammatory and prosurvival transcription factor (Zhi, Leung, & Melendez, 2006). In turn, in cancer cells, SK1 activation has been shown to be mediated by inflammatory cytokines, for example, interleukin 6 (IL-6) through PI3K and MAPK pathways (Pchejetski, Nunes, et al., 2011).

Extracellular S1P was shown to induce COX2 overexpression and PGE2 production in L929 fibrosarcoma and in A549 lung adenocarcinoma cells (Pettus et al., 2003), which may imply the autocrine and paracrine regulatory abilities. Indeed, similar to cytokines secreted by cancer cells to attract macrophages, it was found that S1P secreted from apoptotic tumor cells can induce macrophage polarization (Weigert et al., 2007). This secreted S1P was found to stimulate chemotaxis of primary monocytes and macrophages and therefore attract scavenger cells to the tumor site (Gude et al., 2008) and promote COX2 mRNA stabilization and protein expression (Johann et al., 2008). Conversely, S1P antibody was shown to abrogate macrophage invasion to ischemic areas (Xie et al., 2009).

3.5. S1P: A marker for disease and cancer

The potential of S1P to have a role as a plasma marker of human cancer has been demonstrated in a study which showed that S1P levels in preoperative ovarian cancer patients are higher than in controls, although they were not correlated to clinical stage (Sutphen et al., 2004). Further, Kawamori et al. (2006) have shown that circulating levels of S1P were higher in mice with colon cancers than in those without cancers. In breast cancer patients, modulation of circulating S1P was recently associated with aromatase inhibitors-induced arthralgia (Henry et al., 2010) and chemotherapy-induced weight gain (Pchejetski, Nunes, Sauer, et al., 2010). However, the value of S1P as a disease marker may be confounded by observations that erythrocytes (Hanel, Andreani, & Graler, 2007; Ohkawa et al., 2008) add platelets (English et al., 2000; Tokumura et al., 2009) is the major source of circulating S1P. Indeed, Nunes et al. (2012) have shown that circulating S1P was significantly lower in prostate cancer patients than in healthy controls, which correlated with a downregulation of erythrocyte SK1 activity. Elevated circulating S1P was also proposed to be a plasma marker of inflammatory coronary artery disease (Deutschman et al., 2003), diabetes (Fox et al., 2011), and atherosclerosis (Sugiura et al., 2012).

4. THERAPEUTIC POTENTIAL OF SK1 INHIBITION

4.1. Development of SK1 inhibitors

The clear implication of SK1 in cancer progression, resistance, and metastasis has highlighted its potential as a therapy target. Sphingosine derivatives dihydrosphingospine (DHS) and dimethylsphingosine (DMS) were initially used to shift the ceramide/S1P biostat. Both compounds elicit growth inhibition and provoke apoptosis in a number of tumor cells including acute myeloid leukemia (AML), chronic myeloid leukemia, acute lymphoid leukemia, cervix carcinoma, pheochromocytoma, prostate adenocarcinoma, gastric cancer, lung cancer, colon cancer, melanoma, epidermoid carcinoma, hepatoma, neuroblastoma, and breast adenocarcinoma (reviewed in Cuvillier, 2008). In particular, DMS has been found to dose-dependently inhibit *in vivo* growth of lung and gastric carcinoma tumors in athymic mice (Endo, Igarashi, Nisar, Zhou, & Hakomori, 1991) and severely decrease lung metastasis of melanoma cells in syngeneic mice (Okoshi et al., 1991). DMS was the first reported SK inhibitor (targeting both SK1 and SK2) to induce cancer, but not normal cell apoptosis (Edsall, Van Brocklyn, Cuvillier, Kleuser, & Spiegel, 1998; Sweeney et al., 1996). It, however, lacked specificity being also an inhibitor of protein kinase C, phospholipase A2, and phospholipase D (Merrill et al., 1997). L-threo-dihydrosphingosine (Safingol) was also demonstrated to have sphingosine kinase-inhibiting properties (Banno, Kato, Hara, & Nozawa, 1998).

More specific SK inhibitors F-12509a and B-5354c are synthesized by Kono, Sugiura, and Kohama (2002) based on extracts from marine fungi *Trichopezizella barbata*. Similar to DMS, F-12509a appears to inhibit SK1 competitively which suggests that the sesquiterpene moiety of F-12509a may mimic the sphingosine-conformation when binding to SK1's active site, whereas B-5345c demonstrated noncompetitive inhibition (Kono et al., 2002). Notably, neither F-12509a nor B-5354c appears to inhibit PKC unlike dihydro- and dimethylsphingosine. F-12509a facilitates simultaneous blocking of S1P biosynthesis and also ceramide accumulation in chemisensitive HL-60 and chemoresistant HL-60/Doxo and HL-60/VP16 cells which overexpress MRP1 and MDR1, respectively, inducing cancer cell apoptosis leading to nuclear fragmentation, caspase-3 cleavage, downregulation of XIAP, cytochrome C, and SMAC/Diablo release (Bonhoure et al., 2006). It also promotes apoptosis of Imatinib-resistant cells by shifting the ceramide/S1P biostat toward ceramide (Bonhoure et al., 2008). The administration of B-5354c triggers dose-dependent apoptosis in LnCaP and PC-3 prostate cancers and this may be reversed by upregulation of SK1 (Pchejetski et al., 2008).

Another group of SK inhibitors (SKI-I–V) also possessing anticancer properties was reported by French et al. (2003), and confirmed by other groups (Leroux et al., 2007; Pchejetski et al., 2005). SK-selective inhibitors (SKI-I–V) include 2-(p-hydroxyanilo)-4-(p-chlorophenyl)thiazole (SKI-II), which exhibits strong cytotoxicity toward T24 bladder carcinoma and MCF-7 and MCF-7/VP breast cancer cells (French et al., 2003) and shifts the ceramide/S1P biostat toward ceramide production resulting in death of androgen-sensitive (LnCaP) and hormone-resistant human prostate cancer cells (PC-3) irrespective of their p53 status (Pchejetski et al., 2005). SKI-II and a synthetic variant, compound V, have been then tested in mice. Intraperitoneal administration of either substances resulted in decreased tumor size and an inhibition of tumor growth of 50–80% with no obvious toxicity (French et al., 2006).

FTY720, a sphingosine analog (later shown to be an SK1 inhibitor, Pchejetski, Bohler, et al., 2010; Tonelli et al., 2010; Vessey et al., 2007, both *in situ* and *in vitro*), has been previously shown to induce caspase-3-mediated prostate cancer cell apoptosis (Permpongkosol et al., 2002; Wang, Takahara, et al., 1999), while normal human prostate stromal cells were resistant to FTY720. This compound has potential applications in treating numerous cancer cell and animal models (described in detail below).

Recently, a selective SK1 inhibitor (SK1-I) ([(2-hydroxy-1-naphthyl) methylene]-3-(2-napythyl)-1H-pyrazole-5-carbohydrazide) has been demonstrated to efficiently induce apoptosis of leukemia cells, but not normal bone marrow derived cells (Paugh et al., 2008). Xiang et al. (2009) have developed further SK1-specific inhibitors (6ag, 9ab, and 12aa) through a series of modifications of sphingosine. Amidine-based subtype-selective SK1 inhibitors were reported to induce reduction of endogenous S1P levels in human leukemia cells at nanomolar concentrations (Kennedy et al., 2011). Recent research investigating the structure–activity relationship of various analogs of SK1-I has demonstrated new inhibitors with optimized selectivity and activity (Hengst et al., 2010), noting that the napthyl rings were unnecessary for SK1 inhibition. One such discovery was small molecule SK1-178 which is active both *in vitro* and *in vivo* and could be useful in determining the exact functions of SK1 and SK2 isoforms in the development and progression of disease (Hengst et al., 2010).

Lately, (S)-FTY720 vinylphosphonate (Tonelli et al., 2010) and sphingoguanidines (LCL146 and LCL351) (Sharma, 2011) were also shown to induce SK1 inhibition in breast and prostate cancer cells and decrease the migration rate of human prostate (DU145) cells.

Finally, a recent Phase I clinical trial has reported a safe use of Safingol, a putative inhibitor of SK in combination with cisplatin in 43 cancer patients. Safingol administration was associated with a dose-dependent reduction in S1P in plasma and one patient with adrenal cortical cancer was reported to have a significant regression of liver and lung metastases and another had prolonged stable disease (Dickson et al., 2011).

4.2. SK2 inhibitors

These findings have highlighted the potential of SK1 as a therapy target for cancer treatment, while the role of another SK isoform SK2 remained still debated (Hait et al., 2005). Indeed, SK2 was shown to induce apoptosis via regulation of the cyclin-dependent kinase inhibitor p21 (Sankala et al., 2007) and enhanced histone H3 acetylation and p21 transcription (Hait et al., 2009). However, later studies showed a critical role of SK2 for EGF-stimulated migration of breast cancer cells (Hait et al., 2005) and growth of tumor xenografts (Weigert et al., 2009). Schnitzer et al. (2009) have further shown that SK2 is increased in response to hypoxia and in turn mediates chemoresistance in A549 lung cancer cells.

Recently, a new selective inhibitor of SK2 (ABC294640) was identified and reported to have an antitumor activity (French et al., 2010). One of the proposed mechanisms for its anticancer activity was induction of autophagy (Beljanski, Knaak, & Smith, 2010). It was further reported to have antiestrogenic effects in breast cancer cells (Antoon et al., 2010) to block NF-kappaB-mediated breast cancer chemoresistance (Antoon et al., 2011) and to suppress colitis-driven colon cancer in mice reducing the tumor incidence by 2.6-fold (Chumanevich et al., 2010). Finally, in hepatocellular carcinoma, ABC294640 was shown to synergize with sorafenib proposing its potential use in clinical trials in HCC patients (Beljanski et al., 2011). Taking into consideration a potential role of SK2 in cancer progression and its recently described functions in mitochondria (Strub et al., 2011), these data highlight a possibility for targeting SK2 for cancer therapy.

4.3. Role of SK1 in resistance to therapies and chemosensitization

In addition to promoting cancer cell growth, migration, and secretion of angiogenic and inflammatory factors (as described above), SK1 overexpression was shown to markedly inhibit apoptosis induced by chemotherapies such as: anthracyclines in MCF-7 breast cancer cells (Nava et al., 2002), doxorubicin and etoposide in HL-60 acute myeloid cells (Bonhoure et al., 2006), and camptothecin and docetaxel in PC3 and LNCaP prostate cancer cells (Pchejetski et al., 2005). In AML model, sustained activation of SK1 was associated with MDR-associated chemoresistance (Bonhoure et al., 2006).

In an orthotopic model of prostate cancer, SK1-overexpressing PC3 cells developed larger tumors and exhibited resistance to docetaxel treatment (Pchejetski et al., 2005). This was later confirmed by another study indicating that in prostate cancer cell and animal models, indirect SK1 inhibition was a valid chemotherapeutic strategy (Brizuela et al., 2010).

Therefore, targeting of sphingolipid metabolism (notably the ceramide/S1P biostat through inhibition of SK1) was proposed as a new strategy for cancer treatment. SK1 inhibition was reported to restore endocrine response in breast cancer cells (Sukocheva, Wang, Verrier, Vadas, & Xia, 2009), decrease colony formation (Huwiler et al., 2011), cell motility, and chemotaxis (Jung et al., 2007; Melendez & Ibrahim, 2004). In concordance with these findings, several studies showed that in spite of sensitivity or insensitivity to different therapies, many cancer cell lines were equally sensitive to treatment with either siRNA to SK1 or pharmacological inhibitors of this enzyme (Bonhoure et al., 2006; French et al., 2003; Pchejetski et al., 2005). Interestingly, SK1-specific siRNA or pharmacological inhibitors were shown to induce cancer cell apoptosis independently of p53 mutation (Pchejetski et al., 2005) or Bcl-2 status (Leroux et al., 2007), which implies that SK1 inhibition can override these important pathways of tumor progression.

DHS and DMS were first shown to potentiate apoptotic mechanisms and overcoming radio and chemoresistance in AML (Cuvillier & Levade, 2001; Jendiroba, Klostergaard, Keyhani, Pagliaro, & Freireich, 2002; Klostergaard, Auzenne, & Leroux, 1998) or solid tumors such as breast adenocarcinoma (Maurer, Melton, Billups, Cabot, & Reynolds, 2000; Sachs, Safa, Harrison, & Fine, 1995), prostate adenocarcinoma (Nava et al., 2000; Sachs et al., 1995), cervix carcinoma (Shirahama et al., 1997), gastric adenocarcinoma (Schwartz et al., 1995), neuroblastoma (Maurer et al., 2000), melanoma (Maurer et al., 2000), lung (Maurer et al., 2000), colon (Maurer et al., 2000), and pancreas (Maurer et al., 2000).

A concept of SK1 acting as a "sensor" to anticancer therapies was, therefore, introduced, where SK1 inhibition was proposed to correlate to therapy efficiency (Pchejetski et al., 2005). It was further proposed that pharmacological SK1 inhibition may be a successful sensitizer to therapies that are unable to modulate sphingolipid metabolism. Indeed, several studies indicate that pharmacological inhibition of SK1 results in resensitization to anticancer therapies. In various cancer cell lines, SK1 inhibition and/or shifting ceramide/S1P rheostat toward ceramide were shown to promote sensitivity to cancer therapies (Bonhoure et al., 2006; Bruno et al., 1998; Chmura et al., 1997). Safingol administered independently or in combination with doxorubicin resulted in tumor suppression and showed no apparent toxicity or alteration in pharmacokinetics of doxorubicin (Schwartz et al., 1997).

Modulation of SK or S1P lyase has been suggested to contribute to altered sensitivity to cisplatin (Min et al., 2005). *In vitro* and *in vivo* models of prostate cancer have demonstrated that the regulation of the SK1/S1P pathway has the potential to synergize with the effects of camptothecin chemotherapy (Pchejetski et al., 2008), docetaxel chemotherapy (Sauer et al., 2009), and radiotherapy (Pchejetski, Bohler, et al., 2010).

5. FTY720: A NOVEL CLINICALLY APPLICABLE SK1 INHIBITOR

5.1. Targeting S1PRs and immunosuppression

FTY720 (Fingolimod) is a sphingosine analog and a potent immunosuppressive drug that induces lymphopenia via an inhibition of lymphocytes' egress from lymphoid organs through its antagonist function on the lymphocytes' S1PRs (Budde et al., 2006; Chiba, 2005). The principal mechanism of action of FTY720 was shown to be through internalization of S1P1 on lymphocytes and inhibition of S1P/S1P1-dependent lymphocyte egress from secondary lymphoid tissues and thymus (Lo, Xu, Proia, & Cyster, 2005; Matloubian et al., 2004). While S1P was reported to induce S1P1 internalization and recycling, FTY720-P was shown to induce S1P1 internalization and degradation (Jo et al., 2005) and, therefore, was termed a functional antagonist (Sanna et al., 2006). Two mechanisms of FTY720-mediated inhibition of lymphocyte egress were proposed: (a) down-modulation of S1P1 on T cells and (b) activation of S1P1 on the sinus lining endothelium to increase barrier function and to reduce transmigration of lymphocytes (Baumruker, Billich, & Brinkmann, 2007; Brinkmann, Cyster, & Hla, 2004). Several other mechanisms were proposed to mediate FTY720-induced immunosuppression including antagonism of S1P4 (Wang, Graeler, & Goetzl, 2005), inhibition of S1P lyase, an enzyme responsible for the degradation of S1P (Bandhuvula, Tam, Oskouian, & Saba, 2005), inhibition of cytosolic phospholipase A2 independently of S1PRs (Payne et al., 2007), and interaction with the CB1 cannabinoid receptor (Paugh et al., 2006).

The major clinical outcome of FTY720-mediated lymphocyte sequestration has been recently reported in two Phase III trials demonstrating significant potential of oral FTY720 in multiple sclerosis (MS) patients (Cohen et al., 2010; Kappos et al., 2010).

5.2. Anticancer effects of FTY720

Surprisingly, at higher doses, FTY720 has been shown to be a potent apoptosis inducer in prostate, liver, and bladder cancer cell lines (Ho et al., 2005; Ubai et al., 2007; Wang, Takahara, et al., 1999). In a mouse model of melanoma, FTY720 inhibits tumor growth and metastasis (LaMontagne et al., 2006) without causing detectable toxicity in vital organs.

A bulk of preclinical *in vitro* and *in vivo* studies had shown potent efficacy of FTY720 in various different cancers: for example, prostate cancer (and metastasis) (Chua et al., 2005; Pchejetski, Bohler, et al., 2010; Permpongkosol et al., 2002; Tonelli et al., 2010; Zhou et al., 2006), breast cancer (and metastasis) (Azuma et al., 2002; Lim et al., 2011; Tonelli et al., 2010), different forms of leukemia (Liu, Alinari, et al., 2010; Roberts et al., 2010; Yasui et al., 2005), lung cancer (Lucas da Silva, Ribeiro, Cury, Cordeiro, & Bueno, 2008; Salinas, Lopes, Palma, Oshima, & Bueno, 2009), liver cancer (Ng et al., 2007), pancreatic cancer (Shen et al., 2007; Shen et al., 2008), bladder cancer (Azuma, Takahara, et al., 2003), and renal cancer (Ubai et al., 2007). Interestingly, while in prostate cancer cells, FTY720 has been shown to be a potent inducer of caspase-3 activation and apoptosis (Wang, Takahara, 1999), normal prostate cells exhibited resistance to the drug.

Various potential mechanisms have been proposed to mediate FTY720-induced cancer cell growth arrest and apoptosis: direct mitochondria damage (Nagahara, Ikekita, & Shinomiya, 2000), activation of caspases (Ng et al., 2007; Permpongkosol et al., 2002; Wang, Takahara, et al., 1999), dephosphorylation of Akt (Azuma, Horie, et al., 2003; Liu, Alinari, et al., 2010; Ng et al., 2007; Roberts et al., 2010), G(1) arrest, a decrease in CDK2 and CDK4, and induction of Cip1/p21 (Permpongkosol et al., 2002). Other molecular targets of FTY720 signaling that may mediate cancer cell growth and/or metastasis include: Bcl-2 (Chua et al., 2005; Liu, Alinari, et al., 2010; Shen et al., 2007; Ubai et al., 2007), Stat5 (Roberts et al., 2010), ERK (Liu, Alinari, et al., 2010; Permpongkosol et al., 2002), PP2A (Roberts et al., 2010), p38MAPK (Permpongkosol et al., 2002), FAK (Permpongkosol et al., 2002), RhoA-GTPase (Zhou et al., 2006), and integrins (Shen et al., 2007). Administration of FTY720 at 10 mg/kg/day reduced the growth of prostate CWR22R xenografts in castrated nude mice (Chua et al., 2005), and it has been reported that FTY720 can inhibit tumor angiogenesis (Schmid et al., 2007). The proapoptotic effects of FTY720 were reported be both S1PRs dependent (LaMontagne et al., 2006) and independent (Payne et al., 2007).

5.3. FTY720 inhibits SK1

In 2007, assays using chromatographic fractions containing SK showed that FTY720 possess SK1-, but not SK2-inhibiting properties (Vessey et al., 2007). While a possibility remained that FTY720 inhibitory action could be mediated by modulation of SK1 regulatory protein(s) also present in these chromatographic fractions, a definitive proof that FTY720 is a *bone fide* SK1 inhibitor came from Tonelli et al. (2010), who have used purified enzyme. FTY720 was further shown to be a potent *in vitro* SK1 inhibitor in prostate cancer cells and this inhibition was critical for induction of prostate cancer cell apoptosis (Pchejetski, Bohler, et al., 2010). Sublethal concentrations FTY720 were shown to acts as a radiosensitiser in prostate cancer cell lines and in implanted tumors, reducing primary tumor size and inhibiting cancer metastasis (Pchejetski, Bohler, et al., 2010). As S1PR signaling is also critical to cancer progression (reviewed in Takabe et al., 2008), it may be hypothesized that both S1PR- and SK1-mediated modes of action of FTY720 may contribute to inhibition of cancer growth. Indeed, FTY720 has been shown to induce a downregulation of S1P1 in hepatocellular carcinoma (Ushitora et al., 2009). In contrast, FTY720-phosphate has been demonstrated to induce a growth of breast and colon cancer cells (Nagaoka, Otsuki, Fujita, & Uesato, 2008). It may be therefore a competing action between S1P1 internalization and activation of other S1PRs by phosphorylated FTY720 and SK1 inhibition by nonphosphorylated FTY720 that determines the physiopathological effect of FTY720. This effect may therefore be dependent on tissue type (relative expression of S1PRs) and on local FTY720 concentration (higher concentrations of FTY720 are required to induce SK1 inhibition). Moreover, the release of phosphorylated FTY720 after its intracellular phosphorylation may be dependent on the expression of sphingolipid transporters (Hisano et al., 2011), which may determine the balance between the presence of phosphorylated/nonphosphorylated forms of the drug.

In addition to its direct antitumor effect, FTY720 has strong immunosuppressive properties, specifically against regulatory T cells (Wolf et al., 2009), which can contribute to tolerance of malignant tumor cells (Nishikawa & Sakaguchi, 2010). Based on FTY720's immunosuppressive and anticancer activity, it has therefore been suggested to evaluate its use in organ-transplanted patients with posttransplant malignancies (Domhan, Zeier, & Abdollahi, 2009).

These findings point to the need for clinical testing to establish whether FTY720 therapy might provide an advantage in terms of increased local tumor control to patients with solid cancers treated with radiotherapy. In a view of a significant synergy between SK1-targeting therapies and docetaxel, it is reasonable to propose an extensive testing of FTY720 or other novel selective SK1 inhibitors in combination with taxanes for improved chemosensitivity/tolerance/clinical outcome for solid cancers.

5.4. Pharmacological assessment of FTY720 applicability for future clinical trials in cancer patients

FTY720 ($C_{19}H_{33}NO_2$; MW 344 Da) is the first therapeutically active compound targeting S1P pathway that has been successfully translated into clinical practice (reviewed in Pchejetski, Bohler, et al., 2011). Due to its lymphopenic properties, FTY720 was investigated for the prevention of renal allograft rejection. However, due to the lack of benefits compared with mycophenolate mofetil (MMF, CellCept®) in a large Phase III trial, the clinical development for this indication was stopped (Tedesco-Silva et al., 2007). FTY, however, plays a key protective role in several autoimmune disease models (Graler, 2010). Recently, Gilenya™ (FTY720, Fingolimod) was successfully approved by the FDA as an oral therapeutic drug for the treatment of MS (Cohen et al., 2010; Kappos et al., 2010).

In humans, after an oral intake of a single dose of 0.25–3.5 mg FTY720 circulating drug concentration peaked after 8–36 h and remained elevated up to 36 h with a linear correlation with the drug dose, and then declined monoexponentially (Skerjanec et al., 2005). This elimination was independent of dose with a mean value of 108 h with a low intersubject variability (12–24% coefficient of variation) with the exception of the lowest concentration of 0.25 mg. The volume of distribution of FTY720 was proportional to body mass. The clearance (9.48–10.8 L/h) was comparable to other drugs used in clinical practice. The clearance of FTY720 was not affected by age, gender, or ethnicity (Kovarik et al., 2007), indices of hepatic or renal function, and history of diabetes (Kovarik, Schmouder, Barilla, Wang, & Kraus, 2004). FTY720 is metabolized by ß-oxidation by a cytochrome P450-dependent pathway to inactive metabolites which are excreted in urine and feces. On the basis of the structural and biochemical properties of FTY720, it may be assumed that it is integrated into cell membranes after intake. The advantage of a long half-life is a possibility for a single (once a day) drug administration. The potential disadvantage might lie in a less accurate control and a risk of accumulation (Sensken, Bode, & Graler, 2009).

Clinical data are available for more than 4000 patients/years on FTY treatment, where some patients were treated up to 7 years, and overall tolerance was rated as very good. The major reported complication of FTY720 is a transient and mostly asymptomatic bradycardia (about 25% occurrence)

(Budde et al., 2002), which was dose dependent especially in patients with already existing borderline bradycardia and no toxic effects had been observed in healthy volunteers treated for seven consecutive days with 5 mg/day oral FTY720 (Kovarik, Schmouder, Barilla, Riviere, et al., 2004; Koyrakh, Roman, Brinkmann, & Wickman, 2005). Usually, bradycardia appears as a first dosing effect and disappears approximately after 7–10 days. Heart rate reaches nadir after 4–8 h after start with FTY720 therapy and is sensitive for atropine treatment (Kovarik et al., 2008). A recent TRANSFORMS study has indicated a slight increase in herpes infection, transient macula oedema, skin cancer, and elevated liver enzymes in MS patients receiving FTY720 0.5–1.25 mg/day for 12 months (Cohen et al., 2010).

Preclinical studies suggest that higher FTY720 doses in a shorter time frame are required for the treatment of prostate cancer (Chua et al., 2005; Pchejetski, Bohler, et al., 2010). While local application of the water and cell membrane soluble FTY720 may be considered as a strategy to avoid currently known adverse effects, these side effects in cancer patients may be different to patients with end stage renal disease or MS because of a different profile of co-morbidities. Currently, no clinical data are available on the safety of high FTY720 doses, a question that remains to be addressed by future clinical trials.

6. CONCLUSIONS AND CLINICAL PERSPECTIVE

Overall these data show a significant potential for SK1 as a molecular target for cancer therapy. Considering a noteworthy complexity of the SK1/S1P signaling and specifically intracellular or S1P-independent functions of SK1, it is becoming more and more evident that targeting S1P alone is insufficient to control cancer growth and spread, and treatments targeting SK1 may be required to obtain a complete response. As the result, there is an urgent requirement to develop a new generation of clinically applicable pharmacological SK1 inhibitors.

The current generation of SK1 inhibitors is in most cases neither specific nor selective. FTY720, which is an agonist of four S1PRs (Schwab & Cyster, 2007), has been shown to promote proteasomal degradation of SK1 in several cell lines including the androgen-independent LNCaP-AI prostate cancer cells (Tonelli et al., 2010). There are similar observations reported for the widely preclinically used SK inhibitor SKI-II (Loveridge et al., 2010), while recently described specific SK2 inhibitor ABC294640, which possesses antitumor properties, was shown to be a potent inhibitor of estrogen receptor alpha (Antoon et al., 2010; Table 6.5).

Table 6.5 Structures of SK inhibitors

Name	Structure	IC50	References
DMS	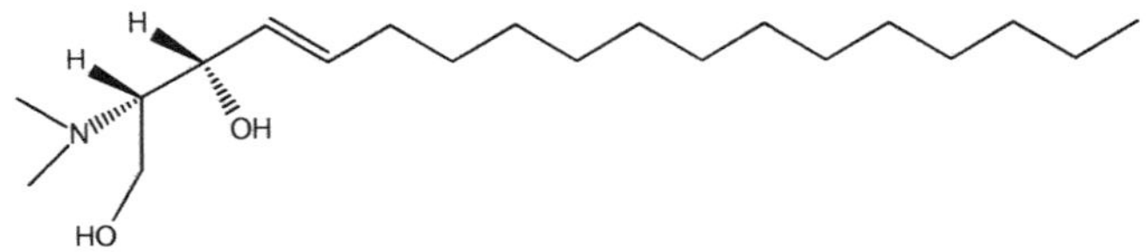	5.7 μM (SK1)	Igarashi and Hakomori (1989)
DHS	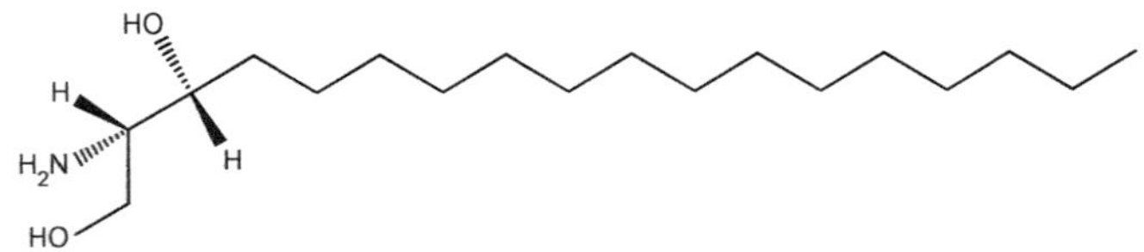	6 μM (SK1)	Darges, Robinson, and Adams (1997)
F-12509a	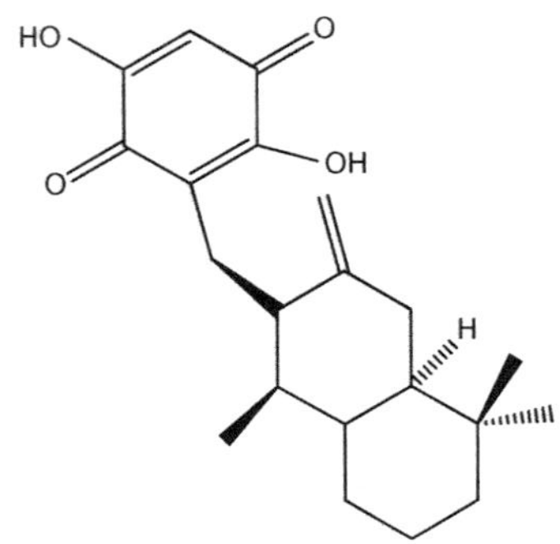	41 μM (SK1) 11 μM (SK2)	Kono et al. (2002)

Name	Structure	IC50	References
B5354c	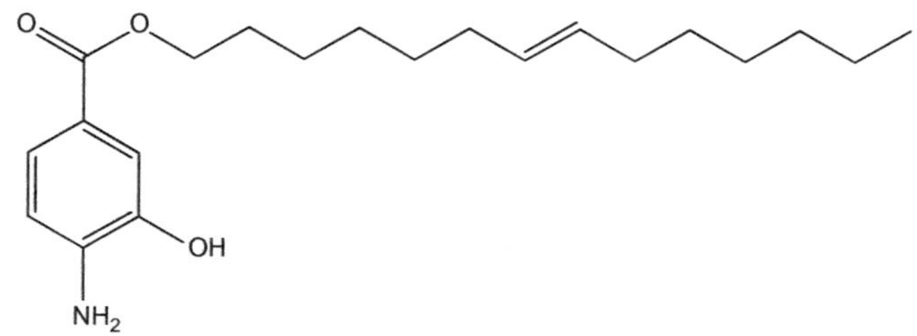	7.8 μM (SK1) 5.0 μM (SK2)	Kono et al. (2002)
SK1-I	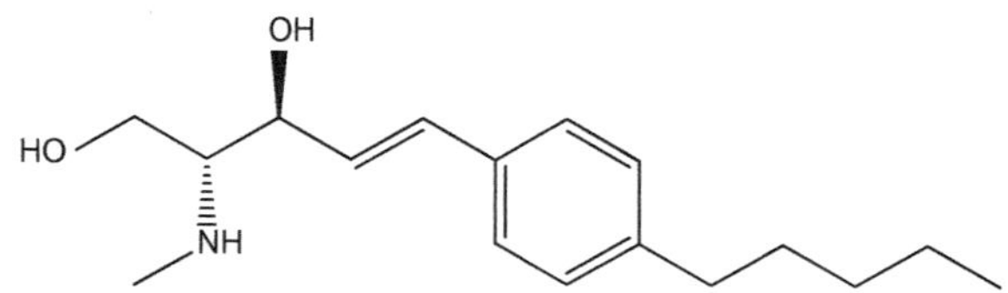	Ki 10 μM (SK1)	Paugh et al. (2008)
SK1-II	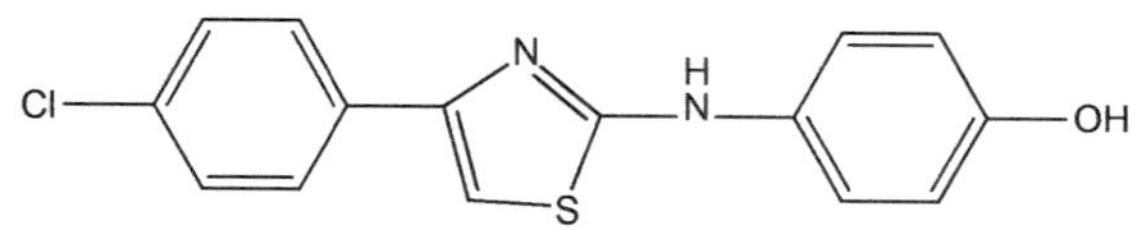	0.5 μM (SK1)	French et al. (2003)
FTY720	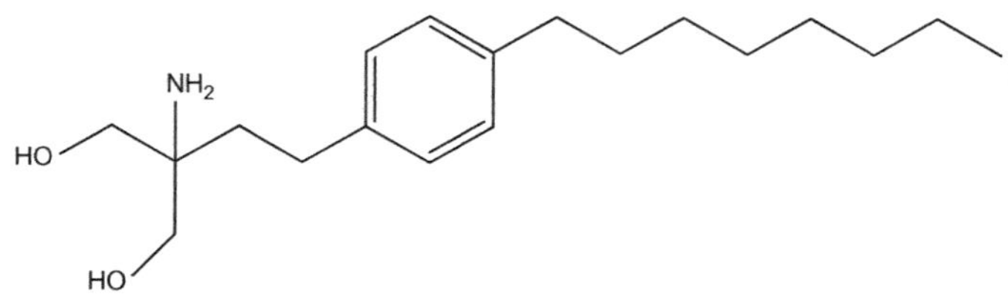	10 μM (SK1) Ki 2.15 μM (SK1)	Vessey et al. (2007) and Tonelli et al. (2010)

Continued

Table 6.5 Structures of SK inhibitors—cont'd

Name	Structure	IC50	References
SKI-178	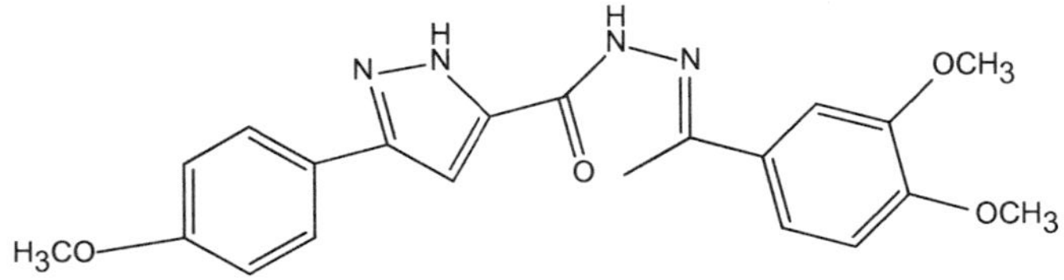	1–3 μM (SK1) 1–3 μM (SK2)	Hengst et al. (2010)
Compound V	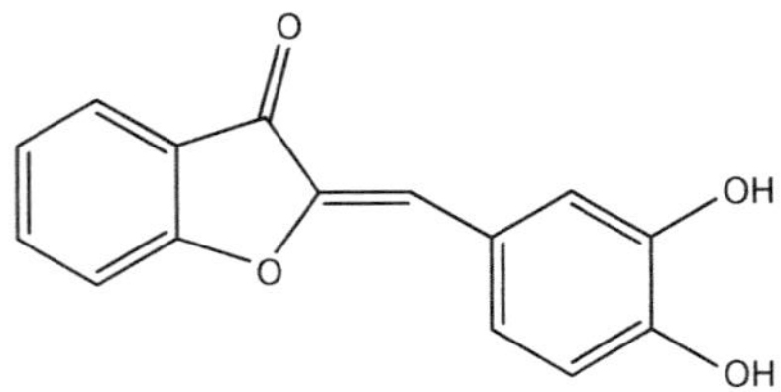	2 μM (SK1)	French et al. (2003)
(S)-FTY720 vinylphosphonate	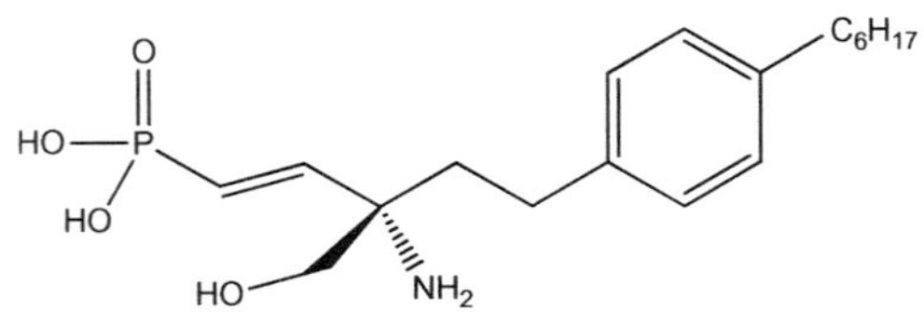	20 μM (SK1)	Lim et al. (2011)
Safingol	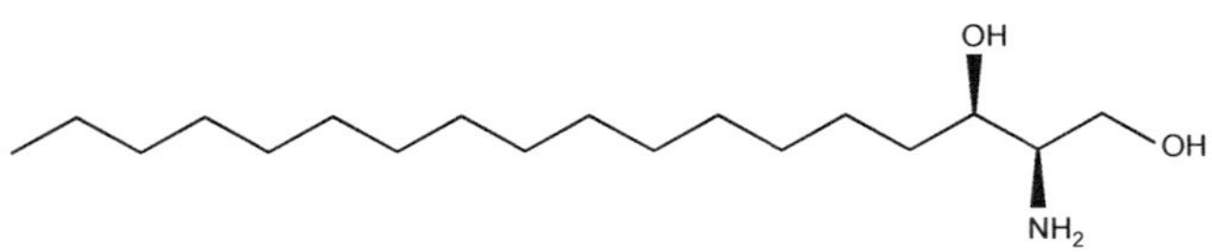	14 μM (SK1)	Coward et al. (2009)

Name	Structure	IC50	References
LCL 146	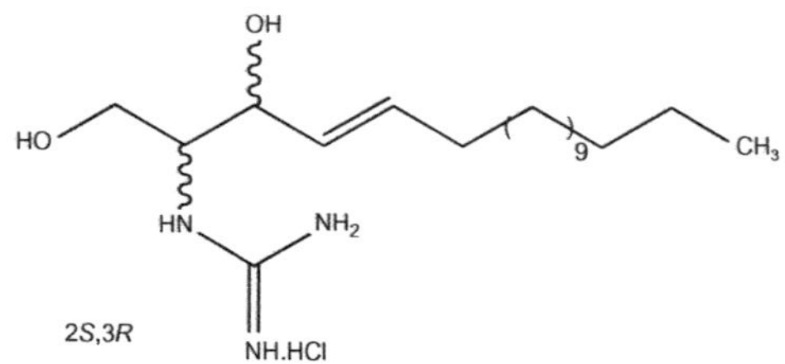	~1 μM (SK1) ~1 μM (SK2)	Sharma, (2011)
LCL 351	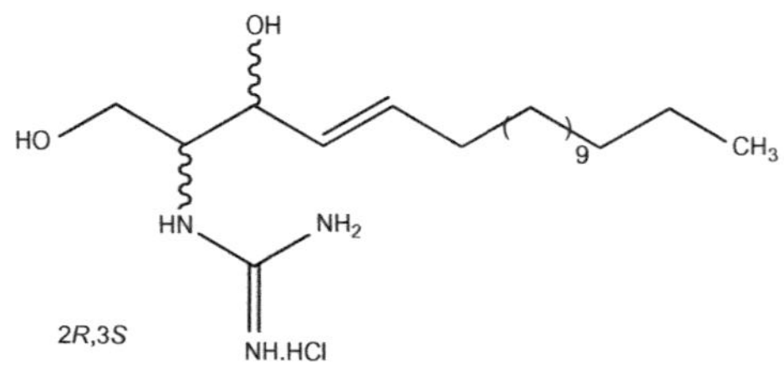	~1 μM (SK1) ~1 μM (SK2)	Sharma, (2011)
6ag	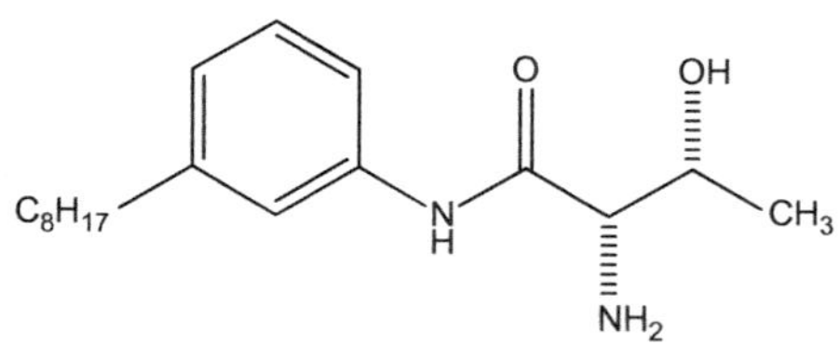	0.65 μM (SK1)	Xiang et al. (2009)

Continued

Table 6.5 Structures of SK inhibitors—cont'd

Name	Structure	IC50	References
9ab	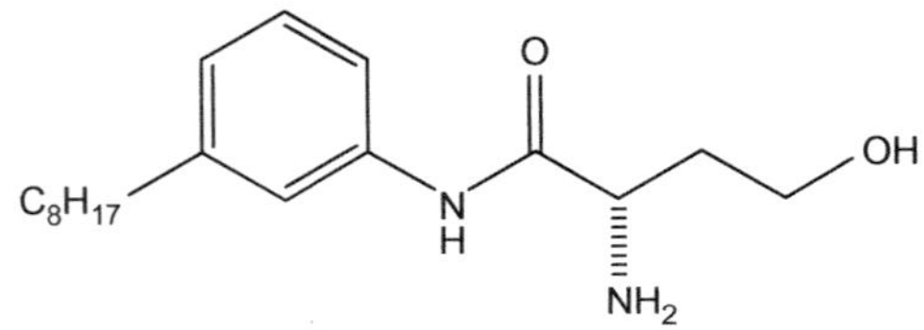	0.05 (SK1)	Xiang et al. (2009)
12aa	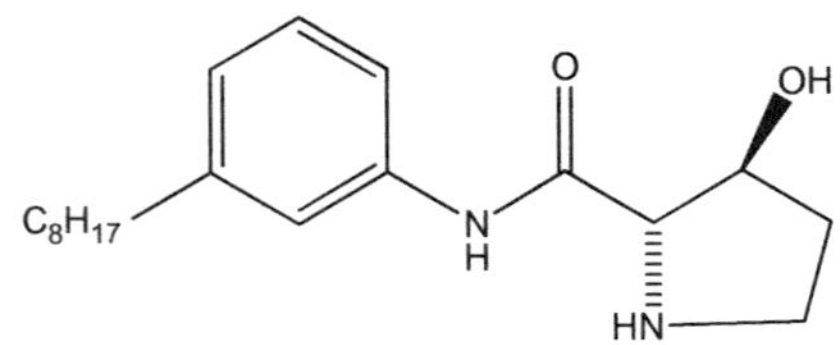	0.062 (SK1)	Xiang et al. (2009)

It should be noted that SK1 is ubiquitously expressed and its enzymatic activity is required for cell survival and organ function (Table 6.2). As a result, the systemic inhibition of SK1 may have toxicities that will limit its use in the clinic. This could be circumvented by a local delivery of SK1-targeting therapy. Similar problems may arise from using nonspecific dual SK1 and SK2 inhibitors. It has been shown that dual SK1/SK2 knockouts are lethal (Mizugishi et al., 2005), while single SK1 or SK2 knockout mice do not have any physiological abnormalities (Zemann et al., 2007); therefore, there is reasonable evidence that the toxicity from selective inhibitors will be limited.

A particular hope rises in the field of combination therapies, where synergistic treatment with SK1 inhibitors and conventional therapies may provide advantage over monotherapies in terms of efficacy and reduction of unspecific toxicity. As an example, in prostate cancer patients cabazitaxel is the first chemotherapy agent to have been shown to have a survival benefit in the setting of recurrence after docetaxel chemotherapy (de Bono et al., 2010), which was previously shown to synergise with SK1 inhibitors (Pchejetski et al., 2008; Sauer et al., 2009). It is possible that the use of SK1 inhibitors in combination with this new agent might lead to further survival benefits and laboratory studies in this area of research are currently under way to provide new hope for the treatment of prostate cancer.

REFERENCES

Ader, I., Brizuela, L., Bouquerel, P., Malavaud, B., & Cuvillier, O. (2008). Sphingosine kinase 1: A new modulator of hypoxia inducible factor 1alpha during hypoxia in human cancer cells. *Cancer Research*, *68*, 8635–8642.

Akao, Y., Banno, Y., Nakagawa, Y., Hasegawa, N., Kim, T. J., Murate, T., et al. (2006). High expression of sphingosine kinase 1 and S1P receptors in chemotherapy-resistant prostate cancer PC3 cells and their camptothecin-induced up-regulation. *Biochemical and Biophysical Research Communications*, *342*, 1284–1290.

Alemany, R., van Koppen, C. J., Danneberg, K., Ter Braak, M., & Meyer Zu Heringdorf, D. (2007). Regulation and functional roles of sphingosine kinases. *Naunyn-Schmiedeberg's Archives of Pharmacology*, *374*, 413–428.

Alvarez, S. E., Harikumar, K. B., Hait, N. C., Allegood, J., Strub, G. M., Kim, E. Y., et al. (2010). Sphingosine-1-phosphate is a missing cofactor for the E3 ubiquitin ligase TRAF2. *Nature*, *465*, 1084–1088.

Alvarez, S. E., Milstien, S., & Spiegel, S. (2007). Autocrine and paracrine roles of sphingosine-1-phosphate. *Trends in Endocrinology and Metabolism*, *18*, 300–307.

Anelli, V., Gault, C. R., Snider, A. J., & Obeid, L. M. (2010). Role of sphingosine kinase-1 in paracrine/transcellular angiogenesis and lymphangiogenesis in vitro. *The FASEB Journal*, *24*, 2727–2738.

Antoon, J. W., White, M. D., Meacham, W. D., Slaughter, E. M., Muir, S. E., Elliott, S., et al. (2010). Antiestrogenic effects of the novel sphingosine kinase-2 inhibitor ABC294640. *Endocrinology*, *151*, 5124–5135.

Antoon, J. W., White, M. D., Slaughter, E. M., Davis, J. L., Khalili, H. S., Elliott, S., et al. (2011). Targeting NF-kB mediated breast cancer chemoresistance through selective inhibition of sphingosine kinase-2. *Cancer Biology & Therapy, 11*, 678–689.

Arikawa, K., Takuwa, N., Yamaguchi, H., Sugimoto, N., Kitayama, J., Nagawa, H., et al. (2003). Ligand-dependent inhibition of B16 melanoma cell migration and invasion via endogenous S1P2 G protein-coupled receptor. Requirement of inhibition of cellular RAC activity. *Journal of Biological Chemistry, 278*, 32841–32851.

Azuma, H., Horie, S., Muto, S., Otsuki, Y., Matsumoto, K., Morimoto, J., et al. (2003). Selective cancer cell apoptosis induced by FTY720; evidence for a Bcl-dependent pathway and impairment in ERK activity. *Anticancer Research, 23*, 3183–3193.

Azuma, H., Takahara, S., Horie, S., Muto, S., Otsuki, Y., & Katsuoka, Y. (2003). Induction of apoptosis in human bladder cancer cells in vitro and in vivo caused by FTY720 treatment. *Journal of Urology, 169*, 2372–2377.

Azuma, H., Takahara, S., Ichimaru, N., Wang, J. D., Itoh, Y., Otsuki, Y., et al. (2002). Marked prevention of tumor growth and metastasis by a novel immunosuppressive agent, FTY720, in mouse breast cancer models. *Cancer Research, 62*, 1410–1419.

Balkwill, F. (2004). Cancer and the chemokine network. *Nature Reviews. Cancer, 4*, 540–550.

Bandhuvula, P., Tam, Y. Y., Oskouian, B., & Saba, J. D. (2005). The immune modulator FTY720 inhibits sphingosine-1-phosphate lyase activity. *Journal of Biological Chemistry, 280*, 33697–33700.

Banno, Y., Kato, M., Hara, A., & Nozawa, Y. (1998). Evidence for the presence of multiple forms of Sph kinase in human platelets. *Biochemical Journal, 335*, 301.

Barr, R. K., Lynn, H. E., Moretti, P. A., Khew-Goodall, Y., & Pitson, S. M. (2008). Deactivation of sphingosine kinase 1 by protein phosphatase 2A. *Journal of Biological Chemistry, 283*, 34994–35002.

Baumruker, T., Billich, A., & Brinkmann, V. (2007). FTY720, an immunomodulatory sphingolipid mimetic: Translation of a novel mechanism into clinical benefit in multiple sclerosis. *Expert Opinion on Investigational Drugs, 16*, 283–289.

Bayerl, M. G., Bruggeman, R. D., Conroy, E. J., Hengst, J. A., King, T. S., Jimenez, M., et al. (2008). Sphingosine kinase 1 protein and mRNA are overexpressed in non-Hodgkin lymphomas and are attractive targets for novel pharmacological interventions. *Leukemia & Lymphoma, 49*, 948–954.

Bektas, M., Jolly, P. S., Muller, C., Eberle, J., Spiegel, S., & Geilen, C. C. (2005). Sphingosine kinase activity counteracts ceramide-mediated cell death in human melanoma cells: Role of Bcl-2 expression. *Oncogene, 24*, 178–187.

Beljanski, V., Knaak, C., & Smith, C. D. (2010). A novel sphingosine kinase inhibitor induces autophagy in tumor cells. *Journal of Pharmacology and Experimental Therapeutics, 333*, 454–464.

Beljanski, V., Lewis, C. S., & Smith, C. D. (2011). Antitumor activity of Sphingosine Kinase 2 inhibitor ABC294640 and sorafenib in hepatocellular carcinoma xenografts. *Cancer Biology & Therapy, 11*, 524–534.

Berdyshev, E. V., Gorshkova, I., Usatyuk, P., Kalari, S., Zhao, Y., Pyne, N. J., et al. (2011). Intracellular S1P generation is essential for S1P-induced motility of human lung endothelial cells: Role of sphingosine kinase 1 and S1P lyase. *PLoS One, 6*, e16571.

Bergelin, N., Blom, T., Heikkila, J., Lof, C., Alam, C., Balthasar, S., et al. (2009). Sphingosine kinase as an oncogene: Autocrine sphingosine 1-phosphate modulates ML-1 thyroid carcinoma cell migration by a mechanism dependent on protein kinase C-alpha and ERK1/2. *Endocrinology, 150*, 2055–2063.

Billich, A., Bornancin, F., Mechtcheriakova, D., Natt, F., Huesken, D., & Baumruker, T. (2005). Basal and induced sphingosine kinase 1 activity in A549 carcinoma cells: Function in cell survival and IL-1beta and TNF-alpha induced production of inflammatory mediators. *Cellular Signalling, 17*, 1203–1217.

Blom, T., Slotte, J. P., Pitson, S. M., & Tornquist, K. (2005). Enhancement of intracellular sphingosine-1-phosphate production by inositol 1,4,5-trisphosphate-evoked calcium mobilisation in HEK-293 cells: Endogenous sphingosine-1-phosphate as a modulator of the calcium response. *Cellular Signalling, 17*, 827–836.

Bonhoure, E., Lauret, A., Barnes, D., Martin, C., Malavaud, B., Kohama, T., et al. (2008). Sphingosine kinase-1 is a downstream regulator of imatinib-induced apoptosis in chronic myeloid leukemia cells. *Leukemia, 22*, 971–979.

Bonhoure, E., Pchejetski, D., Aouali, N., Morjani, H., Levade, T., Kohama, T., et al. (2006). Overcoming MDR-associated chemoresistance in HL-60 acute myeloid leukemia cells by targeting sphingosine kinase-1. *Leukemia, 20*, 95–102.

Bornfeldt, K. E., Graves, L. M., Raines, E. W., Igarashi, Y., Wayman, G., Yamamura, S., et al. (1995). Sphingosine-1-phosphate inhibits PDGF-induced chemotaxis of human arterial smooth muscle cells: Spatial and temporal modulation of PDGF chemotactic signal transduction. *The Journal of Cell Biology, 130*, 193–206.

Boujaoude, L. C., Bradshaw-Wilder, C., Mao, C., Cohn, J., Ogretmen, B., Hannun, Y. A., et al. (2001). Cystic fibrosis transmembrane regulator regulates uptake of sphingoid base phosphates and lysophosphatidic acid: Modulation of cellular activity of sphingosine 1-phosphate. *Journal of Biological Chemistry, 276*, 35258–35264.

Brailoiu, E., Cooper, R. L., & Dun, N. J. (2002). Sphingosine 1-phosphate enhances spontaneous transmitter release at the frog neuromuscular junction. *British Journal of Pharmacology, 136*, 1093–1097.

Brinkmann, V., Cyster, J. G., & Hla, T. (2004). FTY720: Sphingosine 1-phosphate receptor-1 in the control of lymphocyte egress and endothelial barrier function. *American Journal of Transplantation, 4*, 1019–1025.

Brizuela, L., Dayon, A., Doumerc, N., Ader, I., Golzio, M., Izard, J. C., et al. (2010). The sphingosine kinase-1 survival pathway is a molecular target for the tumor-suppressive tea and wine polyphenols in prostate cancer. *The FASEB Journal, 24*, 3882–3894.

Bruno, A. P., Laurent, G., Averbeck, D., Demur, C., Bonnet, J., Bettaieb, A., et al. (1998). Lack of ceramide generation in TF-1 human myeloid leukemic cells resistant to ionizing radiation. *Cell Death and Differentiation, 5*, 172–182.

Bryan, L., Paugh, B. S., Kapitonov, D., Wilczynska, K. M., Alvarez, S. M., Singh, S. K., et al. (2008). Sphingosine-1-phosphate and interleukin-1 independently regulate plasminogen activator inhibitor-1 and urokinase-type plasminogen activator receptor expression in glioblastoma cells: Implications for invasiveness. *Molecular Cancer Research, 6*, 1469–1477.

Budde, K., Schmouder, R. L., Brunkhorst, R., Nashan, B., Lucker, P. W., Mayer, T., et al. (2002). First human trial of FTY720, a novel immunomodulator, in stable renal transplant patients. *Journal of the American Society of Nephrology, 13*, 1073–1083.

Budde, K., Schutz, M., Glander, P., Peters, H., Waiser, J., Liefeldt, L., et al. (2006). FTY720 (fingolimod) in renal transplantation. *Clinical Transplantation, 20*(Suppl 17), 17–24.

Candelore, M. R., Wright, M. J., Tota, L. M., Milligan, J., Shei, G. J., Bergstrom, J. D., et al. (2002). Phytosphingosine 1-phosphate: A high affinity ligand for the S1P(4)/Edg-6 receptor. *Biochemical and Biophysical Research Communications, 297*, 600–606.

Carpio, L. C., Stephan, E., Kamer, A., & Dziak, R. (1999). Sphingolipids stimulate cell growth via MAP kinase activation in osteoblastic cells. *Prostaglandins, Leukotrienes, and Essential Fatty Acids, 61*, 267–273.

Cencetti, F., Bernacchioni, C., Nincheri, P., Donati, C., & Bruni, P. (2010). Transforming growth factor-beta1 induces transdifferentiation of myoblasts into myofibroblasts via up-regulation of sphingosine kinase-1/S1P3 axis. *Molecular Biology of the Cell, 21*, 1111–1124.

Chae, S. S., Paik, J. H., Furneaux, H., & Hla, T. (2004). Requirement for sphingosine 1-phosphate receptor-1 in tumor angiogenesis demonstrated by in vivo RNA interference. *Journal of Clinical Investigation, 114*, 1082–1089.

Chiba, K. (2005). FTY720, a new class of immunomodulator, inhibits lymphocyte egress from secondary lymphoid tissues and thymus by agonistic activity at sphingosine 1-phosphate receptors. *Pharmacology and Therapeutics*, *108*, 308–319.

Chmura, S. J., Nodzenski, E., Beckett, M. A., Kufe, D. W., Quintans, J., & Weichselbaum, R. R. (1997). Loss of ceramide production confers resistance to radiation-induced apoptosis. *Cancer Research*, *57*, 1270–1275.

Chua, C. W., Lee, D. T., Ling, M. T., Zhou, C., Man, K., Ho, J., et al. (2005). FTY720, a fungus metabolite, inhibits in vivo growth of androgen-independent prostate cancer. *International Journal of Cancer*, *117*, 1039–1048.

Chumanevich, A. A., Poudyal, D., Cui, X., Davis, T., Wood, P. A., Smith, C. D., et al. (2010). Suppression of colitis-driven colon cancer in mice by a novel small molecule inhibitor of sphingosine kinase. *Carcinogenesis*, *31*, 1787–1793.

Cohen, J. A., Barkhof, F., Comi, G., Hartung, H. P., Khatri, B. O., Montalban, X., et al. (2010). Oral fingolimod or intramuscular interferon for relapsing multiple sclerosis. *The New England Journal of Medicine*, *362*, 402–415.

Condeelis, J., & Pollard, J. W. (2006). Macrophages: Obligate partners for tumor cell migration, invasion, and metastasis. *Cell*, *124*, 263–266.

Coward, J., Ambrosini, G., Musi, E., Truman, J. P., Haimovitz-Friedman, A., Allegood, J. C., et al. (2009). Safingol (L-threo-sphinganine) induces autophagy in solid tumor cells through inhibition of PKC and the PI3-kinase pathway. *Autophagy*, *5*, 184–193.

Cuvillier, O. (2008). Downregulating sphingosine kinase-1 for cancer therapy. *Expert Opinion on Therapeutic Targets*, *12*, 1009–1020.

Cuvillier, O., & Levade, T. (2001). Sphingosine 1-phosphate antagonizes apoptosis of human leukemia cells by inhibiting release of cytochrome c and Smac/DIABLO from mitochondria. *Blood*, *98*, 2828.

Cuvillier, O., Pirianov, G., Kleuser, B., Vanek, P. G., Coso, O. A., Gutkind, S., et al. (1996). Suppression of ceramide-mediated programmed cell death by sphingosine-1-phosphate. *Nature*, *381*, 800–803.

Darges, J. W., Robinson, S. P., & Adams, L. M. (1997). Inhibition of leukotriene B4 (LTB4) in human neutrophils by L-threo-dihydrosphingosine. *Advances in Experimental Medicine and Biology*, *400A*, 387–392.

Dayon, A., Brizuela, L., Martin, C., Mazerolles, C., Pirot, N., Doumerc, N., et al. (2009). Sphingosine kinase-1 is central to androgen-regulated prostate cancer growth and survival. *PLoS One*, *4*, e8048.

de Bono, J. S., Oudard, S., Ozguroglu, M., Hansen, S., Machiels, J. P., Kocak, I., et al. (2010). Prednisone plus cabazitaxel or mitoxantrone for metastatic castration-resistant prostate cancer progressing after docetaxel treatment: A randomised open-label trial. *Lancet*, *376*, 1147–1154.

Delon, C., Manifava, M., Wood, E., Thompson, D., Krugmann, S., Pyne, S., et al. (2004). Sphingosine kinase 1 is an intracellular effector of phosphatidic acid. *Journal of Biological Chemistry*, *279*, 44763–44774.

Deutschman, D. H., Carstens, J. S., Klepper, R. L., Smith, W. S., Page, M. T., Young, T. R., et al. (2003). Predicting obstructive coronary artery disease with serum sphingosine-1-phosphate. *American Heart Journal*, *146*, 62–68.

Dickson, M. A., Carvajal, R. D., Merrill, A. H., Jr., Gonen, M., Cane, L. M., & Schwartz, G. K. (2011). A phase I clinical trial of safingol in combination with cisplatin in advanced solid tumors. *Clinical Cancer Research*, *17*, 2484–2492.

Ding, G., Sonoda, H., Yu, H., Kajimoto, T., Goparaju, S. K., Jahangeer, S., et al. (2007). Protein kinase D-mediated phosphorylation and nuclear export of sphingosine kinase 2. *Journal of Biological Chemistry*, *282*, 27493–27502.

Doll, F., Pfeilschifter, J., & Huwiler, A. (2007). Prolactin upregulates sphingosine kinase-1 expression and activity in the human breast cancer cell line MCF7 and triggers enhanced proliferation and migration. *Endocrine-Related Cancer, 14,* 325–335.

Domhan, S., Zeier, M., & Abdollahi, A. (2009). Immunosuppressive therapy and post-transplant malignancy. *Nephrology, Dialysis, Transplantation, 24,* 1097–1103.

Duan, H. F., Qu, C. K., Zhang, Q. W., Yu, W. M., Wang, H., Wu, C. T., et al. (2006). Shp-2 tyrosine phosphatase is required for hepatocyte growth factor-induced activation of sphingosine kinase and migration in embryonic fibroblasts. *Cellular Signalling, 18,* 2049–2055.

Duan, H. F., Wu, C. T., Lu, Y., Wang, H., Liu, H. J., Zhang, Q. W., et al. (2004). Sphingosine kinase activation regulates hepatocyte growth factor induced migration of endothelial cells. *Experimental Cell Research, 298,* 593–601.

Edsall, L. C., Van Brocklyn, J. R., Cuvillier, O., Kleuser, B., & Spiegel, S. (1998). N, N-Dimethylsphingosine is a potent competitive inhibitor of sphingosine kinase but not of protein kinase C: Modulation of cellular levels of sphingosine 1-phosphate and ceramide. *Biochemistry, 37,* 12892–12898.

Endo, K., Igarashi, Y., Nisar, M., Zhou, Q., & Hakomori, S. (1991). Cell membrane signaling as target in cancer therapy: Inhibitory effect of N, N-dimethyl and N, N, N-trimethyl sphingosine derivatives on in vitro and in vivo growth of human tumor cells in nude mice. *Cancer Research, 51,* 1613.

English, D., Welch, Z., Kovala, A. T., Harvey, K., Volpert, O. V., Brindley, D. N., et al. (2000). Sphingosine 1-phosphate released from platelets during clotting accounts for the potent endothelial cell chemotactic activity of blood serum and provides a novel link between hemostasis and angiogenesis. *The FASEB Journal, 14,* 2255–2265.

Estrada, R., Wang, L., Jala, V. R., Lee, J. F., Lin, C. Y., Gray, R. D., et al. (2009). Ligand-induced nuclear translocation of S1P(1) receptors mediates Cyr61 and CTGF transcription in endothelial cells. *Histochemistry and Cell Biology, 131,* 239–249.

Facchinetti, M. M., Gandini, N. A., Fermento, M. E., Sterin-Speziale, N. B., Ji, Y., Patel, V., et al. (2010). The Expression of Sphingosine Kinase-1 in Head and Neck Carcinoma. *Cells, Tissues, Organs, 192,* 314–324.

Fox, T. E., Bewley, M. C., Unrath, K. A., Pedersen, M. M., Anderson, R. E., Jung, D. Y., et al. (2011). Circulating sphingolipid biomarkers in models of type 1 diabetes. *Journal of Lipid Research, 52,* 509–517.

French, K. J., Schrecengost, R. S., Lee, B. D., Zhuang, Y., Smith, S. N., Eberly, J. L., et al. (2003). Discovery and evaluation of inhibitors of human sphingosine kinase. *Cancer Research, 63,* 5962–5969.

French, K., Upson, J., Keller, S., Zhuang, Y., Yun, J., & Smith, C. (2006). Antitumor activity of sphingosine kinase inhibitors. *Journal of Pharmacology and Experimental Therapeutics, 318,* 596.

French, K. J., Zhuang, Y., Maines, L. W., Gao, P., Wang, W., Beljanski, V., et al. (2010). Pharmacology and antitumor activity of ABC294640, a selective inhibitor of sphingosine kinase-2. *Journal of Pharmacology and Experimental Therapeutics, 333,* 129–139.

Fujita, T., Okada, T., Hayashi, S., Jahangeer, S., Miwa, N., & Nakamura, S. (2004). Delta-catenin/NPRAP (neural plakophilin-related armadillo repeat protein) interacts with and activates sphingosine kinase 1. *Biochemical Journal, 382,* 717–723.

Fukuda, Y., Aoyama, Y., Wada, A., & Igarashi, Y. (2004). Identification of PECAM-1 association with sphingosine kinase 1 and its regulation by agonist-induced phosphorylation. *Biochimica et Biophysica Acta, 1636,* 12–21.

Gellings Lowe, N., Swaney, J. S., Moreno, K. M., & Sabbadini, R. A. (2009). Sphingosine-1-phosphate and sphingosine kinase are critical for transforming growth factor-beta-stimulated collagen production by cardiac fibroblasts. *Cardiovascular Research, 82,* 303–312.

Gillies, L., Lee, S. C., Long, J. S., Ktistakis, N., Pyne, N. J., & Pyne, S. (2009). The sphingosine 1-phosphate receptor 5 and sphingosine kinases 1 and 2 are localised in centrosomes: Possible role in regulating cell division. *Cellular Signalling, 21*, 675–684.

Gomez-Brouchet, A., Pchejetski, D., Brizuela, L., Garcia, V., Altie, M. F., Maddelein, M. L., et al. (2007). Critical role for sphingosine kinase-1 in regulating survival of neuroblastoma cells exposed to amyloid-beta peptide. *Molecular Pharmacology, 72*, 341–349.

Gonda, K., Okamoto, H., Takuwa, N., Yatomi, Y., Okazaki, H., Sakurai, T., et al. (1999). The novel sphingosine 1-phosphate receptor AGR16 is coupled via pertussis toxin-sensitive and -insensitive G-proteins to multiple signalling pathways. *Biochemical Journal, 337*(Pt 1), 67–75.

Goparaju, S. K., Jolly, P. S., Watterson, K. R., Bektas, M., Alvarez, S., Sarkar, S., et al. (2005). The S1P2 receptor negatively regulates platelet-derived growth factor-induced motility and proliferation. *Molecular and Cellular Biology, 25*, 4237–4249.

Graler, M. H. (2010). Targeting sphingosine 1-phosphate (S1P) levels and S1P receptor functions for therapeutic immune interventions. *Cellular Physiology and Biochemistry, 26*, 79–86.

Graler, M. H., Grosse, R., Kusch, A., Kremmer, E., Gudermann, T., & Lipp, M. (2003). The sphingosine 1-phosphate receptor S1P4 regulates cell shape and motility via coupling to Gi and G12/13. *Journal of Cellular Biochemistry, 89*, 507–519.

Grey, A., Chen, Q., Callon, K., Xu, X., Reid, I. R., & Cornish, J. (2002). The phospholipids sphingosine-1-phosphate and lysophosphatidic acid prevent apoptosis in osteoblastic cells via a signaling pathway involving Gi proteins and phosphatidylinositol-3 kinase. *Endocrinology, 143*, 4755–4763.

Grey, A., Xu, X., Hill, B., Watson, M., Callon, K., Reid, I. R., et al. (2004). Osteoblastic cells express phospholipid receptors and phosphatases and proliferate in response to sphingosine-1-phosphate. *Calcified Tissue International, 74*, 542–550.

Gude, D. R., Alvarez, S. E., Paugh, S. W., Mitra, P., Yu, J., Griffiths, R., et al. (2008). Apoptosis induces expression of sphingosine kinase 1 to release sphingosine-1-phosphate as a "come-and-get-me" signal. *The FASEB Journal, 22*, 2629–2638.

Hait, N. C., Allegood, J., Maceyka, M., Strub, G. M., Harikumar, K. B., Singh, S. K., et al. (2009). Regulation of histone acetylation in the nucleus by sphingosine-1-phosphate. *Science (New York, N.Y), 325*, 1254–1257.

Hait, N. C., Bellamy, A., Milstien, S., Kordula, T., & Spiegel, S. (2007). Sphingosine kinase type 2 activation by ERK-mediated phosphorylation. *Journal of Biological Chemistry, 282*, 12058–12065.

Hait, N. C., Sarkar, S., Le Stunff, H., Mikami, A., Maceyka, M., Milstien, S., et al. (2005). Role of sphingosine kinase 2 in cell migration toward epidermal growth factor. *Journal of Biological Chemistry, 280*, 29462–29469.

Hall, A. (1998). Rho GTPases and the actin cytoskeleton. *Science, 279*, 509–514.

Hanel, P., Andreani, P., & Graler, M. H. (2007). Erythrocytes store and release sphingosine 1-phosphate in blood. *The FASEB Journal, 21*, 1202–1209.

Hannun, Y. A., & Obeid, L. M. (2008). Principles of bioactive lipid signalling: Lessons from sphingolipids. *Nature Reviews. Molecular Cell Biology, 9*, 139–150.

Hashimoto, M., Wang, X., Mao, L., Kobayashi, T., Kawasaki, S., Mori, N., et al. (2008). Sphingosine 1-phosphate potentiates human lung fibroblast chemotaxis through the S1P2 receptor. *American Journal of Respiratory Cell and Molecular Biology, 39*, 356–363.

Hayashi, S., Okada, T., Igarashi, N., Fujita, T., Jahangeer, S., & Nakamura, S. (2002). Identification and characterization of RPK118, a novel sphingosine kinase-1-binding protein. *Journal of Biological Chemistry, 277*, 33319–33324.

Hengst, J., Wang, X., Sk, U., Sharma, A., Amin, S., & Yun, J. (2010). Development of a Sphingosine Kinase 1 Specific Small-Molecule Inhibitor. *Bioorganic & Medicinal Chemistry Letters, 20*, 7498–7502.

Henry, N. L., Pchejetski, D., A'Hern, R., Nguyen, A. T., Charles, P., Waxman, J., et al. (2010). Inflammatory cytokines and aromatase inhibitor-associated musculoskeletal syndrome: A case-control study. *British Journal of Cancer, 103*, 291–296.

Himmel, H. M., Meyer zu Heringdorf, D., Graf, E., Dobrev, D., Kortner, A., Schüler, S., et al. (2000). Evidence for Edg-3 receptor-mediated activation of I K.ACh by sphingosine-1-phosphate in human atrial cardiomyocytes. *Molecular Pharmacology, 58*, 449–454.

Hisano, Y., Kobayashi, N., Kawahara, A., Yamaguchi, A., & Nishi, T. (2011). The sphingosine 1-phosphate transporter, SPNS2, functions as a transporter of the phosphorylated form of the immunomodulating agent FTY720. *Journal of Biological Chemistry, 286*, 1758–1766.

Hla, T. (2003). Signaling and biological actions of sphingosine 1-phosphate. *Pharmacological Research, 47*, 401–407.

Hla, T., & Brinkmann, V. (2011). Sphingosine 1-phosphate (S1P): Physiology and the effects of S1P receptor modulation. *Neurology, 76*, S3–S8.

Ho, J. W., Man, K., Sun, C. K., Lee, T. K., Poon, R. T., & Fan, S. T. (2005). Effects of a novel immunomodulating agent, FTY720, on tumor growth and angiogenesis in hepatocellular carcinoma. *Molecular Cancer Therapeutics, 4*, 1430–1438.

Hobson, J. P., Rosenfeldt, H. M., Barak, L. S., Olivera, A., Poulton, S., Caron, M. G., et al. (2001). Role of the sphingosine-1-phosphate receptor EDG-1 in PDGF-induced cell motility. *Science (New York, N.Y), 291*, 1800–1803.

Huwiler, A., Kotelevets, N., Xin, C., Pastukhov, O., Pfeilschifter, J., & Zangemeister-Wittke, U. (2011). Loss of sphingosine kinase-1 in carcinoma cells increases formation of reactive oxygen species and sensitivity to doxorubicin-induced DNA damage. *British Journal of Pharmacology, 162*, 532–543.

Igarashi, Y., & Hakomori, S. (1989). Enzymatic synthesis of N, N-dimethyl-sphingosine: Demonstration of the sphingosine: N-methyltransferase in mouse brain. *Biochemical and Biophysical Research Communications, 164*, 1411–1416.

Igarashi, N., Okada, T., Hayashi, S., Fujita, T., Jahangeer, S., & Nakamura, S. (2003). Sphingosine kinase 2 is a nuclear protein and inhibits DNA synthesis. *Journal of Biological Chemistry, 278*, 46832–46839.

Iyer, S. S., & Kusner, D. J. (2009). Coordinate regulation of sphingosine kinase and actin dynamics. *Methods in Molecular Biology, 531*, 347–361 Clifton, N.J.

Jarman, K. E., Moretti, P. A., Zebol, J. R., & Pitson, S. M. (2010). Translocation of sphingosine kinase 1 to the plasma membrane is mediated by calcium- and integrin-binding protein 1. *Journal of Biological Chemistry, 285*, 483–492.

Jendiroba, D., Klostergaard, J., Keyhani, A., Pagliaro, L., & Freireich, E. (2002). Effective cytotoxicity against human leukemias and chemotherapy-resistant leukemia cell lines by NN-dimethylsphingosine. *Leukemia Research, 26*, 301.

Jin, Z. Q., Goetzl, E. J., & Karliner, J. S. (2004). Sphingosine kinase activation mediates ischemic preconditioning in murine heart. *Circulation, 110*, 1980–1989.

Jo, E., Sanna, M. G., Gonzalez-Cabrera, P. J., Thangada, S., Tigyi, G., Osborne, D. A., et al. (2005). S1P1-selective in vivo-active agonists from high-throughput screening: Off-the-shelf chemical probes of receptor interactions, signaling, and fate. *Chemical Biology, 12*, 703–715.

Johann, A. M., Weigert, A., Eberhardt, W., Kuhn, A. M., Barra, V., von Knethen, A., et al. (2008). Apoptotic cell-derived sphingosine-1-phosphate promotes HuR-dependent cyclooxygenase-2 mRNA stabilization and protein expression. *Journal of Immunology, 180*, 1239–1248.

Johnson, K. R., Becker, K. P., Facchinetti, M. M., Hannun, Y. A., & Obeid, L. M. (2002). PKC-dependent activation of sphingosine kinase 1 and translocation to the plasma membrane. Extracellular release of sphingosine-1-phosphate induced by phorbol 12-myristate 13-acetate (PMA). *Journal of Biological Chemistry, 277*, 35257–35262.

Johnson, K. R., Johnson, K. Y., Crellin, H. G., Ogretmen, B., Boylan, A. M., Harley, R. A., et al. (2005). Immunohistochemical distribution of sphingosine kinase 1 in normal and tumor lung tissue. *Journal of Histochemistry and Cytochemistry, 53*, 1159–1166.

Jung, I. D., Lee, J. S., Kim, Y. J., Jeong, Y. I., Lee, C. M., Baumruker, T., et al. (2007). Sphingosine kinase inhibitor suppresses a Th1 polarization via the inhibition of immunostimulatory activity in murine bone marrow-derived dendritic cells. *International Immunology, 19*, 411–426.

Kajimoto, T., Okada, T., Yu, H., Goparaju, S. K., Jahangeer, S., & Nakamura, S. (2007). Involvement of sphingosine-1-phosphate in glutamate secretion in hippocampal neurons. *Molecular and Cellular Biology, 27*, 3429–3440.

Kaneider, N. C., Djanani, A., Fischer-Colbrie, R., & Wiedermann, C. J. (2002). Sphingosine kinase-dependent directional migration of leukocytes in response to phorbol ester. *Biochemical and Biophysical Research Communications, 297*, 806–810.

Kang, Y. C., Kim, K. M., Lee, K. S., Namkoong, S., Lee, S. J., Han, J. A., et al. (2004). Serum bioactive lysophospholipids prevent TRAIL-induced apoptosis via PI3K/Akt-dependent cFLIP expression and Bad phosphorylation. *Cell Death and Differentiation, 11*, 1287–1298.

Kappos, L., Radue, E. W., O'Connor, P., Polman, C., Hohlfeld, R., Calabresi, P., et al. (2010). A placebo-controlled trial of oral fingolimod in relapsing multiple sclerosis. *The New England Journal of Medicine, 362*, 387–401.

Karin, M., & Greten, F. R. (2005). NF-kappaB: Linking inflammation and immunity to cancer development and progression. *Nature Reviews, 5*, 749–759.

Kawa, S., Kimura, S., Hakomori, S., & Igarashi, Y. (1997). Inhibition of chemotactic motility and trans-endothelial migration of human neutrophils by sphingosine 1-phosphate. *FEBS Letters, 420*, 196–200.

Kawahara, A., Nishi, T., Hisano, Y., Fukui, H., Yamaguchi, A., & Mochizuki, N. (2009). The sphingolipid transporter spns2 functions in migration of zebrafish myocardial precursors. *Science (New York, N.Y), 323*, 524–527.

Kawamori, T., Osta, W., Johnson, K. R., Pettus, B. J., Bielawski, J., Tanaka, T., et al. (2006). Sphingosine kinase 1 is up-regulated in colon carcinogenesis. *The FASEB Journal, 20*, 386–388.

Kennedy, A. J., Mathews, T. P., Kharel, Y., Field, S. D., Moyer, M. L., East, J. E., et al. (2011). Development of amidine-based sphingosine kinase 1 nanomolar inhibitors and reduction of sphingosine 1-phosphate in human leukemia cells. *Journal of Medicinal Chemistry, 54*, 3524–3548.

Klostergaard, J., Auzenne, E., & Leroux, E. (1998). Characterization of cytotoxicity induced by sphingolipids in multidrug-resistant leukemia cells. *Leukemia Research, 22*, 1049–1056.

Kluk, M. J., & Hla, T. (2002). Signaling of sphingosine-1-phosphate via the S1P/EDG-family of G-protein-coupled receptors. *Biochimica et Biophysica Acta, 1582*, 72–80.

Kobayashi, N., Nishi, T., Hirata, T., Kihara, A., Sano, T., Igarashi, Y., et al. (2006). Sphingosine 1-phosphate is released from the cytosol of rat platelets in a carrier-mediated manner. *Journal of Lipid Research, 47*, 614–621.

Kohno, M., Momoi, M., Oo, M. L., Paik, J. H., Lee, Y. M., Venkataraman, K., et al. (2006). Intracellular role for sphingosine kinase 1 in intestinal adenoma cell proliferation. *Molecular and Cellular Biology, 26*, 7211–7223.

Kono, M., Mi, Y., Liu, Y., Sasaki, T., Allende, M. L., Wu, Y. P., et al. (2004). The sphingosine-1-phosphate receptors S1P1, S1P2, and S1P3 function coordinately during embryonic angiogenesis. *Journal of Biological Chemistry, 279*, 29367–29373.

Kono, K., Sugiura, M., & Kohama, T. (2002). Inhibition of recombinant sphingosine kinases by novel inhibitors of microbial origin, F-12509A and B-5354c. *The Journal of Antibiotics (Tokyo), 55*, 99–103.

Kovanich, D., van der Heyden, M. A., Aye, T. T., van Veen, T. A., Heck, A. J., & Scholten, A. (2010). Sphingosine kinase interacting protein is an A-kinase anchoring protein specific for type I cAMP-dependent protein kinase. *ChemBioChem, 11*, 963–971.

Kovarik, J. M., Schmouder, R., Barilla, D., Riviere, G. J., Wang, Y., & Hunt, T. (2004). Multiple-dose FTY720: Tolerability, pharmacokinetics, and lymphocyte responses in healthy subjects. *Journal of Clinical Pharmacology, 44*, 532–537.

Kovarik, J. M., Schmouder, R., Barilla, D., Wang, Y., & Kraus, G. (2004). Single-dose FTY720 pharmacokinetics, food effect, and pharmacological responses in healthy subjects. *British Journal of Clinical Pharmacology, 57*, 586–591.

Kovarik, J. M., Slade, A., Riviere, G. J., Neddermann, D., Maton, S., Hunt, T. L., et al. (2008). The ability of atropine to prevent and reverse the negative chronotropic effect of fingolimod in healthy subjects. *British Journal of Clinical Pharmacology, 66*, 199–206.

Kovarik, J. M., Slade, A., Voss, B., Schmidli, H., Riviere, G. J., Picard, F., et al. (2007). Ethnic sensitivity study of fingolimod in white and Asian subjects. *International Journal of Clinical Pharmacology and Therapeutics, 45*, 98–109.

Koyrakh, L., Roman, M. I., Brinkmann, V., & Wickman, K. (2005). The heart rate decrease caused by acute FTY720 administration is mediated by the G protein-gated potassium channel I. *American Journal of Transplantation, 5*, 529–536.

Kozawa, O., Kawamura, H., Matsuno, H., & Uematsu, T. (2000). p38 MAP kinase is involved in the signalling of sphingosine in osteoblasts: Sphingosine inhibits prostaglandin F2[alpha]-induced phosphoinositide hydrolysis. *Cellular Signalling, 12*, 447–450.

Kozawa, O., Tokuda, H., Matsuno, H., & Uematsu, T. (1997). Activation of mitogen-activated protein kinase is involved in sphingosine 1-phosphate-stimulated interleukin-6 synthesis in osteoblasts. *FEBS Letters, 418*, 149–151.

Kusner, D. J., Thompson, C. R., Melrose, N. A., Pitson, S. M., Obeid, L. M., & Iyer, S. S. (2007). The localization and activity of sphingosine kinase 1 are coordinately regulated with actin cytoskeletal dynamics in macrophages. *Journal of Biological Chemistry, 282*, 23147–23162.

Lacana, E., Maceyka, M., Milstien, S., & Spiegel, S. (2002). Cloning and characterization of a protein kinase A anchoring protein (AKAP)-related protein that interacts with and regulates sphingosine kinase 1 activity. *Journal of Biological Chemistry, 277*, 32947–32953.

LaMontagne, K., Littlewood-Evans, A., Schnell, C., O'Reilly, T., Wyder, L., Sanchez, T., et al. (2006). Antagonism of sphingosine-1-phosphate receptors by FTY720 inhibits angiogenesis and tumor vascularization. *Cancer Research, 66*, 221–231.

Lampasso, J. D., Kamer, A., Margarone, J., & Dziak, R. (2001). Sphingosine-1-phosphate effects on PKC isoform expression in human osteoblastic cells. *Prostaglandins, Leukotrienes, and Essential Fatty Acids, 65*, 139–146.

Leclercq, T. M., Moretti, P. A., & Pitson, S. M. (2011). Guanine nucleotides regulate sphingosine kinase 1 activation by eukaryotic elongation factor 1A and provide a mechanism for eEF1A-associated oncogenesis. *Oncogene, 30*, 372–378.

Leclercq, T. M., & Pitson, S. M. (2006). Cellular signalling by sphingosine kinase and sphingosine 1-phosphate. *IUBMB Life, 58*, 467–472.

Lee, O. H., Kim, Y. M., Lee, Y. M., Moon, E. J., Lee, D. J., Kim, J. H., et al. (1999). Sphingosine 1-phosphate induces angiogenesis: Its angiogenic action and signaling mechanism in human umbilical vein endothelial cells. *Biochemical and Biophysical Research Communications, 264*, 743–750.

Lee, J. F., Ozaki, H., Zhan, X., Wang, E., Hla, T., & Lee, M. J. (2006). Sphingosine-1-phosphate signaling regulates lamellipodia localization of cortactin complexes in endothelial cells. *Histochemistry and Cell Biology, 126*, 297–304.

Lee, M. J., Thangada, S., Claffey, K. P., Ancellin, N., Liu, C. H., Kluk, M., et al. (1999). Vascular endothelial cell adherens junction assembly and morphogenesis induced by sphingosine-1-phosphate. *Cell, 99*, 301–312.

Lee, M. J., Thangada, S., Paik, J. H., Sapkota, G. P., Ancellin, N., Chae, S. S., et al. (2001). Akt-mediated phosphorylation of the G protein-coupled receptor EDG-1 is required for endothelial cell chemotaxis. *Molecular Cell, 8*, 693–704.

Lee, Y. M., Venkataraman, K., Hwang, S. I., Han, D. K., & Hla, T. (2007). A novel method to quantify sphingosine 1-phosphate by immobilized metal affinity chromatography (IMAC). *Prostaglandins & Other Lipid Mediators, 84*, 154–162.

Leroux, M. E., Auzenne, E., Evans, R., Hail, N., Jr., Spohn, W., Ghosh, S. C., et al. (2007). Sphingolipids and the sphingosine kinase inhibitor, SKI II, induce BCL-2-independent apoptosis in human prostatic adenocarcinoma cells. *The Prostate, 67*, 1699–1717.

Le Scolan, E., Pchejetski, D., Banno, Y., Denis, N., Mayeux, P., Vainchenker, W., et al. (2005). Overexpression of sphingosine kinase 1 is an oncogenic event in erythroleukemic progression. *Blood, 106*, 1808–1816.

Li, J., Guan, H. Y., Gong, L. Y., Song, L. B., Zhang, N., Wu, J., et al. (2008). Clinical significance of sphingosine kinase-1 expression in human astrocytomas progression and overall patient survival. *Clinical Cancer Research, 14*, 6996–7003.

Li, W., Yu, C. P., Xia, J. T., Zhang, L., Weng, G. X., Zheng, H. Q., et al. (2009). Sphingosine kinase 1 is associated with gastric cancer progression and poor survival of patients. *Clinical Cancer Research, 15*, 1393–1399.

Licht, T., Tsirulnikov, L., Reuveni, H., Yarnitzky, T., & Ben-Sasson, S. A. (2003). Induction of pro-angiogenic signaling by a synthetic peptide derived from the second intracellular loop of S1P3 (EDG3). *Blood, 102*, 2099–2107.

Lim, K. G., Tonelli, F., Li, Z., Lu, X., Bittman, R., Pyne, S., et al. (2011). FTY720 analogues as sphingosine kinase 1 inhibitors: Enzyme inhibition kinetics, allosterism, proteasomal degradation, and actin rearrangement in MCF-7 breast cancer cells. *Journal of Biological Chemistry, 286*, 18633–18640.

Lin, C. I., Chen, C. N., Chen, J. H., & Lee, H. (2006). Lysophospholipids increase IL-8 and MCP-1 expressions in human umbilical cord vein endothelial cells through an IL-1-dependent mechanism. *Journal of Cellular Biochemistry, 99*, 1216–1232.

Liu, Q., Alinari, L., Chen, C. S., Yan, F., Dalton, J. T., Lapalombella, R., et al. (2010). FTY720 shows promising in vitro and in vivo preclinical activity by downmodulating Cyclin D1 and phospho-Akt in mantle cell lymphoma. *Clinical Cancer Research, 16*, 3182–3192.

Liu, H., Sugiura, M., Nava, V. E., Edsall, L. C., Kono, K., Poulton, S., et al. (2000). Molecular cloning and functional characterization of a novel mammalian sphingosine kinase type 2 isoform. *Journal of Biological Chemistry, 275*, 19513–19520.

Liu, F., Verin, A. D., Wang, P., Day, R., Wersto, R. P., Chrest, F. J., et al. (2001). Differential regulation of sphingosine-1-phosphate- and VEGF-induced endothelial cell chemotaxis. Involvement of G(ialpha2)-linked Rho kinase activity. *American Journal of Respiratory Cell and Molecular Biology, 24*, 711–719.

Liu, Y., Wada, R., Yamashita, T., Mi, Y., Deng, C. X., Hobson, J. P., et al. (2000). Edg-1, the G protein-coupled receptor for sphingosine-1-phosphate, is essential for vascular maturation. *Journal of Clinical Investigation, 106*, 951–961.

Liu, G., Zheng, H., Zhang, Z., Wu, Z., Xiong, H., Li, J., et al. (2010). Overexpression of sphingosine kinase 1 is associated with salivary gland carcinoma progression and might be a novel predictive marker for adjuvant therapy. *BMC Cancer, 10*, 495.

Lo, C. G., Xu, Y., Proia, R. L., & Cyster, J. G. (2005). Cyclical modulation of sphingosine-1-phosphate receptor 1 surface expression during lymphocyte recirculation and relationship to lymphoid organ transit. *The Journal of Experimental Medicine, 201*, 291–301.

Long, J. S., Edwards, J., Watson, C., Tovey, S., Mair, K. M., Schiff, R., et al. (2010). Sphingosine kinase 1 induces tolerance to human epidermal growth factor receptor 2 and prevents formation of a migratory phenotype in response to sphingosine 1-phosphate in estrogen receptor positive breast cancer cells. *Molecular and Cellular Biology, 30*, 3827–3841.

Loveridge, C., Tonelli, F., Leclercq, T., Lim, K. G., Long, J. S., Berdyshev, E., et al. (2010). The sphingosine kinase 1 inhibitor 2-(p-hydroxyanilino)-4-(p-chlorophenyl)thiazole induces proteasomal degradation of sphingosine kinase 1 in mammalian cells. *Journal of Biological Chemistry*, *285*, 38841–38852.

Lucas da Silva, L. B., Ribeiro, D. A., Cury, P. M., Cordeiro, J. A., & Bueno, V. (2008). FTY720 treatment in experimentally urethane-induced lung tumors. *Journal of Experimental Therapeutics and Oncology*, *7*, 9–15.

Maceyka, M., Alvarez, S. E., Milstien, S., & Spiegel, S. (2008). Filamin A links sphingosine kinase 1 and sphingosine-1-phosphate receptor 1 at lamellipodia to orchestrate cell migration. *Molecular and Cellular Biology*, *28*, 5687–5697.

Maceyka, M., Harikumar, K. B., Milstien, S., & Spiegel, S. (2012). Sphingosine-1-phosphate signaling and its role in disease. *Trends in Cell Biology*, *22*, 50–60.

Maceyka, M., Nava, V. E., Milstien, S., & Spiegel, S. (2004). Aminoacylase 1 is a sphingosine kinase 1-interacting protein. *FEBS Letters*, *568*, 30–34.

Maceyka, M., Sankala, H., Hait, N. C., Le Stunff, H., Liu, H., Toman, R., et al. (2005). SphK1 and SphK2, sphingosine kinase isoenzymes with opposing functions in sphingolipid metabolism. *Journal of Biological Chemistry*, *280*, 37118–37129.

Maeda, Y., Matsuyuki, H., Shimano, K., Kataoka, H., Sugahara, K., & Chiba, K. (2007). Migration of CD4 T cells and dendritic cells toward sphingosine 1-phosphate (S1P) is mediated by different receptor subtypes: S1P regulates the functions of murine mature dendritic cells via S1P receptor type 3. *Journal of Immunology*, *178*, 3437–3446.

Malavaud, B., Pchejetski, D., Mazerolles, C., de Paiva, G. R., Calvet, C., Doumerc, N., et al. (2010). Sphingosine kinase-1 activity and expression in human prostate cancer resection specimens. *European Journal of Cancer*, *46*, 3417–3424.

Malek, R. L., Toman, R. E., Edsall, L. C., Wong, S., Chiu, J., Letterle, C. A., et al. (2001). Nrg-1 belongs to the endothelial differentiation gene family of G protein-coupled sphingosine-1-phosphate receptors. *Journal of Biological Chemistry*, *276*, 5692–5699.

Martin, C., Lafosse, J. M., Malavaud, B., & Cuvillier, O. (2010). Sphingosine kinase-1 mediates androgen-induced osteoblast cell growth. *Biochemical and Biophysical Research Communications*, *391*, 669–673.

Matloubian, M., Lo, C. G., Cinamon, G., Lesneski, M. J., Xu, Y., Brinkmann, V., et al. (2004). Lymphocyte egress from thymus and peripheral lymphoid organs is dependent on S1P receptor 1. *Nature*, *427*, 355–360.

Maurer, B., Melton, L., Billups, C., Cabot, M., & Reynolds, C. (2000). Synergistic cytotoxicity in solid tumor cell lines between N-(4-hydroxyphenyl) retinamide and modulators of ceramide metabolism. *Journal of the National Cancer Institute*, *92*, 1897.

Mehta, D., Konstantoulaki, M., Ahmmed, G. U., & Malik, A. B. (2005). Sphingosine 1-phosphate-induced mobilization of intracellular Ca2+ mediates rac activation and adherens junction assembly in endothelial cells. *Journal of Biological Chemistry*, *280*, 17320–17328.

Melendez, A. J., & Ibrahim, F. B. (2004). Antisense knockdown of sphingosine kinase 1 in human macrophages inhibits C5a receptor-dependent signal transduction, Ca2+ signals, enzyme release, cytokine production, and chemotaxis. *Journal of Immunology*, *173*, 1596–1603.

Merrill, A. H., Jr., Schmelz, E. M., Dillehay, D. L., Spiegel, S., Shayman, J. A., Schroeder, J. J., et al. (1997). Sphingolipids—The enigmatic lipid class: Biochemistry, physiology, and pathophysiology. *Toxicology and Applied Pharmacology*, *142*, 208–225.

Meyer zu Heringdorf, D., Lass, H., Kuchar, I., Lipinski, M., Alemany, R., Rumenapp, U., et al. (2001). Stimulation of intracellular sphingosine-1-phosphate production by G-protein-coupled sphingosine-1-phosphate receptors. *European Journal of Pharmacology*, *414*, 145–154.

Min, J., Van Veldhoven, P. P., Zhang, L., Hanigan, M. H., Alexander, H., & Alexander, S. (2005). Sphingosine-1-phosphate lyase regulates sensitivity of human cells to select chemotherapy drugs in a p38-dependent manner. *Molecular Cancer Research, 3*, 287–296.

Mitra, P., Oskeritzian, C. A., Payne, S. G., Beaven, M. A., Milstien, S., & Spiegel, S. (2006). Role of ABCC1 in export of sphingosine-1-phosphate from mast cells. *Proceedings of the National Academy of Sciences of the United States of America, 103*, 16394–16399.

Mizugishi, K., Yamashita, T., Olivera, A., Miller, G. F., Spiegel, S., & Proia, R. L. (2005). Essential role for sphingosine kinases in neural and vascular development. *Molecular and Cellular Biology, 25*, 11113–11121.

Mullen, T. D., Hannun, Y. A., & Obeid, L. M. (2012). Ceramide synthases at the centre of sphingolipid metabolism and biology. *Biochemical Journal, 441*, 789–802.

Nagahara, Y., Ikekita, M., & Shinomiya, T. (2000). Immunosuppressant FTY720 induces apoptosis by direct induction of permeability transition and release of cytochrome c from mitochondria. *Journal of Immunology, 165*, 3250–3259.

Nagaoka, Y., Otsuki, K., Fujita, T., & Uesato, S. (2008). Effects of phosphorylation of immunomodulatory agent FTY720 (fingolimod) on antiproliferative activity against breast and colon cancer cells. *Biological and Pharmaceutical Bulletin, 31*, 1177–1181.

Nava, V., Cuvillier, O., Edsall, L., Kimura, K., Milstien, S., Gelmann, E., et al. (2000). Sphingosine enhances apoptosis of radiation-resistant prostate cancer cells. *Cancer Research, 60*, 4468.

Nava, V. E., Hobson, J. P., Murthy, S., Milstien, S., & Spiegel, S. (2002). Sphingosine kinase type 1 promotes estrogen-dependent tumorigenesis of breast cancer MCF-7 cells. *Experimental Cell Research, 281*, 115–127.

Ng, K. T., Man, K., Ho, J. W., Sun, C. K., Lee, T. K., Zhao, Y., et al. (2007). Marked suppression of tumor growth by FTY720 in a rat liver tumor model: The significance of down-regulation of cell survival Akt pathway. *International Journal of Oncology, 30*, 375–380.

Niedernberg, A., Scherer, C. R., Busch, A. E., & Kostenis, E. (2002). Comparative analysis of human and rat S1P(5) (edg8): Differential expression profiles and sensitivities to antagonists. *Biochemical Pharmacology, 64*, 1243–1250.

Nishikawa, H., & Sakaguchi, S. (2010). Regulatory T cells in tumor immunity. *International Journal of Cancer, 127*, 759–767.

Novgorodov, A. S., El-Alwani, M., Bielawski, J., Obeid, L. M., & Gudz, T. I. (2007). Activation of sphingosine-1-phosphate receptor S1P5 inhibits oligodendrocyte progenitor migration. *The FASEB Journal, 21*, 1503–1514.

Nunes, J., Naymark, M., Sauer, L., Muhammad, A., Keun, H., Sturge, J., et al. (2012). Circulating sphingosine-1-phosphate and erythrocyte sphingosine kinase-1 activity as novel biomarkers for early prostate cancer detection. *British Journal of Cancer, 106*, 909–915.

Ogretmen, B., & Hannun, Y. A. (2004). Biologically active sphingolipids in cancer pathogenesis and treatment. *Nature Reviews. Cancer, 4*, 604–616.

Ohkawa, R., Nakamura, K., Okubo, S., Hosogaya, S., Ozaki, Y., Tozuka, M., et al. (2008). Plasma sphingosine-1-phosphate measurement in healthy subjects: Close correlation with red blood cell parameters. *Annals of Clinical Biochemistry, 45*, 356–363.

Ohotski, J., Long, J. S., Orange, C., Elsberger, B., Mallon, E., Doughty, J., et al. (2012). Expression of sphingosine 1-phosphate receptor 4 and sphingosine kinase 1 is associated with outcome in oestrogen receptor-negative breast cancer. *British Journal of Cancer, 106*, 1453–1459.

Okada, T., Kajimoto, T., Jahangeer, S., & Nakamura, S. (2009). Sphingosine kinase/sphingosine 1-phosphate signalling in central nervous system. *Cellular Signalling, 21*, 7–13.

Okamoto, H., Takuwa, N., Gonda, K., Okazaki, H., Chang, K., Yatomi, Y., et al. (1998). EDG1 is a functional sphingosine-1-phosphate receptor that is linked via a Gi/o to multiple signaling pathways, including phospholipase C activation, $Ca2+$ mobilization,

Ras-mitogen-activated protein kinase activation, and adenylate cyclase inhibition. *Journal of Biological Chemistry, 273*, 27104–27110.

Okamoto, H., Takuwa, N., Yokomizo, T., Sugimoto, N., Sakurada, S., Shigematsu, H., et al. (2000). Inhibitory regulation of Rac activation, membrane ruffling, and cell migration by the G protein-coupled sphingosine-1-phosphate receptor EDG5 but not EDG1 or EDG3. *Molecular and Cellular Biology, 20*, 9247–9261.

Okoshi, H., Hakomori, S., Nisar, M., Zhou, Q., Kimura, S., Tashiro, K., et al. (1991). Cell membrane signaling as target in cancer therapy II: Inhibitory effect of N, N, N-trimethylsphingosine on metastatic potential of murine B16 melanoma cell line through blocking of tumor cell-dependent platelet aggregation. *Cancer Research, 51*, 6019.

Olivera, A., Kohama, T., Edsall, L., Nava, V., Cuvillier, O., Poulton, S., et al. (1999). Sphingosine kinase expression increases intracellular sphingosine-1-phosphate and promotes cell growth and survival. *The Journal of Cell Biology, 147*, 545–558.

Oo, M. L., Thangada, S., Wu, M. T., Liu, C. H., Macdonald, T. L., Lynch, K. R., et al. (2007). Immunosuppressive and anti-angiogenic sphingosine 1-phosphate receptor-1 agonists induce ubiquitinylation and proteasomal degradation of the receptor. *Journal of Biological Chemistry, 282*, 9082–9089.

Osamu, K., Haruhiko, T., Hiroyuki, M., & Toshihiko, U. (1998). Sphingosine modulates interleukin-6 synthesis in osteoblasts. *Journal of Cellular Biochemistry, 70*, 338–345.

Osawa, Y., Uchinami, H., Bielawski, J., Schwabe, R. F., Hannun, Y. A., & Brenner, D. A. (2005). Roles for C16-ceramide and sphingosine 1-phosphate in regulating hepatocyte apoptosis in response to tumor necrosis factor-alpha. *Journal of Biological Chemistry, 280*, 27879–27887.

Osinde, M., Mullershausen, F., & Dev, K. K. (2007). Phosphorylated FTY720 stimulates ERK phosphorylation in astrocytes via S1P receptors. *Neuropharmacology, 52*, 1210–1218.

Paik, J. H., Chae, S., Lee, M. J., Thangada, S., & Hla, T. (2001). Sphingosine 1-phosphate-induced endothelial cell migration requires the expression of EDG-1 and EDG-3 receptors and Rho-dependent activation of alpha vbeta3- and beta1-containing integrins. *Journal of Biological Chemistry, 276*, 11830–11837.

Pappu, R., Schwab, S. R., Cornelissen, I., Pereira, J. P., Regard, J. B., Xu, Y., et al. (2007). Promotion of lymphocyte egress into blood and lymph by distinct sources of sphingosine-1-phosphate. *Science (New York, N.Y), 316*, 295–298.

Paugh, S., Cassidy, M., He, H., Milstien, S., Sim-Selley, L., Spiegel, S., et al. (2006). Sphingosine and its analog, the immunosuppressant 2-amino-2-(2-[4-octylphenyl] ethyl)-1, 3-propanediol, interact with the CB1 cannabinoid receptor. *Molecular Pharmacology, 70*, 41.

Paugh, S. W., Paugh, B. S., Rahmani, M., Kapitonov, D., Almenara, J. A., Kordula, T., et al. (2008). A selective sphingosine kinase 1 inhibitor integrates multiple molecular therapeutic targets in human leukemia. *Blood, 112*, 1382–1391.

Payne, S. G., Milstien, S., & Spiegel, S. (2002). Sphingosine-1-phosphate: Dual messenger functions. *FEBS Letters, 531*, 54–57.

Payne, S. G., Oskeritzian, C. A., Griffiths, R., Subramanian, P., Barbour, S. E., Chalfant, C. E., et al. (2007). The immunosuppressant drug FTY720 inhibits cytosolic phospholipase A2 independently of sphingosine-1-phosphate receptors. *Blood, 109*, 1077–1085.

Pchejetski, D., Bohler, T., Brizuela, L., Sauer, L., Doumerc, N., Golzio, M., et al. (2010). FTY720 (Fingolimod) sensitizes prostate cancer cells to radiotherapy by inhibition of sphingosine kinase-1. *Cancer Research, 70*, 8651–8661.

Pchejetski, D., Bohler, T., Stebbing, J., & Waxman, J. (2011). Therapeutic potential of targeting sphingosine kinase 1 in prostate cancer. *Nature Reviews. Urology, 8*, 569–678.

Pchejetski, D., Doumerc, N., Golzio, M., Naymark, M., Teissie, J., Kohama, T., et al. (2008). Chemosensitizing effects of sphingosine kinase-1 inhibition in prostate cancer cell and animal models. *Molecular Cancer Therapeutics*, 7, 1836–1845.

Pchejetski, D., Golzio, M., Bonhoure, E., Calvet, C., Doumerc, N., Garcia, V., et al. (2005). Sphingosine kinase-1 as a chemotherapy sensor in prostate adenocarcinoma cell and mouse models. *Cancer Research*, *65*, 11667–11675.

Pchejetski, D., Kunduzova, O., Dayon, A., Calise, D., Seguelas, M. H., Leducq, N., et al. (2007). Oxidative stress-dependent sphingosine kinase-1 inhibition mediates monoamine oxidase A-associated cardiac cell apoptosis. *Circulation Research*, *100*, 41–49.

Pchejetski, D., Nunes, J., Coughlan, K., Lall, H., Pitson, S. M., Waxman, J., et al. (2010). The involvement of sphingosine kinase 1 in LPS-induced Toll-like receptor 4-mediated accumulation of HIF-1alpha protein, activation of ASK1 and production of the pro-inflammatory cytokine IL-6. *Immunology and Cell Biology*, *89*, 268–274.

Pchejetski, D., Nunes, J., Coughlan, K., Lall, H., Pitson, S. M., Waxman, J., et al. (2011). The involvement of sphingosine kinase 1 in LPS-induced Toll-like receptor 4-mediated accumulation of HIF-1alpha protein, activation of ASK1 and production of the pro-inflammatory cytokine IL-6. *Immunology and Cell Biology*, *89*, 268–274.

Pchejetski, D., Nunes, J., Sauer, L., Sidhu, J., Sharma, A., Keun, H. C., et al. (2010). Circulating sphingosine-1-phosphate inversely correlates with chemotherapy-induced weight gain during early breast cancer. *Breast Cancer Research and Treatment*, *124*, 543–549.

Permpongkosol, S., Wang, J. D., Takahara, S., Matsumiya, K., Nonomura, N., Nishimura, K., et al. (2002). Anticarcinogenic effect of FTY720 in human prostate carcinoma DU145 cells: Modulation of mitogenic signaling, FAK, cell-cycle entry and apoptosis. *International Journal of Cancer*, *98*, 167–172.

Pettus, B. J., Bielawski, J., Porcelli, A. M., Reames, D. L., Johnson, K. R., Morrow, J., et al. (2003). The sphingosine kinase 1/sphingosine-1-phosphate pathway mediates COX-2 induction and PGE2 production in response to TNF-alpha. *The FASEB Journal*, *17*, 1411–1421.

Pitson, S. M. (2010). Regulation of sphingosine kinase and sphingolipid signaling. *Trends in Biochemical Sciences*, *36*, 97–107.

Pitson, S. M. (2011). Regulation of sphingosine kinase and sphingolipid signaling. *Trends in Biochemical Sciences*, *36*, 97–107.

Pitson, S. M., Moretti, P. A., Zebol, J. R., Lynn, H. E., Xia, P., Vadas, M. A., et al. (2003). Activation of sphingosine kinase 1 by ERK1/2-mediated phosphorylation. *EMBO Journal*, *22*, 5491–5500.

Pitson, S. M., Moretti, P. A., Zebol, J. R., Xia, P., Gamble, J. R., Vadas, M. A., et al. (2000). Expression of a catalytically inactive sphingosine kinase mutant blocks agonist-induced sphingosine kinase activation. A dominant-negative sphingosine kinase. *Journal of Biological Chemistry*, *275*, 33945–33950.

Pitson, S. M., Xia, P., Leclercq, T. M., Moretti, P. A., Zebol, J. R., Lynn, H. E., et al. (2005). Phosphorylation-dependent translocation of sphingosine kinase to the plasma membrane drives its oncogenic signalling. *The Journal of Experimental Medicine*, *201*, 49–54.

Pyne, S., Lee, S. C., Long, J., & Pyne, N. J. (2009). Role of sphingosine kinases and lipid phosphate phosphatases in regulating spatial sphingosine 1-phosphate signalling in health and disease. *Cellular Signalling*, *21*, 14–21.

Pyne, S., & Pyne, N. J. (2002). Sphingosine 1-phosphate signalling and termination at lipid phosphate receptors. *Biochimica et Biophysica Acta*, *1582*, 121–131.

Radeff-Huang, J., Seasholtz, T. M., Chang, J. W., Smith, J. M., & Walsh, C. T. (2006). TNF-alpha -stimulated cell proliferation is mediated through sphingosine kinase-dependent AKT activation and cyclin D expression. *Journal of Biological Chemistry*, *282*, 863–870.

Rigogliuso, S., Donati, C., Cassara, D., Taverna, S., Salamone, M., Bruni, P., et al. (2010). An active form of sphingosine kinase-1 is released in the extracellular medium as component of membrane vesicles shed by two human tumor cell lines. *Journal of Oncology*, *2010*, 509329.

Robert, P., Tsui, P., Laville, M. P., Livi, G. P., Sarau, H. M., Bril, A., et al. (2001). EDG1 receptor stimulation leads to cardiac hypertrophy in rat neonatal myocytes. *Journal of Molecular and Cellular Cardiology*, *33*, 1589–1606.

Roberts, J. L., Moretti, P. A., Darrow, A. L., Derian, C. K., Vadas, M. A., & Pitson, S. M. (2004). An assay for sphingosine kinase activity using biotinylated sphingosine and streptavidin-coated membranes. *Analytical Biochemistry*, *331*, 122–129.

Roberts, K. G., Smith, A. M., McDougall, F., Carpenter, H., Horan, M., Neviani, P., et al. (2010). Essential requirement for PP2A inhibition by the oncogenic receptor c-KIT suggests PP2A reactivation as a strategy to treat c-KIT+ cancers. *Cancer Research*, *70*, 5438–5447.

Rosen, H., Gonzalez-Cabrera, P. J., Sanna, M. G., & Brown, S. (2009). Sphingosine 1-phosphate receptor signaling. *Annual Review of Biochemistry*, *78*, 743–768.

Rosenfeldt, H. M., Amrani, Y., Watterson, K. R., Murthy, K. S., Panettieri, R. A., Jr., & Spiegel, S. (2003). Sphingosine-1-phosphate stimulates contraction of human airway smooth muscle cells. *The FASEB Journal*, *17*, 1789–1799.

Ruckhaberle, E., Rody, A., Engels, K., Gaetje, R., von Minckwitz, G., Schiffmann, S., et al. (2007). Microarray analysis of altered sphingolipid metabolism reveals prognostic significance of sphingosine kinase 1 in breast cancer. *Breast Cancer Research and Treatment*, *112*, 41–52.

Ruckhaberle, E., Rody, A., Engels, K., Gaetje, R., von Minckwitz, G., Schiffmann, S., et al. (2008). Microarray analysis of altered sphingolipid metabolism reveals prognostic significance of sphingosine kinase 1 in breast cancer. *Breast Cancer Research and Treatment*, *112*, 41–52.

Ryu, J., Kim, H. J., Chang, E. J., Huang, H., Banno, Y., & Kim, H. H. (2006). Sphingosine 1-phosphate as a regulator of osteoclast differentiation and osteoclast-osteoblast coupling. *EMBO Journal*, *25*, 5840–5851.

Sachs, C., Safa, A., Harrison, S., & Fine, R. (1995). Partial inhibition of multidrug resistance by safingol is independent of modulation of P-glycoprotein substrate activities and correlated with inhibition of protein kinase C. *Journal of Biological Chemistry*, *270*, 26639.

Sadahira, Y., Ruan, F., Hakomori, S., & Igarashi, Y. (1992). Sphingosine 1-phosphate, a specific endogenous signaling molecule controlling cell motility and tumor cell invasiveness. *Proceedings of the National Academy of Sciences of the United States of America*, *89*, 9686–9690.

Salinas, N. R., Lopes, C. T., Palma, P. V., Oshima, C. T., & Bueno, V. (2009). Lung tumor development in the presence of sphingosine 1-phosphate agonist FTY720. *Pathology and Oncology Research*, *15*, 549–554.

Sanchez, T., & Hla, T. (2004). Structural and functional characteristics of S1P receptors. *Journal of Cellular Biochemistry*, *92*, 913–922.

Sanchez, T., Thangada, S., Wu, M. T., Kontos, C. D., Wu, D., Wu, H., et al. (2005). PTEN as an effector in the signaling of antimigratory G protein-coupled receptor. *Proceedings of the National Academy of Sciences of the United States of America*, *102*, 4312–4317.

Sankala, H. M., Hait, N. C., Paugh, S. W., Shida, D., Lepine, S., Elmore, L. W., et al. (2007). Involvement of sphingosine kinase 2 in p53-independent induction of p21 by the chemotherapeutic drug doxorubicin. *Cancer Research*, *67*, 10466–10474.

Sanna, M. G., Wang, S. K., Gonzalez-Cabrera, P. J., Don, A., Marsolais, D., Matheu, M. P., et al. (2006). Enhancement of capillary leakage and restoration of lymphocyte egress by a chiral S1P1 antagonist in vivo. *Nature Chemical Biology*, *2*, 434–441.

Sarkar, S., Maceyka, M., Hait, N. C., Paugh, S. W., Sankala, H., Milstien, S., et al. (2005). Sphingosine kinase 1 is required for migration, proliferation and survival of MCF-7 human breast cancer cells. *FEBS Letters*, *579*, 5313–5317.

Sato, K., Malchinkhuu, E., Horiuchi, Y., Mogi, C., Tomura, H., Tosaka, M., et al. (2007). Critical role of ABCA1 transporter in sphingosine 1-phosphate release from astrocytes. *Journal of Neurochemistry*, *103*, 2610–2619.

Sauer, L., Nunes, J., Salunkhe, V., Skalska, L., Kohama, T., Cuvillier, O., et al. (2009). Sphingosine kinase 1 inhibition sensitizes hormone-resistant prostate cancer to docetaxel. *International Journal of Cancer, 125*, 2728–2736.

Schmid, G., Guba, M., Ischenko, I., Papyan, A., Joka, M., Schrepfer, S., et al. (2007). The immunosuppressant FTY720 inhibits tumor angiogenesis via the sphingosine 1-phosphate receptor 1. *Journal of Cellular Biochemistry, 101*, 259–270.

Schnitzer, S. E., Weigert, A., Zhou, J., & Brune, B. (2009). Hypoxia enhances sphingosine kinase 2 activity and provokes sphingosine-1-phosphate-mediated chemoresistance in A549 lung cancer cells. *Molecular Cancer Research, 7*, 393–401.

Schwab, S. R., & Cyster, J. G. (2007). Finding a way out: Lymphocyte egress from lymphoid organs. *Nature Immunology, 8*, 1295–1301.

Schwalm, S., Doll, F., Romer, I., Bubnova, S., Pfeilschifter, J., & Huwiler, A. (2008). Sphingosine kinase-1 is a hypoxia-regulated gene that stimulates migration of human endothelial cells. *Biochemical and Biophysical Research Communications, 368*, 1020–1025.

Schwartz, G., Haimovitz-Friedman, A., Dhupar, S., Ehleiter, D., Maslak, P., Lai, L., et al. (1995). Potentiation of apoptosis by treatment with the protein kinase C-specific inhibitor safingol in mitomycin C-treated gastric cancer cells. *Journal of the National Cancer Institute, 87*, 1394.

Schwartz, G., Ward, D., Saltz, L., Casper, E., Spiess, T., Mullen, E., et al. (1997). A pilot clinical/pharmacological study of the protein kinase C-specific inhibitor safingol alone and in combination with doxorubicin. *Clinical Cancer Research, 3*, 537.

Sensken, S. C., Bode, C., & Graler, M. H. (2009). Accumulation of fingolimod (FTY720) in lymphoid tissues contributes to prolonged efficacy. *Journal of Pharmacology and Experimental Therapeutics, 328*, 963–969.

Serrano-Sanchez, M., Tanfin, Z., & Leiber, D. (2008). Signaling pathways involved in sphingosine kinase activation and sphingosine-1-phosphate release in rat myometrium in late pregnancy: Role in the induction of cyclooxygenase 2. *Endocrinology, 149*, 4669–4679.

Serriere-Lanneau, V., Teixeira-Clerc, F., Li, L., Schippers, M., de Wries, W., Julien, B., et al. (2007). The sphingosine 1-phosphate receptor S1P2 triggers hepatic wound healing. *The FASEB Journal, 21*, 2005–2013.

Sharma, A. K. (2011). Sphingo-guanidines and their use as inhibitors of sphingosine kinase (WO2010078247). *Expert Opinion on Therapeutic Patents, 21*, 807–812.

Shen, Y., Cai, M., Xia, W., Liu, J., Zhang, Q., Xie, H., et al. (2007). FTY720, a synthetic compound from Isaria sinclairii, inhibits proliferation and induces apoptosis in pancreatic cancer cells. *Cancer Letters, 254*, 288–297.

Shen, Y., Wang, X., Xia, W., Wang, C., Cai, M., Xie, H., et al. (2008). Antiproliferative and overadditive effects of rapamycin and FTY720 in pancreatic cancer cells in vitro. *Transplantation Proceedings, 40*, 1727–1733.

Shida, D., Fang, X., Kordula, T., Takabe, K., Lepine, S., Alvarez, S. E., et al. (2008). Cross-talk between LPA1 and epidermal growth factor receptors mediates up-regulation of sphingosine kinase 1 to promote gastric cancer cell motility and invasion. *Cancer Research, 68*, 6569–6577.

Shirahama, T., Sweeney, E., Sakakura, C., Singhal, A., Nishiyama, K., Akiyama, S., et al. (1997). In vitro and in vivo induction of apoptosis by sphingosine and N, N-dimethylsphingosine in human epidermoid carcinoma KB-3-1 and its multidrug-resistant cells. *Clinical Cancer Research, 3*, 257.

Shu, X., Wu, W., Mosteller, R. D., & Broek, D. (2002). Sphingosine kinase mediates vascular endothelial growth factor-induced activation of ras and mitogen-activated protein kinases. *Molecular and Cellular Biology, 22*, 7758–7768.

Sinha, U. K., Schorn, V. J., Hochstim, C., Chinn, S. B., Zhu, S., & Masood, R. (2010). Increased radiation sensitivity of head and neck squamous cell carcinoma with sphingosine kinase 1 inhibition. *Head & Neck, 33*, 178–188.

Skerjanec, A., Tedesco, H., Neumayer, H. H., Cole, E., Budde, K., Hsu, C. H., et al. (2005). FTY720, a novel immunomodulator in de novo kidney transplant patients: Pharmacokinetics and exposure-response relationship. *Journal of Clinical Pharmacology, 45*, 1268–1278.
Soldi, R., Mandinova, A., Venkataraman, K., Hla, T., Vadas, M., Pitson, S., et al. (2007). Sphingosine kinase 1 is a critical component of the copper-dependent FGF1 export pathway. *Experimental Cell Research, 313*, 3308–3318.
Song, L., Xiong, H., Li, M., Liao, W. T., Wang, L., & Wu, J. (2011). Sphingosine kinase-1 enhances resistance to apoptosis through activation of PI3K/Akt/NF-κB pathway in human non-small cell lung cancer. *Clinical Cancer Research, 17*, 1839–1849.
Spiegel, S., & Kolesnick, R. (2002). Sphingosine 1-phosphate as a therapeutic agent. *Leukemia, 16*, 1596–1602.
Spiegel, S., & Milstien, S. (2003). Sphingosine-1-phosphate: An enigmatic signalling lipid. *Nature Reviews. Molecular Cell Biology, 4*, 397–407.
Stahelin, R. V., Hwang, J. H., Kim, J. H., Park, Z. Y., Johnson, K. R., Obeid, L. M., et al. (2005). The mechanism of membrane targeting of human sphingosine kinase 1. *Journal of Biological Chemistry, 280*, 43030–43038.
Stam, J. C., Michiels, F., van der Kammen, R. A., Moolenaar, W. H., & Collard, J. G. (1998). Invasion of T-lymphoma cells: Cooperation between Rho family GTPases and lysophospholipid receptor signaling. *EMBO Journal, 17*, 4066–4074.
Strub, G. M., Maceyka, M., Hait, N. C., Milstien, S., & Spiegel, S. (2010). Extracellular and intracellular actions of sphingosine-1-phosphate. *Advances in Experimental Medicine and Biology, 688*, 141–155.
Strub, G. M., Paillard, M., Liang, J., Gomez, L., Allegood, J. C., Hait, N. C., et al. (2011). Sphingosine-1-phosphate produced by sphingosine kinase 2 in mitochondria interacts with prohibitin 2 to regulate complex IV assembly and respiration. *The FASEB Journal, 25*, 600–612.
Sugimoto, N., Takuwa, N., Okamoto, H., Sakurada, S., & Takuwa, Y. (2003). Inhibitory and stimulatory regulation of Rac and cell motility by the G12/13-Rho and Gi pathways integrated downstream of a single G protein-coupled sphingosine-1-phosphate receptor isoform. *Molecular and Cellular Biology, 23*, 1534–1545.
Sugiura, T., Dohi, Y., Yamashita, S., Ohte, N., Ito, S., Iwaki, S., et al. (2012). Analytical evaluation of plasma serotonin and sphingosine 1-phosphate and their clinical assessment in early atherosclerosis. *Coronary Artery Disease, 23*, 234–238.
Sukocheva, O. A., Wang, L., Albanese, N., Pitson, S. M., Vadas, M. A., & Xia, P. (2003). Sphingosine kinase transmits estrogen signaling in human breast cancer cells. *Molecular Endocrinology (Baltimore MD), 17*, 2002–2012.
Sukocheva, O., Wang, L., Verrier, E., Vadas, M. A., & Xia, P. (2009). Restoring endocrine response in breast cancer cells by inhibition of the sphingosine kinase-1 signaling pathway. *Endocrinology, 150*, 4484–4492.
Sun, Y., Lin, J., Katz, A. E., & Fisher, P. B. (1997). Human prostatic carcinoma oncogene PTI-1 is expressed in human tumor cell lines and prostate carcinoma patient blood samples. *Cancer Research, 57*, 18–23.
Sutherland, C. M., Moretti, P. A., Hewitt, N. M., Bagley, C. J., Vadas, M. A., & Pitson, S. M. (2006). The calmodulin-binding site of sphingosine kinase and its role in agonist-dependent translocation of sphingosine kinase 1 to the plasma membrane. *Journal of Biological Chemistry, 281*, 11693–11701.
Sutphen, R., Xu, Y., Wilbanks, G. D., Fiorica, J., Grendys, E. C., Jr., LaPolla, J. P., et al. (2004). Lysophospholipids are potential biomarkers of ovarian cancer. *Cancer Epidemiology, Biomarkers & Prevention, 13*, 1185–1191.
Sweeney, E. A., Sakakura, C., Shirahama, T., Masamune, A., Ohta, H., Hakomori, S., et al. (1996). Sphingosine and its methylated derivative N, N-dimethylsphingosine (DMS) induce apoptosis in a variety of human cancer cell lines. *International Journal of Cancer, 66*, 358–366.

Taha, T. A., Hannun, Y. A., & Obeid, L. M. (2006). Sphingosine kinase: Biochemical and cellular regulation and role in disease. *Journal of Biochemistry and Molecular Biology*, *39*, 113–131.

Takabe, K., Kim, R. H., Allegood, J. C., Mitra, P., Ramachandran, S., Nagahashi, M., et al. (2010). Estradiol induces export of sphingosine-1-phosphate from breast cancer cells via ABCC1 and ABCG2. *Journal of Biological Chemistry*, *285*, 10477–10486.

Takabe, K., Paugh, S. W., Milstien, S., & Spiegel, S. (2008). "Inside-out" signaling of sphingosine-1-phosphate: Therapeutic targets. *Pharmacological Reviews*, *60*, 181–195.

Takuwa, Y. (2002). Subtype-specific differential regulation of Rho family G proteins and cell migration by the Edg family sphingosine-1-phosphate receptors. *Biochimica et Biophysica Acta*, *1582*, 112–120.

Tani, M., Ito, M., & Igarashi, Y. (2007). Ceramide/sphingosine/sphingosine 1-phosphate metabolism on the cell surface and in the extracellular space. *Cellular Signalling*, *19*, 229–237.

Tedesco-Silva, H., Szakaly, P., Shoker, A., Sommerer, C., Yoshimura, N., Schena, F. P., et al. (2007). FTY720 versus mycophenolate mofetil in de novo renal transplantation: Six-month results of a double-blind study. *Transplantation*, *84*, 885–892.

Tokumura, A., Carbone, L. D., Yoshioka, Y., Morishige, J., Kikuchi, M., Postlethwaite, A., et al. (2009). Elevated serum levels of arachidonoyl-lysophosphatidic acid and sphingosine 1-phosphate in systemic sclerosis. *International Journal of Medical Sciences*, *6*, 168–176.

Tonelli, F., Lim, K. G., Loveridge, C., Long, J., Pitson, S. M., Tigyi, G., et al. (2010). FTY720 and (S)-FTY720 vinylphosphonate inhibit sphingosine kinase 1 and promote its proteasomal degradation in human pulmonary artery smooth muscle, breast cancer and androgen-independent prostate cancer cells. *Cellular Signalling*, *22*, 1536–1542.

Ubai, T., Azuma, H., Kotake, Y., Inamoto, T., Takahara, K., Ito, Y., et al. (2007). FTY720 induced Bcl-associated and Fas-independent apoptosis in human renal cancer cells in vitro and significantly reduced in vivo tumor growth in mouse xenograft. *Anticancer Research*, *27*, 75–88.

Ushitora, Y., Tashiro, H., Ogawa, T., Tanimoto, Y., Kuroda, S., Kobayashi, T., et al. (2009). Suppression of hepatocellular carcinoma recurrence after rat liver transplantation by FTY720, a sphingosine-1-phosphate analog. *Transplantation*, *88*, 980–986.

Usui, S., Sugimoto, N., Takuwa, N., Sakagami, S., Takata, S., Kaneko, S., et al. (2004). Blood lipid mediator sphingosine 1-phosphate potently stimulates platelet-derived growth factor-A and -B chain expression through S1P1-Gi-Ras-MAPK-dependent induction of Kruppel-like factor 5. *Journal of Biological Chemistry*, *279*, 12300–12311.

Van Brocklyn, J. R., Graler, M. H., Bernhardt, G., Hobson, J. P., Lipp, M., & Spiegel, S. (2000). Sphingosine-1-phosphate is a ligand for the G protein-coupled receptor EDG-6. *Blood*, *95*, 2624–2629.

Van Brocklyn, J. R., Jackson, C. A., Pearl, D. K., Kotur, M. S., Snyder, P. J., & Prior, T. W. (2005). Sphingosine kinase-1 expression correlates with poor survival of patients with glioblastoma multiforme: Roles of sphingosine kinase isoforms in growth of glioblastoma cell lines. *Journal of Neuropathology and Experimental Neurology*, *64*, 695–705.

Van Brocklyn, J. R., Young, N., & Roof, R. (2003). Sphingosine-1-phosphate stimulates motility and invasiveness of human glioblastoma multiforme cells. *Cancer Letters*, *199*, 53–60.

Venkataraman, K., Thangada, S., Michaud, J., Oo, M. L., Ai, Y., Lee, Y. M., et al. (2006). Extracellular export of sphingosine kinase-1a contributes to the vascular S1P gradient. *Biochemical Journal*, *397*, 461–471.

Verzijl, D., Peters, S. L., & Alewijnse, A. E. (2010). Sphingosine-1-phosphate receptors: Zooming in on ligand-induced intracellular trafficking and its functional implications. *Molecules and Cells*, *29*, 99–104.

Vessey, D. A., Kelley, M., Zhang, J., Li, L., Tao, R., & Karliner, J. S. (2007). Dimethylsphingosine and FTY720 inhibit the SK1 form but activate the SK2 form of sphingosine kinase from rat heart. *Journal of Biochemical and Molecular Toxicology*, *21*, 273–279.

Visentin, B., Vekich, J. A., Sibbald, B. J., Cavalli, A. L., Moreno, K. M., Matteo, R. G., et al. (2006). Validation of an anti-sphingosine-1-phosphate antibody as a potential therapeutic in reducing growth, invasion, and angiogenesis in multiple tumor lineages. *Cancer Cell, 9*, 225–238.

Wang, W., Graeler, M. H., & Goetzl, E. J. (2005). Type 4 sphingosine 1-phosphate G protein-coupled receptor (S1P4) transduces S1P effects on T cell proliferation and cytokine secretion without signaling migration. *The FASEB Journal, 19*, 1731–1733.

Wang, L., Lee, J. F., Lin, C. Y., & Lee, M. J. (2008). Rho GTPases mediated integrin alpha v beta 3 activation in sphingosine-1-phosphate stimulated chemotaxis of endothelial cells. *Histochemistry and Cell Biology, 129*, 579–588.

Wang, J. D., Takahara, S., Nonomura, N., Ichimaru, N., Toki, K., Azuma, H., et al. (1999). Early induction of apoptosis in androgen-independent prostate cancer cell line by FTY720 requires caspase-3 activation. *The Prostate, 40*, 50–55.

Wang, F., Van Brocklyn, J. R., Edsall, L., Nava, V. E., & Spiegel, S. (1999). Sphingosine-1-phosphate inhibits motility of human breast cancer cells independently of cell surface receptors. *Cancer Research, 59*, 6185–6191.

Wang, F., Van Brocklyn, J. R., Hobson, J. P., Movafagh, S., Zukowska-Grojec, Z., Milstien, S., et al. (1999). Sphingosine 1-phosphate stimulates cell migration through a G(i)-coupled cell surface receptor. Potential involvement in angiogenesis. *Journal of Biological Chemistry, 274*, 35343–35350.

Watson, C., Long, J. S., Orange, C., Tannahill, C. L., Mallon, E., McGlynn, L. M., et al. (2010). High expression of sphingosine 1-phosphate receptors, S1P1 and S1P3, sphingosine kinase 1, and extracellular signal-regulated kinase-1/2 is associated with development of tamoxifen resistance in estrogen receptor-positive breast cancer patients. *American Journal of Pathology, 177*, 2205–2215.

Wei, E. K., Wolin, K. Y., & Colditz, G. A. (2010). Time course of risk factors in cancer etiology and progression. *Journal of Clinical Oncology, 28*, 4052–4057.

Weigert, A., Schiffmann, S., Sekar, D., Ley, S., Menrad, H., Werno, C., et al. (2009). Sphingosine kinase 2 deficient tumor xenografts show impaired growth and fail to polarize macrophages towards an anti-inflammatory phenotype. *International Journal of Cancer, 125*, 2114–2121.

Weigert, A., Tzieply, N., von Knethen, A., Johann, A. M., Schmidt, H., Geisslinger, G., et al. (2007). Tumor cell apoptosis polarizes macrophages role of sphingosine-1-phosphate. *Molecular Biology of the Cell, 18*, 3810–3819.

Windh, R. T., Lee, M. J., Hla, T., An, S., Barr, A. J., & Manning, D. R. (1999). Differential coupling of the sphingosine 1-phosphate receptors Edg-1, Edg-3, and H218/Edg-5 to the G(i), G(q), and G(12) families of heterotrimeric G proteins. *Journal of Biological Chemistry, 274*, 27351–27358.

Wolf, A. M., Eller, K., Zeiser, R., Durr, C., Gerlach, U. V., Sixt, M., et al. (2009). The sphingosine 1-phosphate receptor agonist FTY720 potently inhibits regulatory T cell proliferation in vitro and in vivo. *Journal of Immunology, 183*, 3751–3760.

Wu, W., Shu, X., Hovsepyan, H., Mosteller, R. D., & Broek, D. (2003). VEGF receptor expression and signaling in human bladder tumors. *Oncogene, 22*, 3361–3370.

Xia, P., Gamble, J. R., Rye, K. A., Wang, L., Hii, C. S., Cockerill, P., et al. (1998). Tumor necrosis factor-alpha induces adhesion molecule expression through the sphingosine kinase pathway. *Proceedings of the National Academy of Sciences of the United States of America, 95*, 14196–14201.

Xia, P., Gamble, J. R., Wang, L., Pitson, S. M., Moretti, P. A., Wattenberg, B. W., et al. (2000). An oncogenic role of sphingosine kinase. *Current Biology, 10*, 1527–1530.

Xia, P., Wang, L., Gamble, J. R., & Vadas, M. A. (1999). Activation of sphingosine kinase by tumor necrosis factor-alpha inhibits apoptosis in human endothelial cells. *Journal of Biological Chemistry, 274*, 34499–34505.

Xia, P., Wang, L., Moretti, P. A., Albanese, N., Chai, F., Pitson, S. M., et al. (2002). Sphingosine kinase interacts with TRAF2 and dissects tumor necrosis factor-alpha signaling. *Journal of Biological Chemistry, 277*, 7996–8003.

Xiang, Y., Asmussen, G., Booker, M., Hirth, B., Kane, J. L., Jr., Liao, J., et al. (2009). Discovery of novel sphingosine kinase 1 inhibitors. *Bioorganic & Medicinal Chemistry Letters, 19*, 6119–6121.

Xie, B., Shen, J., Dong, A., Rashid, A., Stoller, G., & Campochiaro, P. A. (2009). Blockade of sphingosine-1-phosphate reduces macrophage influx and retinal and choroidal neovascularization. *Journal of Cellular Physiology, 218*, 192–198.

Yamamura, S., Yatomi, Y., Ruan, F., Sweeney, E. A., Hakomori, S., & Igarashi, Y. (1997). Sphingosine 1-phosphate regulates melanoma cell motility through a receptor-coupled extracellular action and in a pertussis toxin-insensitive manner. *Biochemistry, 36*, 10751–10759.

Yamanaka, M., Shegogue, D., Pei, H., Bu, S., Bielawska, A., Bielawski, J., et al. (2004). Sphingosine kinase 1 (SPHK1) is induced by transforming growth factor-beta and mediates TIMP-1 up-regulation. *Journal of Biological Chemistry, 279*, 53994–54001.

Yamazaki, Y., Kon, J., Sato, K., Tomura, H., Sato, M., Yoneya, T., et al. (2000). Edg-6 as a putative sphingosine 1-phosphate receptor coupling to Ca(2+) signaling pathway. *Biochemical and Biophysical Research Communications, 268*, 583–589.

Yasui, H., Hideshima, T., Raje, N., Roccaro, A. M., Shiraishi, N., Kumar, S., et al. (2005). FTY720 induces apoptosis in multiple myeloma cells and overcomes drug resistance. *Cancer Research, 65*, 7478–7484.

Young, N., Pearl, D. K., & Van Brocklyn, J. R. (2009). Sphingosine-1-phosphate regulates glioblastoma cell invasiveness through the urokinase plasminogen activator system and CCN1/Cyr61. *Molecular Cancer Research, 7*, 23–32.

Young, N., & Van Brocklyn, J. R. (2006). Signal transduction of sphingosine-1-phosphate G protein-coupled receptors. *ScientificWorldJournal, 6*, 946–966.

Young, N., & Van Brocklyn, J. R. (2007). Roles of sphingosine-1-phosphate (S1P) receptors in malignant behavior of glioma cells. Differential effects of S1P2 on cell migration and invasiveness. *Experimental Cell Research, 313*, 1615–1627.

Zemann, B., Urtz, N., Reuschel, R., Mechtcheriakova, D., Bornancin, F., Badegruber, R., et al. (2007). Normal neutrophil functions in sphingosine kinase type 1 and 2 knockout mice. *Immunology Letters, 109*, 56–63.

Zhang, J., Honbo, N., Goetzl, E. J., Chatterjee, K., Karliner, J. S., & Gray, M. O. (2007). Signals from type 1 sphingosine 1-phosphate receptors enhance adult mouse cardiac myocyte survival during hypoxia. *American Journal of Physiology. Heart and Circulatory Physiology, 293*, H3150–H3158.

Zhi, L., Leung, B. P., & Melendez, A. J. (2006). Sphingosine kinase 1 regulates proinflammatory responses triggered by TNFalpha in primary human monocytes. *Journal of Cellular Physiology, 208*, 109–115.

Zhou, C., Ling, M. T., Kin-Wah Lee, T., Man, K., Wang, X., & Wong, Y. C. (2006). FTY720, a fungus metabolite, inhibits invasion ability of androgen-independent prostate cancer cells through inactivation of RhoA-GTPase. *Cancer Letters, 233*, 36–47.

CHAPTER SEVEN

Sphingosine Kinase 1 in Cancer

Linda A. Heffernan-Stroud[*,†,1], **Lina M. Obeid**[*,‡,2,3]
[*]Molecular and Cellular Biology and Pathobiology Program, Medical University of South Carolina, Charleston, South Carolina, USA
[†]College of Medicine, Medical University of South Carolina, Charleston, South Carolina, USA
[‡]Department of Medicine, Medical University of South Carolina, Charleston, South Carolina, USA
[1]Current Address: Department of Internal Medicine, Greenville Memorial Hospital, Greenville, South Carolina, USA
[2]Current Address: Department of Medicine, Stony Brook University, Stony Brook, New York, USA
[3]Corresponding author: e-mail address: lina.obeid@stonybrookmedicine.edu

Contents

Abstract

The role of sphingolipids as bioactive signaling molecules that can regulate cell fate decisions puts them at center stage for cancer treatment and prevention. While ceramide and sphingosine have been established as antigrowth molecules, sphingosine-1-phosphate (S1P) offers a progrowth message to cells. The enzymes responsible for maintaining the balance between these "stop" or "go" signals are the sphingosine kinases (SK), SK1 and SK2. While the relative contribution of SK2 is still being elucidated and may involve an intranuclear role, a substantial amount of evidence suggests that regulation of

Advances in Cancer Research, Volume 117
ISSN 0065-230X
http://dx.doi.org/10.1016/B978-0-12-394274-6.00007-8

sphingolipid levels by SK1 is an important component of carcinogenesis. Here, we review the literature regarding the role of SK1 as an oncogene that can function to enhance cancer cell viability and promote tumor growth and metastasis; highlighting the importance of developing specific SK1 inhibitors to supplement current cancer therapies.

1. INTRODUCTION

While sphingolipids, along with cholesterol and other phospholipids, serve as critical structural components of cell membranes, their most interesting functions involve their roles as bioactive signaling molecules. Although there is still much to be elucidated about specific mechanisms involved in sphingolipid signal transduction, some themes have emerged. Ceramide is known to be involved in apoptosis, cell senescence, differentiation, and cell stress (Hannun and Luberto, 2000; Mathias et al., 1998; Perry and Hannun, 1998). Sphingosine has also been revealed as an antigrowth signaling molecule (Taha et al., 2006b). In contrast, however, sphingosine-1-phosphate (S1P) is known to promote proliferation, survival, and inhibition of apoptosis (Spiegel and Milstien, 2003) as well as having roles in cell migration, vascular development, and inflammation (Hla, 2004). Sphingosine kinase (SK) is a key enzyme in the sphingolipid pathway, as it regulates the levels of all three of the aforementioned lipids to influence cell fate. Furthermore, its activity is necessary for the clearance of sphingolipids, as once it produces S1P, the latter is hydrolyzed by S1P lyase in an irreversible reaction to produce hexadecenal and ethanolamine phosphate which marks the only exit point for the sphingolipid metabolic pathway.

1.1. S1P signaling

Unlike the antiproliferative effects of ceramide and sphingosine, S1P has been shown to play a significant role in proliferation, migration, survival, angiogenesis, inflammation, and lymphocyte egress (Spiegel and Milstien, 2003). S1P can exert its various effects by binding five distinct and differentially expressed cell surface S1P receptors; these G-protein-coupled receptors were formerly referred to as endothelial differentiation gene (Edg) receptors (Taha et al., 2004), further highlighting S1P's important prosurvival role.

S1P is capable of increasing cell survival and inhibiting the apoptotic process in a number of different cell types, including T cells (Goetzl et al., 1999; Kwon et al., 2001), and it has been also shown to induce a proliferative response in endothelial cells (Kimura et al., 2000) and vascular smooth muscle cells (Tamama et al., 2001). S1P, like potent angiogenic peptide growth factors such as VEGF and FGF-2, is now also considered one of the key regulators in angiogenesis. S1P acts on vascular endothelial cells via S1PR1 and S1PR3 receptors to stimulate migration and capillary-like tube formation *in vitro* (Kimura et al., 2000; Lee et al., 1999; Wang et al., 1999). This stimulatory activity of S1P on angiogenesis has recently been demonstrated *in vivo* in ischemic hindlimbs of mice (Oyama et al., 2008). Furthermore, siRNA-induced S1P receptor knockdown (Theilmeier et al., 2006) or FTY720-induced receptor downregulation has been shown to suppress tumor angiogenesis and tumor growth (LaMontagne et al., 2006; Nagaoka et al., 2008; Permpongkosol et al., 2002) as well as inhibit lymphocyte trafficking (Graler and Goetzl, 2004; Morris et al., 2005).

As S1P has been shown to be involved in cell survival, proliferation, and angiogenesis, it is easy to see how these S1P-mediated activities may contribute to the etiology of cancer (Olivera and Spiegel, 1993; Zhang et al., 1991). Interestingly, parenteral administration of S1P-specific antibodies has been shown to markedly slow human cancer xenograft progression and angiogenesis (Visentin et al., 2006). It is also important to note that S1P signaling has been implicated in the development of the drug resistant phenotype in cancer cells (Akao et al., 2006). Taken together, these findings implicate S1P as a very important signaling molecule in cancer biology that requires further study, and its manipulation may play a role in future anticancer therapies.

1.2. Sphingosine kinase

SK1 and SK2 are the enzymes responsible for catalyzing the conversion of sphingosine to S1P using adenosine triphosphate. Despite differences in structure (Okada et al., 2005) and cellular localization (Johnson et al., 2002) *in vivo* studies suggest that SK1 and SK2 may have some overlapping physiologic functions. For instance, although SK1 knockout (KO) mice and SK2 KO mice appear to develop normally, SK1/SK2 double KO mice suffer embryonic lethality due to inadequate angiogenesis and neurogenesis coupled with neuronal apoptosis (Mizugishi et al., 2005). SK1 and SK2 may in fact have some overlapping functions; however, a review of the current literature describes SK1 as being activated by several extracellular agonists and effecting extracellular S1P levels, whereas SK2 is considered to play more of a housekeeping role, and its function may be localized to the nucleus. Interestingly, it was recently shown that S1P produced by SK2 in the nucleus may regulate histone acetylation of the p21 promoter (Hait et al., 2009).

Further insights into the relative roles of SK1 and SK2 in S1P production have been complicated by the development of SK KO mice. Both the SK1 KO and SK2 KO mice show no abnormal gross or histological phenotypes and no compensation in message levels of the other SK isoform. However,

SK1 KO mice show decreased S1P levels in their serum, while this was not detected in the sera of SK2 KO mice. To further investigate this, SK1 +/− mice were crossed with SK2 +/− mice. Interestingly, offspring lacking one to three SK alleles in any combination were indistinguishable from the wild type, but no animals lacking all four alleles were born, indicating that the SK1 KO SK2 KO genotype was embryonic lethal. These embryos had no detectable S1P and suffered hemorrhage and exencephaly due to vascular and neural tube defects (Mizugishi et al., 2005). It was also shown that SK1 KO SK2 +/− female mice were infertile due to defective decidualization with decreased cell mitosis, increased cell death, and defective decidual blood vessel formation (Mizugishi et al., 2007).

Despite the unclear consequences of SK1 KO on sphingolipid homeostasis, these mice have been useful for studying the oncogenic role of SK1. KO of SK1 was able to decrease the size but not the incidence of adenomas in Apc Min/+ mice by attenuating epithelial cell proliferation in the polyps. SK1 KO led to an elevation of sphingosine content and decreased expression of the G1/S cell cycle regulator CDK4 and c-myc in these adenomas, whereas S1P levels were not altered. Furthermore, this protection was not observed when Apc Min/+ mice were crossed with S1P receptor KOs. This highlights that the intracellular function of SK1, independent of the extracellular signaling of its product, S1P, is critical for the growth of intestinal adenomas (Kohno et al., 2006). Other studies have also shown that KO of SK1 can offer protection from chemically induced colon carcinogenesis (Kawamori et al., 2009; Snider et al., 2009).

The cancer preventative affects of SK1 KO are not confined to colon carcinogenesis but also apply to other chemically and genetically induced cancers. While 4-nitroquinoline-1-oxide (NQO) induced head and neck squamous cell carcinoma (HNSCC) in wild-type mice, genetic loss of SK1 prevented 4-NQO-induced HNSCC carcinogenesis with decreased tumor incidence, multiplicity, and volume. The SK1 KO mice showed decreased cell proliferation, increased caspase 3 activation, and decreased activation of AKT when compared with wild-type controls (Shirai et al., 2011). Work from our lab has shown that SK1 KO can protect p53 KO mice from developing their characteristic thymic lymphomas and increase their survival ~30%. Importantly, while KO of SK1 also protects p53 heterozygote mice from the development of lymphomas, it also protected them from the development of other tumor types including osteosarcoma and decreased tumor incidence while prolonging their survival (Heffernan-Stroud et al., 2012). Thus, studies with the SK1 KO mouse have confirmed the oncogenic character of SK1 *in vivo*.

2. SK1 IS AN ONCOGENE

Given its prime position in sphingolipid metabolism, SK1 expression affects the balance between prodeath and prosurvival sphingolipids to determine cell fate, a prime regulatory position for a potential oncogene. Likewise, screening of normal/tumor patient-matched thyroid and non-small-cell-lung cancer tissue, including carcinoid, squamous, and adenocarcinoma, exhibited overwhelmingly positive immunostaining for SK1 as compared with patient-matched normal tissue (Guan et al., 2011a). Moreover, a twofold elevation of SK1 mRNA expression was observed in cancer versus normal tissue for several types of solid tumors including breast, uterus, ovary, colon, small intestine, and rectum in addition to lung (Johnson et al., 2005). Increased SK1 and S1P were also found in a mouse model of colon cancer, and remarkably SK1 KO mice were resistant to the azoxymethane induced colon cancer observed in mice expressing SK1 (Kawamori et al., 2009).

2.1. Increased expression of SK1 in cancer

The potent oncogene, Bcr/Abl, was shown to upregulate SK1 through various pathways leading to myeloid cell leukemia-1 (Mcl-1) expression in chronic myelogenous leukemia (Li et al., 2007a). In multiple myeloma cells, prokineticin-1, an endocrine gland-derived vascular endothelial growth factor, upregulates SK1 and Mcl-1 in addition to its activation of the MAPK, PI3K-AKT, and Jak-STAT3 pathways in order to protect cells from starvation-induced apoptosis (Li et al., 2010). Transcriptional upregulation of SK1 has also been associated with the tumorigenic phenotype of transgenic proerythroblasts in a mouse model of erythroleukemia (Le Scolan et al., 2005). It has also been shown in MKN1 gastric cancer cells that lysophosphatidic acid signaling through ERK-1 activation upregulates SK1 to exert its proliferative effects (Ramachandran et al., 2010). Increased SK1 expression has been shown to correlate with carcinogenicity in several other cancers including acute leukemia (Sobue et al., 2008), endometrial cancer (Knapp et al., 2010), prostate cancer (Malavaud et al., 2010), HNSCC (Facchinetti et al., 2010; Shirai et al., 2011; Sinha et al., 2011), and thyroid cancer (Guan et al., 2011a).

As for the possible mechanisms for this transcriptional upregulation of SK1 in cancer, in human glioblastoma cells, upregulation of SK1 transcription is induced by IL-1 and is mediated by a novel AP-1 element located within the first intron of the SK1 gene, and this upregulation of SK1 can be blocked by inhibition of JNK (Paugh et al., 2009). Rapidly growing tumors often undergo hypoxia, and this results in HIF-1α and HIF-2α production which has been shown to stimulate two putative hypoxia-inducible factor-responsive-elements in the SK1 promoter region (Anelli et al., 2008; Schwalm et al., 2008). Another pathway of SK1 upregulation has been found in rat pheochromocytoma cells in which nerve growth factor is able to upregulate SK1 expression in a pathway involving transcription factor specificity protein 1 (Sobue et al., 2005). Therefore, it seems that various transcription factors can promote SK1 expression for it to enact its oncogenic effects (Table 7.1).

Table 7.1 Consequences of upregulation of SK1 in cancer

Cancer type	Inducer (reference)	Result	Cell line
Breast adenocarcinoma	Prolactin or 17-β estradiol (Doll et al., 2007)	Proliferation and migration	MCF-7
	Estrogen (Sukocheva et al., 2006)	Release of S1P activates S1P3 receptor resulting in transactivation of enhanced growth factor receptor in a MMP-dependent manner	MCF-7
	EGF (Doll et al., 2005)	Proliferation and migration	MCF-7
	HER2 (Long et al., 2010)	Limits p21-activated protein kinase 1 and ERK1/2 signaling and desensitizes the S1P-induced formation of a migratory phenotype through negative feedback	ER+MCF-7 HER2
Esophageal adenocarcinoma	TGF-β (Miller et al., 2008)	ERK1/2 activation and progression of cancer through non-Smad signaling	OE33
Glioma	Hypoxia and HIF-2α (Anelli et al., 2008)	Neoangiogenesis	U87MG
	EGF and EGFRvIII (Estrada-Bernal et al., 2010)	Proliferation	U-1242, U-251, and U-251-E18
Hepatocellular carcinoma	Hepatopoietin Cn (Chang et al., 2010)	Increases cell viability, decreases trichostatin A-induced apoptosis, and upregulates myeloid cell leukemia-1	SMMC7721 and HepG2

Table 7.1 Consequences of upregulation of SK1 in cancer—cont'd

Cancer type	Inducer (reference)	Result	Cell line
Leukemia and lymphoma	PI3K/AKT2/mTOR (Marfe et al., 2011)	Imatinib resistance	K652
	12-Ootetradecanoyl-phorbol-13-acetate (Cuvillier and Levade, 2001)	Inhibits cytochrome *c* and Smac/DIABLO release	Jurkat, U937, HL-60
	BCR/ABL (Li et al., 2007a)	Upregulation of Mcl-1	BaF3
Multiple myeloma	IL-6 (Li et al., 2007b)	Protection from apoptosis	XG-7, SKo-007, U266
Neuroblastoma	17β-Estradiol (Lopez-Tobon et al., 2009)	Protects from glutamate-induced Tau hyperphosphorylation	SH-5Y5Y
Pheochromo-cytoma	Prosaposin Misasi et al., 2001	DNA synthesis and protection from apoptosis	PC12
Prostate adenocarcinoma	Dihydrotestosterone (Dayon et al., 2009)	Reestablishes cell proliferation of androgen-deprived cells	LNCaP

2.2. SK1 as a prognostic indicator

The elevation of SK1 in various cancers implies that it could be a molecule of prognostic value. In fact, SK1 expression is higher in estrogen receptor (ER) negative breast tumors and its high expression correlates with worse survival and a higher chance of metastasis in patients, with carcinoma cells being the major source of SK1 expression in the tumors (Ruckhaberle et al., 2008). Furthermore, expression of SK1, but not SK2, in glioblastoma multiforme tissue correlated with short patient survival. Patients with low SK1 survived a median of 357 days, whereas those with high levels of SK1 survived a median of 102 days (Van Brocklyn et al., 2005). Also in glioblastoma multiforme, it was found that SK is necessary for basal activity of the urokinase plasminogen activator system and glioma cell invasion (Young et al., 2009). High expression of SK1 in non-Hodgkin lymphomas (Bayerl et al., 2008), astrocytomas (Li et al., 2008), gastric cancer (Li et al., 2009), salivary gland carcinoma (Liu et al., 2010a), esophageal carcinoma (Pan et al., 2011), non-small-cell-lung cancer (Song et al., 2011), and head and neck squamous cell carcinoma (Facchinetti et al., 2010) also correlates with grade and shorter survival time. Thus not only is SK1 overexpressed in cancer, but it can also enhance tumor invasiveness and serve as a prognostic indicator.

Patients with ER-positive breast tumors that expressed high levels of both cytoplasmic SK1 and ERK-1/2 had significantly shorter recurrence times than those that expressed low levels of cytoplasmic SK1 and cytoplasmic ERK-1/2, with a difference in recurrence time of 10.5 years (Watson et al., 2010). In contrast, for patients with ER-positive breast cancer, a low HER1-3/SK1 expression ratio is correlated with improved prognosis compared to patients that have a high HER1-3/SK1 expression ratio. This is thought to occur since although the HER2 oncogene increases SK1 expression in these tumors, then SK1 limits HER2 expression in a negative-feedback manner, this phenomenon has been termed "oncogene tolerance" (Long et al., 2010).

Although treatment failure and Gleason score were found to correlate with tumor SK1 activity in prostate cancer patients (Malavaud et al., 2010), patients with prostate cancer were shown to have lower circulating levels of S1P and decreased SK1 activity in their erythrocytes when compared to healthy subjects. They found that decreased circulating S1P was an early marker of prostate cancer progression to hormonal unresponsiveness and correlated with prostate-specific antigen levels and lymph node metastasis as well as mortality (Nunes et al., 2012). Thus, implying that tumor versus serum SK1/S1P levels may have opposite prognostic indications in certain cancer patients.

2.3. Elevated SK1 confers resistance to chemotherapy

With increased SK1 correlated to poor prognosis in several cancers, it seems likely that overexpression of SK1 may confer resistance to current chemotherapeutics. In fact, imatinib-induced apoptosis in K562 human chronic myeloid leukemia (CML) cells involves an increase in C18-ceramide; however, imatinib-resistant cells have increased expression of SK1 which elevates their S1P to C18-ceramide ratio sixfold and prevents apoptosis (Baran et al., 2007). Furthermore, impaired SK1/S1P signaling enhanced the growth-inhibitory effects of nilotinib against 32D/T315I-Bcr–Abl1-derived mouse allografts, demonstrating SK1 inhibition as a potential chemosensitizer in CML (Salas et al., 2011).

Likewise, it has been found that camptothecin-resistant prostate cancer PC3 cells show higher expression of SK1 and S1PR1 and -3 receptors when compared to camptothecin-sensitive, LNCaP cells, and inhibition of SK1 or the S1P receptors can inhibit cell growth in PC3 cells (Akao et al., 2006). Increased SK1 also appears to be involved in cisplatin resistance in lung cancer cells (Min et al., 2005), daunorubicin resistance in leukemia cells (Sobue et al., 2008), *N*-(4-hydroxyphenyl)retinamide resistance in ovarian carcinoma cells (Illuzzi et al., 2010), oxaliplatin resistance in colon cancer cells (Nemoto et al., 2009), and doxorubicin, docetaxel, and tamoxifen resistance in breast cancer cells (Ling et al., 2009; Nava et al., 2002; Sukocheva et al., 2009). Accordingly, inhibition or knockdown of SK1 has been found to increase drug efficacy in acute myeloid leukemia cells (Bonhoure et al., 2006), chronic myelogenous leukemia cells (Li et al., 2007a; Salas et al., 2011), colon cancer cells (Nemoto et al., 2009), and ovarian carcinoma cells (Illuzzi et al., 2010).

2.4. SK1 functions to suppress apoptosis of cancer cells

In vivo cells undergo apoptosis, cell cycle arrest, or senescence when their survival and growth no longer benefit the organism. When aberrant signaling overrides this response, these cells can become tumorigenic. Bioactive sphingolipids have been shown to be important signaling molecules for such cell fate decisions and SK is at crucial point for contributing to this since its product, S1P, is a prosurvival signaling molecule. Here, we review cell studies that illustrate the prosurvival role of the SK1/S1P pathway, in order to show how their functions at the cellular level might contribute to tumor initiation at the organismal level. For instance, in an elegant study by Le Scolan et al., they showed that overexpression of SK1 in erythroleukemic cells increased proliferation, clonogenicity, and resistance to apoptosis in reduced serum by a mechanism involving ERK1/2 activation and the PI3K/Akt pathway leading to increased tumorigenicity when engrafted *in vivo*. However, expression of a dominant-negative mutant of SK1 or treatment with a pharmacological inhibitor was able to reduce both cell growth and apoptosis-resistance in the tumorigenic erythroleukemia cells (Le Scolan et al., 2005). Furthermore, a novel hepatic growth factor that protects hepatocarcinoma cells, hepatopoietin Cn, was recently found to work through SK1 signaling to upregulate Mcl-1 expression to induce its antiapoptotic effects (Chang et al., 2010).

Work from our lab has shown in breast cancer cells, knockdown of SK1 causes cell cycle arrest and induces apoptosis involving effector caspase activation, cytochrome c release, and Bax oligomerization in the mitochondrial membrane, thus placing SK1 knockdown upstream of the mitochondrial pathway of apoptosis (Taha et al., 2006a). Knockdown or downregulation of SK1 has also been shown to lead to ceramide accumulation, decreased S1P, and apoptosis in other cancer cell lines including prostate cancer (Pchejetski et al., 2005), glioblastoma (Van Brocklyn et al., 2005), chronic myelogenous leukemia cell lines (Li et al., 2007a), and neuroblastoma cells (Gomez-Brouchet et al., 2007). (Table 7.2) Likewise, when SK1 is overexpressed in breast cancer cells, it results in enhanced proliferation and resistance to tamoxifen-induced growth arrest and apoptosis (Sukocheva et al., 2009) and it confers resistance to other anticancer drugs, sphingosine, and TNF-α (Nava et al., 2002). Interestingly, other studies in breast cancer cells have suggested that overexpression of SK1 stimulates autophagy. And it was shown that nutrient starvation stimulates SK1 activity and that SK1/S1P-induced autophagy can protect cells from apoptosis under nutrient-deplete conditions (Lavieu et al., 2006). Overexpression of SK1 was also shown to upregulate the Mcl-1 protein in multiple myeloma cells, leading to increased cell proliferation and survival. Furthermore, activation of SK1 was shown to mediate the suppressive effects of IL-6 on multiple myeloma cell apoptosis (Li et al., 2007b). Overexpression or activation of SK1 has been shown to suppress the apoptotic response in a variety of other cell lines including rat pheochromocytoma cells (Edsall et al., 2001; Misasi et al., 2001) and Jurkat, U937, and HL-60 leukemia cells (Cuvillier and Levade, 2001) (Table 7.3).

Table 7.2 Consequences of downregulation of SK1 in cancer

Cancer type	Inhibitor (reference)	Result	Cell line
Cervical adenocarcinoma	NADH (De Luca et al., 2010)	Antiproliferative	HeLa
Hepatocellular carcinoma	KAI1/CD82 overexpression, Sprouty2 (Mu et al., 2008)	Reduces migration induced by hepatocyte growth factor	SMMC-7721
Leukemia	Dasatinib (Gencer et al., 2011)	Apoptosis	K562 Meg-01
	Spred2 (Liu et al., 2010b)	Inhibits proliferation, induces apoptosis, and enhances imatinib-induced cytotoxicity	K652
	AKT2 suppression (Marfe et al., 2011)	Increased sensitivity to imatinib-induced apoptosis	K652

Table 7.2 Consequences of downregulation of SK1 in cancer—cont'd

Cancer type	Inhibitor (reference)	Result	Cell line
Leukemia and breast cancer	Resveratrol (Cakir et al., 2011; Kartal et al., 2011) and its dimers Ampelopsin A and Balanocarpol (Lim et al., 2012)	Apoptosis PARP cleavage	HL60, K562 MCF-7
Melanoma	AKT inhibitor BI-69A11 (Feng et al., 2011)	Antiproliferative	UACC 903 SW1
Neuroblastoma	Amyloid-β peptide (Gomez-Brouchet et al., 2007)	Apoptosis	SH-SY5Y
Pancreatic carcinoma	KAI1 overexpression (Guo et al., 2006)	Inhibits migration	PANC1
Prostate adenocarcinoma	Inhibition of PI3K/Akt pathway (Dayon et al., 2009)	Prevents neuroendocrine differentiation/transformation	LNCaP
	Docetaxel (Pchejetski et al., 2005)	Decreased tumor volume, reduced occurrence and number of metastases in a mouse model	PC-3
	Docetaxel (Sauer et al., 2009)	Apoptosis	PC-3 and DU145
	Gamma irradiation (Nava et al., 2000)	Apoptosis	TSU-Pr1
	Dietary polyphenols (Brizuela et al., 2010)	Apoptosis and decreased tumor growth	PC-3 and C4-2B

SK1 status appears to be tightly linked with caspase-dependent apoptosis. KO of SK1, but not SK2, was shown to enhance caspase-3-mediated apoptosis induced by staurosporine (Hofmann et al., 2008). In several human leukemia cell lines (Jurkat, U937, and HL-60), S1P can inhibit caspase-3 activation by inhibiting translocation of cytochrome *c* and Smac/DIABLO from mitochondria to cytosol (Cuvillier and Levade, 2001). In addition to

Table 7.3 Consequences of SK1 overexpression in cancer

Cancer type (reference)	Result of SK1 overexpression	Cell line
Breast adenocarcinoma (Sukocheva et al., 2003, 2009)	Prevents cell death and enhances proliferation and resistance to tamoxifen-induced growth arrest and apoptosis	MCF-7
Esophageal adenocarcinoma (Pan et al., 2011)	Increases the invasiveness of cells *in vitro* and cell growth and spontaneous metastasis *in vivo*	EC9706
Fibrosarcoma (Xia et al., 2000)	Leads to transformation and tumor formation	NIH3T3
Glioma (Guan et al., 2011b)	Protects cells from UV- or adriamycin-induced apoptosis through PI3K-dependent activation of Akt, inactivation of FOXO3a, and downregulation of Bim	U87MG and LN-382
Leukemia (Marfe et al., 2011)	Imatinib resistance	K562
Multiple myeloma (Li et al., 2007b)	Increases proliferation and survival, inhibits dexamethasone induced apoptosis, and upregulates Mcl-1	XG-7, SKo-007, U266
Non-small-cell-lung cancer (Song et al., 2011)	Inhibits doxorubicin- or docetaxel-induced apoptosis through activation of the PI3K/Akt/NF-κB pathway	95D and A549
Ovarian carcinoma (Illuzzi et al., 2010a, 2010b)	Induces *N*-(4-hydroxyphenyl) retinamide (HPR) resistance	A2780
Pheochromocytoma (Edsall et al., 2001)	Protects from apoptosis in response to trophic factor withdrawl or C2-ceramide treatment	PC12
Prostate adenocarcinoma (Pchejetski et al., 2005)	Increases tumor volume and resistance to docetaxel in a mouse model	PC-3

its prosurvival product S1P, SK1 also regulates the level of its proapoptotic substrate, sphingosine. Recent studies have shown that binding of sphingosine to 14-3-3 proteins renders them phosphorylatable at the dimer interface, an event that abolishes their prosurvival signaling. Thus by reducing the availability of sphingosine for interaction with 14-3-3, SK1 can inhibit cell death (Woodcock et al., 2010), underlining the important position of SK1 in balancing growth and death signaling sphingolipids.

2.5. SK1 affects apoptosis through controlling the cellular ceramide:S1P ratio

To illustrate this balancing act controlled by SK1, in noncancerous human endothelial cells, TNF activates sphingomyelinase and generates ceramide, but these cells are resistant to TNF-induced apoptosis since TNF also activates SK and increases S1P. However, in a transformed endothelial cell line (C11), TNF fails to activate SK and thus induces apoptosis, which treatment with S1P can prevent (Xia et al., 1999). Clearly, with its important role of regulating the balance between prodeath and prosurvival sphingolipids, modifications of SK1 could lead to aberrantly enhanced cell survival leading to tumorigenesis.

In fact, overexpression of SK1 in prostate cancer cells can increase resistance to chemotherapy by decreasing the ceramide:S1P ratio (Pchejetski et al., 2005). Furthermore, a significantly higher ceramide:S1P ratio was found in melanoma cells that are resistant to ceramide- and Fas-induced death than those sensitive to programmed cell death. Overexpression of SK1-reduced sensitivity of A-375 melanoma cells to Fas- and ceramide-mediated apoptosis while siRNA of SK1 decreased resistance of Mel-2a cells to apoptosis. Interestingly, overexpression of prosurvival protein Bcl-2 in A-375 cells stimulated SK1 expression and activity, and downregulation of Bcl-2-reduced SK1 expression (Bektas et al., 2005). Furthermore, it has been shown that Dasatinib (Gencer et al., 2011) and resveratrol (Kartal et al., 2011) induce apoptosis through downregulating SK1 and upregulating ceramide synthase genes to increase the ceramide:S1P ratio in K562 CML cells (Gencer et al., 2011; Kartal et al., 2011). This opposing regulation of ceramide synthases and SK1 has also been observed in breast cancer samples (Erez-Roman et al., 2009). Moreover, our work has shown that through KO of SK1, an increased ceramide:S1P ratio is protective from thymic lymphoma development and may initiate tumor cell senescence *in vivo* (Heffernan-Stroud et al., 2012) (Table 7.4).

2.6. SK1 inhibitors challenge cancer cell viability to enhance chemotherapy

Various studies have employed pharmacological inhibitors to show that inhibition of SK decreases cancer cell viability in neuroblastoma cells (Tavarini et al., 2000), glioblastoma cells (Bektas et al., 2009), and several types of leukemia cells (Cuvillier and Levade, 2001; Paugh et al., 2008;

Table 7.4 Consequences of SK1 knockdown in cancer

Cancer type (reference)	Result of SK1 knockdown	Cell line
Bladder carcinoma (Shu et al., 2002)	Blocks VEGF-induced DNA synthesis and Ras-GTP and p-ERK accumulation	T24
Breast adenocarcinoma (Sukocheva et al., 2009; Sarkar et al., 2005)	Restores cell growth arrest and apoptosis in tamoxifen-resistant MCF-7 cells	MCF-7
	Reduced EGF and serum stimulated growth; enhanced sensitivity to doxorubicin	MCF-7
Colon adenocarcinoma (Nemoto et al., 2009)	Restores chemosensitivity to oxaliplatin including increased C16-ceramide, decreased pAkt, increased p53 and p21 protein levels and PARP cleavage	RKO
Gastric cancer (Fuereder et al., 2011)	Increases apoptosis, inhibits growth by 50%, and increases sensitivity to doxorubicin treatment	MKN28 and N87
Glioma (Radeff-Huang et al., 2007; Van Brocklyn et al., 2005)	Decreases TNF-α-stimulated Akt phosphorylation, cyclin D expression, and DNA synthesis	1321N1
	Decreases proliferation rate and prevents cells from exiting G1	U-1242 MG, U87MG
Glioma and breast adenocarcinoma (Anelli et al., 2010)	Decreased migration and tube formation in cocultured vascular or lymphatic endothelial cells	U87MG, MDA-MB-231
Head and neck squamous cell carcinoma (Shirai et al., 2011; Sinha et al., 2011)	Prevents 4-nitroquinoline-1-oxide-induced head and neck squamous cell carcinoma (HNSCC)	Mouse model
	Reduces xenograft tumor proliferation and volume synergistically with radiation treatment	SCC-15 in Mouse model
Leukemia (Salas et al., 2011; Bonhoure	Reduces Bcr–Abl1 stability	K562/IMA-3 MEFs
	Apoptosis of imatinib-sensitive and resistant cells	Primary CML

Table 7.4 Consequences of SK1 knockdown in cancer—cont'd

Cancer type (reference)	Result of SK1 knockdown	Cell line
et al., 2008; Marfe et al., 2011; Li et al., 2007a; Sobue et al., 2008)	Increases sensitivity to imatinib-induced apoptosis in resistant cells and returns BCR–ABL to baseline levels	K562
	Enhances STI571-induced apoptosis	K562
	Increases daunorubicin sensitivity	K562
Non-small-cell-lung cancer (Song et al., 2011)	Increases doxorubicin- or docetaxel-induced apoptosis through decreased activation of the PI3K/Akt/NF-κB pathway	95D and A549
Prostate cancer (Cho et al., 2011)	Blocked expression of HIF-1α during hypoxia, decreased phospho-Akt and VEGF production	PC-3
Thyroid cancer (Guan et al., 2011a)	Leads to dephosphorylation of protein kinase B and glycogen synthase kinase-3β and subsequent inactivation of β-catenin-T-cell factor/lymphoid enhancing factor transcriptional activity and decreases cell proliferation	WRO, FRO, SW579
Various tumors (Heffernan-Stroud et al., 2012)	Protects p53 KO mice from thymic lymphoma and p53 heterozygote mice from sarcoma development to increase life span ~30%	Mouse model

Zhang et al., 2008) as well as in various solid tumor cell lines (French et al., 2003). SK inhibitors (outlined in Table 7.5). were found to be antiproliferative toward a panel of tumor cell lines, including lines with the multidrug resistance phenotype because of overexpression of either P-glycoprotein or multidrug resistance phenotype 1, and they could induce apoptosis (French et al., 2003). Furthermore, resistance to doxorubicin- or etoposide-induced apoptosis in acute myeloid leukemia cells was attributed to sustained SK1 activity and reduced ceramide accumulation; likewise, apoptosis could be restored in these cells by treatment with an SK inhibitor, F-12509a (Bonhoure et al., 2006). Later results using the same inhibitor

Table 7.5 Effects of SK1 inhibitors in cancer

Inhibitor	Mechanism of action
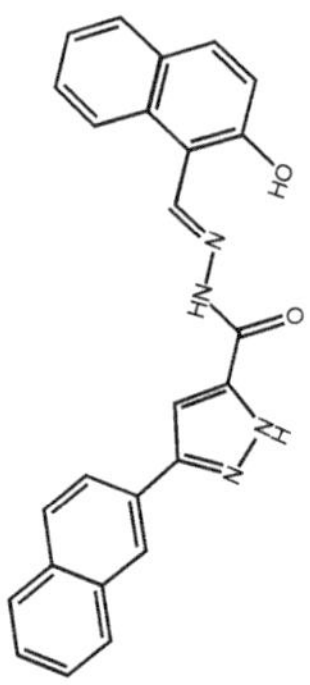 SKI-I (French et al., 2003)	Induces apoptosis and suppresses tumor growth rate and vascularization of glioblastoma xenografts (Kapitonov et al., 2009)
	Sensitizes pancreatic cancer cells to gemcitabine induced death (Guillermet-Guibert et al., 2009)
	Decreases growth and survival while enhancing apoptosis and Bcl-2 cleavage in leukemia cells (Paugh et al., 2008)
	Suppresses lysophosphatidic acid mediated proliferation in gastric cancer cells (Ramachandran et al., 2010)
	Suppressed S1P levels, decreased hemangiogenesis and lymphangiogenesis, reduced metastases to lymph nodes and lungs, and decreased overall tumor burden in murine model of breast cancer (Nagahashi et al., 2012)
	Blocked cell proliferation and induced apoptotic cell death in glioblastoma multiforme neurosphere cell line (Estrada-Bernal et al., 2010)
	Increases oxaliplatin-induced cytotoxicity in colon cancer (Nemoto et al., 2009)
	Increases sensitivity of non-small-cell-lung cancer cells to doxorubicin- or docetaxel-induced apoptosis (Song et al., 2011)
	Rescues daunorubicin sensitivity in leukemia cells (Sobue et al., 2008)
	Enhances ability of imatinib mesylate to inhibit cell growth, survival, and clonogenic potential in leukemia cells (Ricci et al., 2009)
	Sensitizes prostate and breast cancer cells to docetaxel (Sauer et al., 2009) (*as does B-5354c*)
	Reduces proliferation and sensitizes ovarian cancer cells to cytotoxic effect of *N*-(4-hydroxyphenyl) retinamide (Illuzzi et al., 2010a, 2010b)
	Impedes DHT-induced cell proliferation, DNA synthesis, and PSA secretion in prostate cancer cells (Dayon et al., 2009)

Table 7.5 Effects of SK1 inhibitors in cancer—cont'd

Inhibitor	**Mechanism of action**
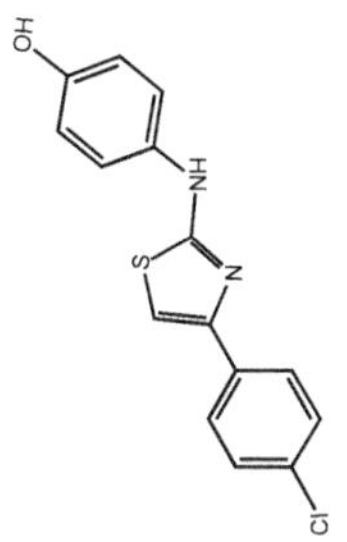 SKI-II (French et al., 2003)	Blocks expression of urokinase plasminogen activator and its receptor to inhibit glioma cell invasion (Young et al., 2009)
	Prevents accumulation and transcriptional activity of HIF-1α in prostate, brain, breast, and lung cancer cells (Ader et al., 2008)
	Reduces cell colony formation and activates of caspase-3 in glioblastoma cells (Bektas et al., 2009)
	Antimigratory and toxic effects on human glioma and glioma stem cells (Mora et al., 2010)
	Blocks breast cancer viability, clonogenic survival, and proliferation and dose-dependently decreases estrogen-stimulated estrogen response element transcriptional activity (Antoon et al., 2011)
	Sensitizes HNSCC to radiation-induced cytotoxicity (Sinha et al., 2011)
	Induces proteasomal degradation of SK1 and apoptosis in androgen-sensitive prostate cancer cells Loveridge et al., 2010
	Triggers lysosomal degradation of SK1 rather than direct inhibition of SK1 in HEK293 cells and lung cancer cells (Ren et al., 2010)
	Increases Tau hyperphosphylation by glutamate in neuroblastoma cells (Lopez-Tobon et al., 2009)
SKI-I and SKI-II	Induce apoptosis of T-LGL leukemia PBMCs but not normal PBMCs (Zhang et al., 2008)
SKI-I, SKI-II, and SKI-V	Reduce rate of solid tumor growth when JC mammary adenocarcinoma cells injected into BALB/c mice (French et al., 2006)
SKI-I, SKI-II, SKI-III, SKI-IV, and SKI-V	Induce apoptosis of bladder cancer cells (French et al., 2003)
SKI-II and DMS	Decreases TNF-α-stimulated DNA synthesis in glioma cells (Radeff-Huang et al., 2007)
DMS	Sensitizes prostate cancer cells to gamma-irradiation (Nava et al., 2000)

Continued

Table 7.5 Effects of SK1 inhibitors in cancer—cont'd

Inhibitor	Mechanism of action
	Enhances cytochrome *c* and Smac/DIABLO release in leukemia cells (Cuvillier and Levade, 2001)
	Inhibits proliferation of Gastric Carcinoma cells (Ren et al., 2002)
	Increases sensitivity to imatinib-induced apoptosis in resistant chronic myeloid leukemia cells and returns BCR–ABL to baseline levels (Marfe et al., 2011)
DMS and DHS	Enhance apoptosis of leiomyoma cells in response to serum starvation by antagonizing the effects of endothelin-1 (Raymond et al., 2006)
DHS	Protects rhabdomyosarcoma cells from from TRAIL-induced apoptosis (Petak et al., 2003)
Phytosphingosine derivatives	Methylphytosphingosine induces apoptosis, decreases phophorylation of ERK, and inhibits daunorubicin-induced ERK activation in leukemia cells (Pewzner-Jung et al., 2010)
	Dimethylphytosphingosine suppresses cell growth and induces apoptosis in human leukemia cells; decreases phophorylation of ERK and inhibits daunorubicin-induced ERK activation to enhance its cytotoxicity in leukemia cells (Park et al., 2010)
F-12509a	Enhances imatinib-induced apoptosis in leukemia cells (Bonhoure et al., 2008)
B-5354c	Induces apoptosis and sensitizes prostate cancer cells to docetaxel and camptothecin to reduce tumor size *in vivo* (Pchejetski et al., 2005, 2008)

suggest that SK1 regulates imatinib-induced apoptosis in primary cells from CML patients, as inhibition of SK1 could kill both imatinib-sensitive and -resistant cells. This work places SK1 downstream of the Bcr–Abl/Ras/ERK pathway inhibited by imatinib and upstream of Bcl-2 family members (Bonhoure et al., 2008).

Inhibition of SK was also shown to reduce cell colony formation and activate caspase-3 in temozolomide-resistant glioblastoma multiforme cells (Bektas et al., 2009). And inhibition of SK1 or the S1P receptors can inhibit cell growth in camptothecin-resistant prostate cancer cells (Akao et al.,

2006). Interestingly, treatment with SK inhibitors SK1-I or SK1-II, can selectively induce apoptosis in T cell large granular lymphocyte leukemia cells but not in normal peripheral blood mononuclear cells (Zhang et al., 2008). Likewise, treatment with SK1-I potently induced apoptosis in leukemic blasts isolated from patients with acute myelogenous leukemia but again was relatively sparing of normal peripheral blood mononuclear leukocytes (Paugh et al., 2008). *In vivo*, treatment with SK1-I induced apoptosis and reduced tumor vascularization leading to a markedly decreased tumor growth rate of glioblastoma xenografts and enhancing the survival of mice harboring LN229 intracranial tumors (Kapitonov et al., 2009). Recently, production of an aspirinyl analog of SK1-I has been shown to increase its half-life for better *in vivo* delivery (Sharma et al., 2010).

Moreover, combining SK1-I and the proteasome inhibitor, bortezomib, synergistically increases apoptosis, decreases colony formation, and induces downregulation of BCR/ABL and Mcl-1 in human leukemia cells and was even effective in imatinib-resistant cells (Li et al., 2011). Whereas, SK1-II on its own has been shown to induce proteasomal degradation of SK1 in human pulmonary artery smooth muscle cells, androgen-sensitive LNCaP prostate cancer cells, MCF-7 and MCF-7 HER2 breast cancer cells (Loveridge et al., 2010). Although FTY720 has been reported to be more proficient at inhibiting SK2, some studies claim that FTY720 and its analogue (S)-FTY720 vinylphosphonate behave as typical SK1 inhibitors and induce proteasomal degradation of SK1 and apoptosis in breast and prostate cancer cells (Tonelli et al., 2010) as well as preventing S1P-stimulated rearrangement of actin in MCF-7 cells (Lim et al., 2011). Of note, it has also been reported that FTY720 inhibits SK1 to increase radiation sensitivity of prostate cancer tumor xenografts to reduce tumor growth and metastasis in mice (Pchejetski et al., 2010).

The first putative-SK inhibitor to enter phase 1 clinical trials for advanced solid tumor therapy was Safingol, 1-threo-dihydrosphingosine, in 2011, which was found to decrease plasma levels of S1P and prolong stable disease/inhibit cancer progression when used in combination with cisplatin in a small cohort of patients (Dickson et al., 2011). However, new SK1 inhibitors with increased specificity are being developed. As the upregulation of SK1 in HNSCC cells correlates with their radioresistance, a group developed gold nanorod-SK1 siRNA nanocomplexes as a method of radiosensitization of head and neck cancer to achieve over 50% greater tumor regression as compared to controls (Masood et al., 2012).

Recently developed amidine-based SK1 nanomolar inhibitors were shown to significantly reduce S1P levels in human leukemia U937 cells

(Kennedy et al., 2011). Two sphingo-guanidine salts named LCL146 and LCL135 have been reported to be cytotoxic to cancer cells and decrease the migration rate of human prostate (DU145) cells through their inhibition of SK (Sharma, 2011). Interestingly, the anticancer agent resveratrol and its dimers, ampelopsin A and balanocarpol, have been shown to downregulate SK1 in MCF-7 human breast cancer cells (Lim et al., 2012). The AKT inhibitor, BI-69A11, was shown to reduce NF-κB signaling through its inhibition of SK1 to inhibit melanoma growth (Feng et al., 2011). All of these results indicate that an SK1 inhibitor could be a beneficial addition to the chemotherapeutic strategies employed for treating various types of tumors, leading to extensive review of the subject (Cuvillier, 2007, 2008; Edmonds et al., 2011; Gault and Obeid, 2011; Pallis, 2002; Pitman and Pitson, 2010; Pyne et al., 2011; Shida et al., 2008b; Uddin and Platanias, 2008).

3. SK1/S1P SIGNALS FOR TUMOR GROWTH AND SPREAD

Not only does SK1 protect cell survival, it also plays an important role in signaling for cell proliferation. It appears that SK1-dependent Akt activation plays a significant role in cell proliferation induced by TNF-α in 1321N1 glioblastoma cells (Radeff-Huang et al., 2007) and by 11,12-epoxyeicosatrienoic acid in human umbilical vein endothelial cells (Yan et al., 2008). SK1 has also been found to act as a proproliferative oncogenic kinase in thyroid cancer where it appears to activate the Akt/glycogen synthase kinase-3β/β-catenin pathway (Guan et al., 2011a). Moreover, various growth factors have been shown to function in part by upregulating SK1 expression and/or activity, these include PDGF (Francy et al., 2007), NGF (Edsall et al., 2001), IGF-BP3 (Martin et al., 2009), VEGF (Shu et al., 2002), and EGF (Doll et al., 2005). VEGF induces DNA synthesis in a pathway which sequentially involves protein kinase C, SK, Ras, Raf, and ERK1/2 (Shu et al., 2002). And epidermal growth factor both acutely and transcriptionally upregulates SK1 in a MAPK-, PKC-, and PI3K-dependent manner. Accordingly, when cells are depleted of SK1, but not SK2, by siRNA, EGF-induced proliferation and migration are drastically reduced (Doll et al., 2005). Also through the ERK1/2 and PI3K signaling pathways, glial cell line-derived neurotrophic factor induces transcription of SK1 which leads to neurite formation, GAP43 expression, and cell growth in TGW human neuroblastoma cells and growth and anchorage-independent colony formation in a model of multiple endocrine neoplasia type 2 tumor (Murakami et al., 2007).

3.1. Hormonal regulation of SK1

Hormones have also been shown to activate SK1. Progesterone induced SK1 expression (30-fold) in the rat uterus during pregnancy in the glandular epithelium, vasculature, and myometrium. Overexpression of SK1 in a rat myometrial leiomyoma cell line (ELT3) resulted in increased levels of the cell cycle regulator cyclin D1 and increased myosin light-chain phosphorylation and increased proliferation rates (Jeng et al., 2007). Prolactin and 17-β estradiol were also shown to activate SK1, but not SK2, in MCF-7 breast adenocarcinoma cells both within minutes and within hours after STAT5-, PKC-, and MAPK-dependent promoter activation. Either inhibition of SK1 activation with glucocorticoids or direct knockdown of SK1 with siRNA could abolish the hormone-induced cell proliferation and migration observed in the breast cancer cells (Doll et al., 2007).

Interestingly, the SK1-II inhibitor binds the ER directly in the antagonist ligand-binding domain. SK1-II was shown to dose-dependently decrease estrogen-stimulated estrogen response element transcriptional activity and diminish mRNA levels of the ER regulated genes progesterone receptor and steroid derived factor-1 to block breast cancer viability, clonogenic survival, and proliferation (Antoon et al., 2011). In addition to hormonal regulation of SK1 in breast cancer, dihydrotestosterone was shown to signal for proliferation through an androgen receptor/PI3K/Akt-dependent stimulation of SK1 in hormone-sensitive prostate cancer cells (Dayon et al., 2009). This hormonal regulation of SK1 has even been shown to be important in neuroblastoma cell lines in which 17β-estradiol upregulates SK1 to protect from Tau hyperphosphorylation and glutamate toxicity (Lopez-Tobon et al., 2009).

3.2. SK1 functions to enhance proliferation of cancer cells

SK1 can signal to a cell to proliferate through its product S1P. For instance, in the presence of retinoic acid receptor alpha, retinoic acid upregulates neutral sphingomyelinase and downregulates SK1 to inhibit cell growth, but in the absence of that receptor, retinoic acid activates SK1 and promotes cell growth through S1P (Somenzi et al., 2007). Addition of S1P increases proliferation in neuroblastoma cells (Tavarini et al., 2000) and extracellular S1P is required for basic fibroblast growth factor-induced growth stimulation of astrocytes (Bassi et al., 2006). In a glioma cell line, S1P induces expression of early growth response-1, an essential transcription factor for fibroblast growth factor 2 (Sato et al., 1999). This was the first report of S1P acting as a transcription factor, and its role as a molecule capable of epigenetic regulation was more recently elucidated when it was shown to regulate histone acetylation (Hait et al., 2009). Clearly, SK1 and its product S1P could employ their potent proliferative signaling to enhance tumor growth.

3.3. SK1 can stimulate angiogenesis to increase blood supply to tumors

In order to maintain rapid growth, tumors require a sufficient vascular supply. As tumor cells rapidly proliferate, they undergo hypoxia which results in their signaling for new blood vessels to be formed. Hypoxia increases SK1 mRNA, protein, and activity in human glioma U87MG cells, followed by intracellular S1P production and S1P release and this could all be inhibited with knockdown of HIF-2α by siRNA (Anelli et al., 2008). Similar results were observed in endothelial cells themselves, where their production of HIF-1α and HIF-2α and their binding to two putative hypoxia-inducible factor and -responsive elements in the SK1 promoter led to increases in SK1 (Schwalm et al., 2008). Furthermore, our lab showed that conditioned medium from hypoxia-treated tumor cells results in neoangiogenesis in human umbilical vein endothelial cells in an S1P receptor-dependent manner (Anelli et al., 2008). Interestingly, inhibition of SK1 activity can prevent the accumulation of HIF-1α and its transcriptional activity in several human cancer cell lineages including prostate, brain, breast, kidney, and lung (Ader et al., 2008). Thus, SK1 signaling appears to play an important role in creating new vasculature in order for oxygen and nutrients to be delivered to oxygen-starved tumors.

In addition to angiogenesis, SK1 has recently been shown to play a role in lymphangiogenesis. SK1 overexpression in HEK cells or its downregulation in glioma or breast cancer cells modulated extracellular S1P levels accordingly, which in turn increased or decreased both migration and tube formation in cocultured vascular or lymphatic endothelial cells. Furthermore, S1P initiated endothelial cell sprouting in three-dimensional collagen matrices (Anelli et al., 2010). Activation of SK1 has been shown to be necessary for the angiogenic effects of 11,12-epoxyeicosatrienoic acid, a product of cytochrome p450 epoxygenases (Yan et al., 2008), TNF-α (Xia et al., 1998), and VEGF (Kolmakova et al., 2009). Likewise, inhibition of SK has been shown to be responsible for the antiangiogenic properties of phenoxodiol (2H-1-benzopyran-7-0, 1,3-[4-hydroxyphenyl]), an anticancer drug undergoing clinical trials for its apoptotic properties (Gamble et al., 2006), and alphastatin, the potent antiangiogenic molecule that has been shown to inhibit gastric cancer angiogenesis (Chen et al., 2006; Li and Chen, 2009). In NIH3T3 cells with dominant negative CBF1, SK1 is required for

FGF1 export and acceleration of cell growth to produce highly angiogenic tumors in nude mice (Kacer et al., 2011). Therefore, the angiogenic signaling properties of SK1 appear to be important for vascularization, and hence expansion, of tumors.

3.4. SK1 may enhance tumor metastasis

In addition to its angiogenic and lymphangiogenic properties, other aspects of SK1 signaling that could potentially contribute to tumor metastasis have been elucidated. For instance, upregulation of SK1 led to a migratory response in endothelial cells (Schwalm et al., 2008) and promotes migration of thyroid follicular carcinoma cells (Bergelin et al., 2009). Whereas, depletion of SK1 can inhibit EGF-induced (Doll et al., 2005; Sarkar et al., 2005), as well as hormone-induced migration of breast cancer cells (Doll et al., 2007), and it can abrogate migration toward EGF in HEK293 cells (Hait et al., 2005). SK1 was also shown to be required for nucleotide-induced migration of renal mesangial cells (Meyer zu Heringdorf and Jakobs, 2007). Moreover, the antimetastasis molecule KAI1 has been shown to suppress migration of pancreatic carcinoma cells through downregulation of SK activity (Guo et al., 2006), and it has been shown to downregulate hepatocyte growth factor induced activation of SK1 in order to impede migration of hepatoma cells (Mu et al., 2008). These results indicate that the regulation of SK1 is a crucial regulator of cell migration.

The ability to migrate as well as invasiveness and the aggressiveness of tumor cells is what distinguishes a benign mass from malignant cancer. Importantly, NIH3T3 fibroblasts overexpressing SK acquire a transformed phenotype, as determined by focus formation, colony growth in soft agar, and the ability to form tumors in NOD/SCID mice, and SK activity appears to be involved in transformation mediated by the potent oncogene, H-Ras (Xia et al., 2000). Furthermore, TGF-β's activation of SK and formation of S1P contribute to non-Smad signaling that could be important for progression of esophageal cancer (Miller et al., 2008), and S1P is able to induce invasion of glioblastoma cells (Bryan et al., 2008). Studies show that hepatocellular carcinoma cell migration and invasion may also be promoted by SK1/S1P signaling through S1PR1 (EDG1) (Bao et al., 2011). Moreover, S1P has been shown to signal through S1PR2 for apoptotic cell extrusion, which because live cells can also be extruded, it is postulated that this may serve as a mechanism for the invasion of cancer cells (Gu et al., 2011). Other studies have suggested that SK1 is a convergence point of multiple cell surface receptors for LPA,

EGF, and S1P, which have all been implicated in the regulation of motility and invasiveness of cancer cells (Shida et al., 2008a). Thus, it appears that SK1 signaling can contribute to the malignant transformation of tumors and their metastasis.

4. CONCLUSION

Through both regulating ceramide and sphingosine accumulation within the cell and producing S1P, increased expression of SK1 results in an oncogenic phenotype. Decreased ceramide and sphingosine levels can confer chemotherapeutic resistance to cancer cells and increased S1P can stimulate the proliferation, vascularization, and metastasis of tumors. Therefore, SK1 represents an important target for chemotherapy.

ACKNOWLEDGMENTS

This work was supported, in part, by NIH/NCI P01 CA097132—project 3 (to L. M. O). NIH MSTP Training Grant (GM08716—to L. A. H. S), MUSC Hollings Cancer Center Abney Foundation Scholarship (to L. A. H. S), NIH/NIEHS TG T32 ES012878, and NIH-NIEHS/NRSA F30ES017379 (to L. A. H. S).

REFERENCES

Ader, I., Brizuela, L., Bouquerel, P., Malavaud, B., & Cuvillier, O. (2008). Sphingosine kinase 1: A new modulator of hypoxia inducible factor 1alpha during hypoxia in human cancer cells. *Cancer Research*, *68*, 8635–8642.

Akao, Y., Banno, Y., Nakagawa, Y., Hasegawa, N., Kim, T.-J., Murate, T., et al. (2006). High expression of sphingosine kinase 1 and S1P receptors in chemotherapy-resistant prostate cancer PC3 cells and their camptothecin-induced up-regulation. *Biochemical and Biophysical Research Communications*, *342*, 1284–1290.

Anelli, V., Gault, C. R., Cheng, A. B., & Obeid, L. M. (2008). Sphingosine kinase 1 is up-regulated during hypoxia in U87MG glioma cells. Role of hypoxia-inducible factors 1 and 2. *The Journal of Biological Chemistry*, *283*, 3365–3375.

Anelli, V., Gault, C. R., Snider, A. J., & Obeid, L. M. (2010). Role of sphingosine kinase-1 in paracrine/transcellular angiogenesis and lymphangiogenesis in vitro. *The FASEB Journal*, *24*, 2727–2738.

Antoon, J. W., Meacham, W. D., Bratton, M. R., Slaughter, E. M., Rhodes, L. V., Ashe, H. B., et al. (2011). Pharmacological inhibition of sphingosine kinase isoforms alters estrogen receptor signaling in human breast cancer. *Journal of Molecular Endocrinology*, *46*, 205–216.

Bao, M., Chen, Z., Xu, Y., Zhao, Y., Zha, R., Huang, S., et al. (2011). Sphingosine kinase 1 promotes tumour cell migration and invasion via the S1P/EDG1 axis in hepatocellular carcinoma. *Liver International*, *32*, 331–338.

Baran, Y., Salas, A., Senkal, C. E., Gunduz, U., Bielawski, J., Obeid, L. M., et al. (2007). Alterations of ceramide/sphingosine 1-phosphate rheostat involved in the regulation of resistance to imatinib-induced apoptosis in K562 human chronic myeloid leukemia cells. *The Journal of Biological Chemistry*, *282*, 10922–10934.

Bassi, R., Anelli, V., Giussani, P., Tettamanti, G., Viani, P., & Riboni, L. (2006). Sphingosine-1-phosphate is released by cerebellar astrocytes in response to bFGF and induces astrocyte proliferation through Gi-protein-coupled receptors. *Glia*, *53*, 621–630.

Bayerl, M. G., Bruggeman, R. D., Conroy, E. J., Hengst, J. A., King, T. S., Jimenez, M., et al. (2008). Sphingosine kinase 1 protein and mRNA are overexpressed in non-Hodgkin lymphomas and are attractive targets for novel pharmacological interventions. *Leukemia & Lymphoma*, *49*, 948–954.

Bektas, M., Jolly, P. S., Muller, C., Eberle, J., Spiegel, S., & Geilen, C. C. (2005). Sphingosine kinase activity counteracts ceramide-mediated cell death in human melanoma cells: Role of Bcl-2 expression. *Oncogene*, *24*, 178–187.

Bektas, M., Johnson, S. P., Poe, W. E., Bigner, D. D., & Friedman, H. S. (2009). A sphingosine kinase inhibitor induces cell death in temozolomide resistant glioblastoma cells. *Cancer Chemotherapy and Pharmacology*, *64*, 1053–1058.

Bergelin, N., Blom, T., Heikkila, J., Lof, C., Alam, C., Balthasar, S., et al. (2009). Sphingosine kinase as an oncogene: Autocrine sphingosine 1-phosphate modulates ML-1 thyroid carcinoma cell migration by a mechanism dependent on protein kinase C-alpha and ERK1/2. *Endocrinology*, *150*, 2055–2063.

Bonhoure, E., Pchejetski, D., Aouali, N., Morjani, H., Levade, T., Kohama, T., et al. (2006). Overcoming MDR-associated chemoresistance in HL-60 acute myeloid leukemia cells by targeting sphingosine kinase-1. *Leukemia*, *20*, 95–102.

Bonhoure, E., Lauret, A., Barnes, D. J., Martin, C., Malavaud, B., Kohama, T., et al. (2008). Sphingosine kinase-1 is a downstream regulator of imatinib-induced apoptosis in chronic myeloid leukemia cells. *Leukemia*, *22*, 971–979.

Brizuela, L., Dayon, A., Doumerc, N., Ader, I., Golzio, M., Izard, J. C., et al. (2010). The sphingosine kinase-1 survival pathway is a molecular target for the tumor-suppressive tea and wine polyphenols in prostate cancer. *The FASEB Journal*, *24*, 3882–3894.

Bryan, L., Kordula, T., Spiegel, S., & Milstien, S. (2008). Regulation and functions of sphingosine kinases in the brain. *Biochimica et Biophysica Acta*, *1781*, 459–466.

Cakir, Z., Saydam, G., Sahin, F., & Baran, Y. (2011). The roles of bioactive sphingolipids in resveratrol-induced apoptosis in HL60: Acute myeloid leukemia cells. *Journal of Cancer Research and Clinical Oncology*, *137*, 279–286.

Chang, J., Liu, Y., Zhang, D. D., Zhang, D. J., Wu, C. T., Wang, L. S., et al. (2010). Hepatopoietin Cn suppresses apoptosis of human hepatocellular carcinoma cells by up-regulating myeloid cell leukemia-1. *World Journal of Gastroenterology*, *16*, 193–200.

Chen, L., Li, T., Li, R., Wei, B., & Peng, Z. (2006). Alphastatin downregulates vascular endothelial cells sphingosine kinase activity and suppresses tumor growth in nude mice bearing human gastric cancer xenografts. *World Journal of Gastroenterology*, *12*, 4130–4136.

Cho, S. Y., Lee, H. J., Jeong, S. J., Kim, H. S., Chen, C. Y., Lee, E. O., et al. (2011). Sphingosine kinase 1 pathway is involved in melatonin-induced HIF-1alpha inactivation in hypoxic PC-3 prostate cancer cells. *Journal of Pineal Research*, *51*, 87–93.

Cuvillier, O. (2007). Sphingosine kinase-1–A potential therapeutic target in cancer. *Anti-cancer Drugs*, *18*, 105–110.

Cuvillier, O. (2008). Downregulating sphingosine kinase-1 for cancer therapy. *Expert Opinion on Therapeutic Targets*, *12*, 1009–1020.

Cuvillier, O., & Levade, T. (2001). Sphingosine 1-phosphate antagonizes apoptosis of human leukemia cells by inhibiting release of cytochrome c and Smac/DIABLO from mitochondria. *Blood*, *98*, 2828–2836.

Dayon, A., Brizuela, L., Martin, C., Mazerolles, C., Pirot, N., Doumerc, N., et al. (2009). Sphingosine kinase-1 is central to androgen-regulated prostate cancer growth and survival. *PLoS One*, *4*, e8048.

De Luca, T., Morre, D. M., & Morre, D. J. (2010). Reciprocal relationship between cytosolic NADH and ENOX2 inhibition triggers sphingolipid-induced apoptosis in HeLa cells. *Journal of Cellular Biochemistry, 110*, 1504–1511.

Dickson, M. A., Carvajal, R. D., Merrill, A. H., Jr., Gonen, M., Cane, L. M., & Schwartz, G. K. (2011). A phase I clinical trial of safingol in combination with cisplatin in advanced solid tumors. *Clinical Cancer Research, 17*, 2484–2492.

Doll, F., Pfeilschifter, J., & Huwiler, A. (2005). The epidermal growth factor stimulates sphingosine kinase-1 expression and activity in the human mammary carcinoma cell line MCF7. *Biochimica et Biophysica Acta, 1738*, 72–81.

Doll, F., Pfeilschifter, J., & Huwiler, A. (2007). Prolactin upregulates sphingosine kinase-1 expression and activity in the human breast cancer cell line MCF7 and triggers enhanced proliferation and migration. *Endocrine-Related Cancer, 14*, 325–335.

Edmonds, Y., Milstien, S., & Spiegel, S. (2011). Development of small-molecule inhibitors of sphingosine-1-phosphate signaling. *Pharmacology & Therapeutics, 132*, 352–360.

Edsall, L. C., Cuvillier, O., Twitty, S., Spiegel, S., & Milstien, S. (2001). Sphingosine kinase expression regulates apoptosis and caspase activation in PC12 cells. *Journal of Neurochemistry, 76*, 1573–1584.

Erez-Roman, R., Pienik, R., & Futerman, A. H. (2009). Increased ceramide synthase 2 and 6 mRNA levels in breast cancer tissues and correlation with sphingosine kinase expression. *Biochemical and Biophysical Research Communications, 391*, 219–223.

Estrada-Bernal, A., Lawler, S. E., Nowicki, M. O., Ray Chaudhury, A., & Van Brocklyn, J. R. (2010). The role of sphingosine kinase-1 in EGFRvIII-regulated growth and survival of glioblastoma cells. *Journal of Neuro-Oncology, 102*, 353–366.

Facchinetti, M. M., Gandini, N. A., Fermento, M. E., Sterin-Speziale, N. B., Ji, Y., Patel, V., et al. (2010). The expression of sphingosine kinase-1 in head and neck carcinoma. *Cells, Tissues, Organs, 192*, 314–324.

Feng, Y., Barile, E., De, S. K., Stebbins, J. L., Cortez, A., Aza-Blanc, P., et al. (2011). Effective inhibition of melanoma by BI-69A11 is mediated by dual targeting of the AKT and NF-kappaB pathways. *Pigment Cell & Melanoma Research, 24*, 703–713.

Francy, J. M., Nag, A., Conroy, E. J., Hengst, J. A., & Yun, J. K. (2007). Sphingosine kinase 1 expression is regulated by signaling through PI3K, AKT2, and mTOR in human coronary artery smooth muscle cells. *Biochimica et Biophysica Acta, 1769*, 253–265.

French, K. J., Schrecengost, R. S., Lee, B. D., Zhuang, Y., Smith, S. N., Eberly, J. L., et al. (2003). Discovery and evaluation of inhibitors of human sphingosine kinase. *Cancer Research, 63*, 5962–5969.

French, K. J., Upson, J. J., Keller, S. N., Zhuang, Y., Yun, J. K., & Smith, C. D. (2006). Antitumor activity of sphingosine kinase inhibitors. *The Journal of Pharmacology and Experimental Therapeutics, 318*, 596–603.

Fuereder, T., Hoeflmayer, D., Jaeger-Lansky, A., Rasin-Streden, D., Strommer, S., Fisker, N., et al. (2011). Sphingosine kinase 1 is a relevant molecular target in gastric cancer. *Anti-cancer Drugs, 22*, 245–252.

Gamble, J. R., Xia, P., Hahn, C. N., Drew, J. J., Drogemuller, C. J., Brown, D., et al. (2006). Phenoxodiol, an experimental anticancer drug, shows potent antiangiogenic properties in addition to its antitumour effects. *International Journal of Cancer, 118*, 2412–2420.

Gault, C. R., & Obeid, L. M. (2011). Still benched on its way to the bedside: Sphingosine kinase 1 as an emerging target in cancer chemotherapy. *Critical Reviews in Biochemistry and Molecular Biology, 46*, 342–351.

Gencer, E. B., Ural, A. U., Avcu, F., & Baran, Y. (2011). A novel mechanism of dasatinib-induced apoptosis in chronic myeloid leukemia; ceramide synthase and ceramide clearance genes. *Annals of Hematology, 90*, 1265–1275.

Goetzl, E. J., Kong, Y., & Mei, B. (1999). Lysophosphatidic acid and sphingosine 1-phosphate protection of T Cells from apoptosis in association with suppression of bax. *The Journal of Immunology, 162*, 2049–2056.

Gomez-Brouchet, A., Pchejetski, D., Brizuela, L., Garcia, V., Altie, M. F., Maddelein, M. L., et al. (2007). Critical role for sphingosine kinase-1 in regulating survival of neuroblastoma cells exposed to amyloid-beta peptide. *Molecular Pharmacology, 72*, 341–349.

Graler, M. H., & Goetzl, E. J. (2004). The immunosuppressant FTY720 down-regulates sphingosine 1-phosphate G-protein-coupled receptors. *The FASEB Journal, 18*, 551–553.

Gu, Y., Forostyan, T., Sabbadini, R., & Rosenblatt, J. (2011). Epithelial cell extrusion requires the sphingosine-1-phosphate receptor 2 pathway. *The Journal of Cell Biology, 193*, 667–676.

Guan, H., Liu, L., Cai, J., Liu, J., Ye, C., Li, M., et al. (2011a). Sphingosine kinase 1 is overexpressed and promotes proliferation in human thyroid cancer. *Molecular Endocrinology, 25*, 1858–1866.

Guan, H., Song, L., Cai, J., Huang, Y., Wu, J., Yuan, J., et al. (2011b). Sphingosine kinase 1 regulates the Akt/FOXO3a/Bim pathway and contributes to apoptosis resistance in glioma cells. *PLoS One, 6*, e19946.

Guillermet-Guibert, J., Davenne, L., Pchejetski, D., Saint-Laurent, N., Brizuela, L., Guilbeau-Frugier, C., et al. (2009). Targeting the sphingolipid metabolism to defeat pancreatic cancer cell resistance to the chemotherapeutic gemcitabine drug. *Molecular Cancer Therapeutics, 8*, 809–820.

Guo, X. Z., Zhang, W. W., Wang, L. S., Lu, Z. Z., Wang, H., Xu, J. H., et al. (2006). Adenovirus-mediated overexpression of KAI1 suppresses sphingosine kinase activation and metastasis of pancreatic carcinoma cells. *Zhonghua Nei Ke Za Zhi, 45*, 752–754.

Hait, N. C., Sarkar, S., Le Stunff, H., Mikami, A., Maceyka, M., Milstien, S., et al. (2005). Role of sphingosine kinase 2 in cell migration toward epidermal growth factor. *The Journal of Biological Chemistry, 280*, 29462–29469.

Hait, N. C., Allegood, J., Maceyka, M., Strub, G. M., Harikumar, K. B., Singh, S. K., et al. (2009). Regulation of histone acetylation in the nucleus by sphingosine-1-phosphate. *Science, 325*, 1254–1257.

Hannun, Y. A., & Luberto, C. (2000). Ceramide in the eukaryotic stress response. *Trends in Cell Biology, 10*, 73–80.

Heffernan-Stroud, L. A., Helke, K. L., Jenkins, R. W., De Costa, A. M., Hannun, Y. A., & Obeid, L. M. (2012). Defining a role for sphingosine kinase 1 in p53-dependent tumors. *Oncogene, 31*, 1166–1175.

Hla, T. (2004). Physiological and pathological actions of sphingosine 1-phosphate. *Seminars in Cell & Developmental Biology, 15*, 513–520.

Hofmann, L. P., Ren, S., Schwalm, S., Pfeilschifter, J., & Huwiler, A. (2008). Sphingosine kinase 1 and 2 regulate the capacity of mesangial cells to resist apoptotic stimuli in an opposing manner. *Biological Chemistry, 389*, 1399–1407.

Illuzzi, G., Bernacchioni, C., Aureli, M., Prioni, S., Frera, G., Donati, C., et al. (2010). Sphingosine kinase mediates resistance to the synthetic retinoid N-(4-hydroxyphenyl) retinamide in human ovarian cancer cells. *The Journal of Biological Chemistry, 285*, 18594–18602.

Jeng, Y. J., Suarez, V. R., Izban, M. G., Wang, H. Q., & Soloff, M. S. (2007). Progesterone-induced sphingosine kinase-1 expression in the rat uterus during pregnancy and signaling consequences. *American Journal of Physiology, Endocrinology and Metabolism, 292*, E1110–E1121.

Johnson, K. R., Becker, K. P., Facchinetti, M. M., Hannun, Y. A., & Obeid, L. M. (2002). PKC-dependent activation of sphingosine kinase 1 and translocation to the plasma membrane. Extracellular release of sphingosine-1-phosphate induced by phorbol 12-myristate 13-acetate (PMA). *The Journal of Biological Chemistry, 277*, 35257–35262.

Johnson, K. R., Johnson, K. Y., Crellin, H. G., Ogretmen, B., Boylan, A. M., Harley, R. A., et al. (2005). Immunohistochemical distribution of sphingosine kinase 1 in normal and tumor lung tissue. *Journal of Histochemistry and Cytochemistry, 53*, 1159–1166.

Kacer, D., McIntire, C., Kirov, A., Kany, E., Roth, J., Liaw, L., et al. (2011). Regulation of non-classical FGF1 release and FGF-dependent cell transformation by CBF1-mediated notch signaling. *Journal of Cellular Physiology, 226*, 3064–3075.

Kapitonov, D., Allegood, J. C., Mitchell, C., Hait, N. C., Almenara, J. A., Adams, J. K., et al. (2009). Targeting sphingosine kinase 1 inhibits Akt signaling, induces apoptosis, and suppresses growth of human glioblastoma cells and xenografts. *Cancer Research, 69*, 6915–6923.

Kartal, M., Saydam, G., Sahin, F., & Baran, Y. (2011). Resveratrol triggers apoptosis through regulating ceramide metabolizing genes in human K562 chronic myeloid leukemia cells. *Nutrition and Cancer, 63*, 637–644.

Kawamori, T., Kaneshiro, T., Okumura, M., Maalouf, S., Uflacker, A., Bielawski, J., et al. (2009). Role for sphingosine kinase 1 in colon carcinogenesis. *The FASEB Journal, 23*, 405–414.

Kennedy, A. J., Mathews, T. P., Kharel, Y., Field, S. D., Moyer, M. L., East, J. E., et al. (2011). Development of amidine-based sphingosine kinase 1 nanomolar inhibitors and reduction of sphingosine 1-phosphate in human leukemia cells. *Journal of Medicinal Chemistry, 54*, 3524–3548.

Kimura, T., Watanabe, T., Sato, K., Kon, J., Tomura, H., Tamama, K., et al. (2000). Sphingosine 1-phosphate stimulates proliferation and migration of human endothelial cells possibly through the lipid receptors, Edg-1 and Edg-3. *The Biochemical Journal, 348*, 71–76.

Knapp, P., Baranowski, M., Knapp, M., Zabielski, P., Blachnio-Zabielska, A. U., & Gorski, J. (2010). Altered sphingolipid metabolism in human endometrial cancer. *Prostaglandins & Other Lipid Mediators, 92*, 62–66.

Kohno, M., Momoi, M., Oo, M. L., Paik, J. H., Lee, Y. M., Venkataraman, K., et al. (2006). Intracellular role for sphingosine kinase 1 in intestinal adenoma cell proliferation. *Molecular and Cellular Biology, 26*, 7211–7223.

Kolmakova, A., Rajesh, M., Zang, D., Pili, R., & Chatterjee, S. (2009). VEGF recruits lactosylceramide to induce endothelial cell adhesion molecule expression and angiogenesis in vitro and in vivo. *Glycoconjugate Journal, 26*, 547–558.

Kwon, Y.-G., Min, J.-K., Kim, K.-M., Lee, D.-J., Billiar, T. R., & Kim, Y.-M. (2001). Sphingosine 1-phosphate protects human umbilical vein endothelial cells from serum-deprived apoptosis by nitric oxide production. *The Journal of Biological Chemistry, 276*, 10627–10633.

LaMontagne, K., Littlewood-Evans, A., Schnell, C., O'Reilly, T., Wyder, L., Sanchez, T., et al. (2006). Antagonism of sphingosine-1-phosphate receptors by FTY720 inhibits angiogenesis and tumor vascularization. *Cancer Research, 66*, 221–231.

Lavieu, G., Scarlatti, F., Sala, G., Carpentier, S., Levade, T., Ghidoni, R., et al. (2006). Regulation of autophagy by sphingosine kinase 1 and its role in cell survival during nutrient starvation. *The Journal of Biological Chemistry, 281*, 8518–8527.

Le Scolan, E., Pchejetski, D., Banno, Y., Denis, N., Mayeux, P., Vainchenker, W., et al. (2005). Overexpression of sphingosine kinase 1 is an oncogenic event in erythroleukemic progression. *Blood, 106*, 1808–1816.

Lee, M.-J., Thangada, S., Claffey, K. P., Ancellin, N., Liu, C. H., Kluk, M., et al. (1999). Vascular endothelial cell adherens junction assembly and morphogenesis induced by sphingosine-1-phosphate. *Cell, 99*, 301–312.

Li, T., & Chen, L. (2009). Alphastatin inhibits tumor angiogenesis in nude mice bearing human gastric cancer xenografts. *Zhonghua Wei Chang Wai Ke Za Zhi, 12*, 61–64.
Li, Q. F., Huang, W. R., Duan, H. F., Wang, H., Wu, C. T., & Wang, L. S. (2007a). Sphingosine kinase-1 mediates BCR/ABL-induced upregulation of Mcl-1 in chronic myeloid leukemia cells. *Oncogene, 26*, 7904–7908.
Li, Q. F., Wu, C. T., Duan, H. F., Sun, H. Y., Wang, H., Lu, Z. Z., et al. (2007b). Activation of sphingosine kinase mediates suppressive effect of interleukin-6 on human multiple myeloma cell apoptosis. *British Journal of Haematology, 138*, 632–639.
Li, J., Guan, H. Y., Gong, L. Y., Song, L. B., Zhang, N., Wu, J., et al. (2008). Clinical significance of sphingosine kinase-1 expression in human astrocytomas progression and overall patient survival. *Clinical Cancer Research, 14*, 6996–7003.
Li, W., Yu, C. P., Xia, J. T., Zhang, L., Weng, G. X., Zheng, H. Q., et al. (2009). Sphingosine kinase 1 is associated with gastric cancer progression and poor survival of patients. *Clinical Cancer Research, 15*, 1393–1399.
Li, Q. F., Zhu, H. Y., Yang, Y. F., Liu, J., Xiao, F. J., Zhang, Q. W., et al. (2010). Prokineticin-1/endocrine gland-derived vascular endothelial growth factor is a survival factor for human multiple myeloma cells. *Leukemia & Lymphoma, 51*, 1902–1912.
Li, Q. F., Yan, J., Zhang, K., Yang, Y. F., Xiao, F. J., Wu, C. T., et al. (2011). Bortezomib and sphingosine kinase inhibitor interact synergistically to induces apoptosis in BCR/ABl+ cells sensitive and resistant to STI571 through down-regulation Mcl-1. *Biochemical and Biophysical Research Communications, 405*, 31–36.
Lim, K. G., Tonelli, F., Li, Z., Lu, X., Bittman, R., Pyne, S., et al. (2011). FTY720 analogues as sphingosine kinase 1 inhibitors: Enzyme inhibition kinetics, allosterism, proteasomal degradation, and actin rearrangement in MCF-7 breast cancer cells. *The Journal of Biological Chemistry, 286*, 18633–18640.
Lim, K. G., Gray, A. I., Pyne, S., & Pyne, N. J. (2012). Resveratrol dimers are novel sphingosine kinase 1 inhibitors and affect sphingosine kinase 1 expression and cancer cell growth and survival. *British Journal of Pharmacology, 166*(5), 1605–1616.
Ling, B., Chen, L., Alcorn, J., Ma, B., & Yang, J. (2011). Sphingosine-1-phosphate: A potential therapeutic agent against human breast cancer. *Investigational New Drugs, 29*(2), 396–399.
Liu, G., Zheng, H., Zhang, Z., Wu, Z., Xiong, H., Li, J., et al. (2010a). Overexpression of sphingosine kinase 1 is associated with salivary gland carcinoma progression and might be a novel predictive marker for adjuvant therapy. *BMC Cancer, 10*, 495.
Liu, X. Y., Yang, Y. F., Wu, C. T., Xiao, F. J., Zhang, Q. W., Ma, X. N., et al. (2010b). Spred2 is involved in imatinib-induced cytotoxicity in chronic myeloid leukemia cells. *Biochemical and Biophysical Research Communications, 393*, 637–642.
Long, J. S., Edwards, J., Watson, C., Tovey, S., Mair, K. M., Schiff, R., et al. (2010). Sphingosine kinase 1 induces tolerance to human epidermal growth factor receptor 2 and prevents formation of a migratory phenotype in response to sphingosine 1-phosphate in estrogen receptor-positive breast cancer cells. *Molecular and Cellular Biology, 30*, 3827–3841.
Lopez-Tobon, A., Cepeda-Prado, E., & Cardona-Gomez, G. P. (2009). Decrease of Tau hyperphosphorylation by 17beta estradiol requires sphingosine kinase in a glutamate toxicity model. *Neurochemical Research, 34*, 2206–2214.
Loveridge, C., Tonelli, F., Leclercq, T., Lim, K. G., Long, J. S., Berdyshev, E., et al. (2010). The sphingosine kinase 1 inhibitor 2-(p-hydroxyanilino)-4-(p-chlorophenyl)thiazole induces proteasomal degradation of sphingosine kinase 1 in mammalian cells. *The Journal of Biological Chemistry, 285*, 38841–38852.
Malavaud, B., Pchejetski, D., Mazerolles, C., de Paiva, G. R., Calvet, C., Doumerc, N., et al. (2010). Sphingosine kinase-1 activity and expression in human prostate cancer resection specimens. *European Journal of Cancer, 46*, 3417–3424.

Marfe, G., Di Stefano, C., Gambacurta, A., Ottone, T., Martini, V., Abruzzese, E., et al. (2011). Sphingosine kinase 1 overexpression is regulated by signaling through PI3K, AKT2, and mTOR in imatinib-resistant chronic myeloid leukemia cells. *Experimental Hematology, 39*(653–665), e6.

Martin, J. L., Lin, M. Z., McGowan, E. M., & Baxter, R. C. (2009). Potentiation of growth factor signaling by insulin-like growth factor-binding protein-3 in breast epithelial cells requires sphingosine kinase activity. *The Journal of Biological Chemistry, 284*, 25542–25552.

Masood, R., Roy, I., Zu, S., Hochstim, C., Yong, K. T., Law, W. C., et al. (2012). Gold nanorod-sphingosine kinase siRNA nanocomplexes: A novel therapeutic tool for potent radiosensitization of head and neck cancer. *Integrative Biology: Quantitative Biosciences from Nano to Macro, 4*, 132–141.

Mathias, S., Peña, L. A., & Kolesnick, R. N. (1998). Signal transduction of stress via ceramide. *The Biochemical Journal, 335*, 465–480.

Meyer zu Heringdorf, D., & Jakobs, K. H. (2007). Renal mesangial cells: Moving on sphingosine kinase-1. *British Journal of Pharmacology, 150*, 255–257.

Miller, A. V., Alvarez, S. E., Spiegel, S., & Lebman, D. A. (2008). Sphingosine kinases and sphingosine-1-phosphate are critical for transforming growth factor beta-induced extracellular signal-regulated kinase 1 and 2 activation and promotion of migration and invasion of esophageal cancer cells. *Molecular and Cellular Biology, 28*, 4142–4151.

Min, J., Van Veldhoven, P. P., Zhang, L., Hanigan, M. H., Alexander, H., & Alexander, S. (2005). Sphingosine-1-phosphate lyase regulates sensitivity of human cells to select chemotherapy drugs in a p38-dependent manner. *Molecular Cancer Research, 3*, 287–296.

Misasi, R., Sorice, M., Di Marzio, L., Campana, W. M., Molinari, S., Cifone, M. G., et al. (2001). Prosaposin treatment induces PC12 entry in the S phase of the cell cycle and prevents apoptosis: Activation of ERKs and sphingosine kinase. *The FASEB Journal, 15*, 467–474.

Mizugishi, K., Yamashita, T., Olivera, A., Miller, G. F., Spiegel, S., & Proia, R. L. (2005). Essential role for sphingosine kinases in neural and vascular development. *Molecular and Cellular Biology, 25*, 11113–11121.

Mizugishi, K., Li, C., Olivera, A., Bielawski, J., Bielawska, A., Deng, C. X., et al. (2007). Maternal disturbance in activated sphingolipid metabolism causes pregnancy loss in mice. *The Journal of Clinical Investigation, 117*, 2993–3006.

Mora, R., Dokic, I., Kees, T., Huber, C. M., Keitel, D., Geibig, R., et al. (2010). Sphingolipid rheostat alterations related to transformation can be exploited for specific induction of lysosomal cell death in murine and human glioma. *Glia, 58*, 1364–1383.

Morris, M. A., Gibb, D. R., Picard, F., Brinkmann, V., Straume, M., & Ley, K. (2005). Transient T cell accumulation in lymph nodes and sustained lymphopenia in mice treated with FTY720. *European Journal of Immunology, 35*, 3570–3580.

Mu, Z., Wang, H., Zhang, J., Li, Q., Wang, L., & Guo, X. (2008). KAI1/CD82 suppresses hepatocyte growth factor-induced migration of hepatoma cells via upregulation of Sprouty2. *Science in China. Series C, Life Sciences, 51*, 648–654.

Murakami, M., Ichihara, M., Sobue, S., Kikuchi, R., Ito, H., Kimura, A., et al. (2007). RET signaling-induced SPHK1 gene expression plays a role in both GDNF-induced differentiation and MEN2-type oncogenesis. *Journal of Neurochemistry, 102*, 1585–1594.

Nagahashi, M., Ramachandran, S., Kim, E. Y., Allegood, J. C., Rashid, O. M., Yamada, A., et al. (2012). Sphingosine-1-phosphate produced by sphingosine kinase 1 promotes breast cancer progression by stimulating angiogenesis and lymphangiogenesis. *Cancer Research, 72*, 726–735.

Nagaoka, Y., Otsuki, K., Fujita, T., & Uesato, S. (2008). Effects of phosphorylation of immunomodulatory agent FTY720 (fingolimod) on antiproliferative activity against breast and colon cancer cells. *Biological & Pharmaceutical Bulletin*, *31*, 1177–1181.

Nava, V. E., Cuvillier, O., Edsall, L. C., Kimura, K., Milstien, S., Gelmann, E. P., et al. (2000). Sphingosine enhances apoptosis of radiation-resistant prostate cancer cells. *Cancer Research*, *60*, 4468–4474.

Nava, V. E., Hobson, J. P., Murthy, S., Milstien, S., & Spiegel, S. (2002). Sphingosine kinase type 1 promotes estrogen-dependent tumorigenesis of breast cancer MCF-7 cells. *Experimental Cell Research*, *281*, 115–127.

Nemoto, S., Nakamura, M., Osawa, Y., Kono, S., Itoh, Y., Okano, Y., et al. (2009). Sphingosine kinase isoforms regulate oxaliplatin sensitivity of human colon cancer cells through ceramide accumulation and Akt activation. *The Journal of Biological Chemistry*, *284*, 10422–10432.

Nunes, J., Naymark, M., Sauer, L., Muhammad, A., Keun, H., Sturge, J., et al. (2012). Circulating sphingosine-1-phosphate and erythrocyte sphingosine kinase-1 activity as novel biomarkers for early prostate cancer detection. *British Journal of Cancer*, *106*(5), 909–915.

Okada, T., Ding, G., Sonoda, H., Kajimoto, T., Haga, Y., Khosrowbeygi, A., et al. (2005). Involvement of N-terminally extended form of sphingosine kinase 2 in serum-dependent regulation of cell proliferation and apoptosis. *Journal of Chemical Biology*, *280*, 36318–36325.

Olivera, A., & Spiegel, S. (1993). Sphingosine-1-phosphate as second messenger in cell proliferation induced by PDGF and FCS mitogens. *Nature*, *365*, 557–560.

Oyama, O., Sugimoto, N., Qi, X., Takuwa, N., Mizugishi, K., Koizumi, J., et al. (2008). The lysophospholipid mediator sphingosine-1-phosphate promotes angiogenesis in vivo in ischaemic hindlimbs of mice. *Cardiovascular Research*, *78*, 301–307.

Pallis, M. (2002). Sphingosine kinase inhibitors in the apoptosis of leukaemia cells. *Leukemia Research*, *26*, 415–416.

Pan, J., Tao, Y. F., Zhou, Z., Cao, B. R., Wu, S. Y., Zhang, Y. L., et al. (2011). An novel role of sphingosine kinase-1 (SPHK1) in the invasion and metastasis of esophageal carcinoma. *Journal of Translational Medicine*, *9*, 157.

Park, S. R., Cho, H. J., Moon, K. J., Chun, K. H., Kong, S. Y., Yoon, S. S., et al. (2010). Cytotoxic effects of novel phytosphingosine derivatives, including N, N-dimethylphytosphingosine and N-monomethylphytosphingosine, in human leukemia cell line HL60. *Leukemia & Lymphoma*, *51*, 132–145.

Paugh, S. W., Paugh, B. S., Rahmani, M., Kapitonov, D., Almenara, J. A., Kordula, T., et al. (2008). A selective sphingosine kinase 1 inhibitor integrates multiple molecular therapeutic targets in human leukemia. *Blood*, *112*, 1382–1391.

Paugh, B. S., Bryan, L., Paugh, S. W., Wilczynska, K. M., Alvarez, S. M., Singh, S. K., et al. (2009). Interleukin-1 regulates the expression of sphingosine kinase 1 in glioblastoma cells. *The Journal of Biological Chemistry*, *284*, 3408–3417.

Pchejetski, D., Golzio, M., Bonhoure, E., Calvet, C., Doumerc, N., Garcia, V., et al. (2005). Sphingosine kinase-1 as a chemotherapy sensor in prostate adenocarcinoma cell and mouse models. *Cancer Research*, *65*, 11667–11675.

Pchejetski, D., Doumerc, N., Golzio, M., Naymark, M., Teissie, J., Kohama, T., et al. (2008). Chemosensitizing effects of sphingosine kinase-1 inhibition in prostate cancer cell and animal models. *Molecular Cancer Therapeutics*, *7*, 1836–1845.

Pchejetski, D., Bohler, T., Brizuela, L., Sauer, L., Doumerc, N., Golzio, M., et al. (2010). FTY720 (fingolimod) sensitizes prostate cancer cells to radiotherapy by inhibition of sphingosine kinase-1. *Cancer Research*, *70*, 8651–8661.

Permpongkosol, S., Wang, J. D., Takahara, S., Matsumiya, K., Nonomura, N., Nishimura, K., et al. (2002). Anticarcinogenic effect of FTY720 in human prostate carcinoma DU145 cells: Modulation of mitogenic signaling, FAK, cell-cycle entry and apoptosis. *International Journal of Cancer, 98*, 167–172.

Perry, D. K., & Hannun, Y. A. (1998). The role of ceramide in cell signaling. *Molecular and Cell Biology of Lipids, 1436*, 233–243.

Petak, I., Vernes, R., Szucs, K. S., Anozie, M., Izeradjene, K., Douglas, L., et al. (2003). A caspase-8-independent component in TRAIL/Apo-2L-induced cell death in human rhabdomyosarcoma cells. *Cell Death and Differentiation, 10*, 729–739.

Pewzner-Jung, Y., Brenner, O., Braun, S., Laviad, E. L., Ben-Dor, S., Feldmesser, E., et al. (2010). A critical role for ceramide synthase 2 in liver homeostasis: II. Insights into molecular changes leading to hepatopathy. *The Journal of Biological Chemistry, 285*, 10911–10923.

Pitman, M. R., & Pitson, S. M. (2010). Inhibitors of the sphingosine kinase pathway as potential therapeutics. *Current Cancer Drug Targets, 10*, 354–367.

Pyne, S., Bittman, R., & Pyne, N. J. (2011). Sphingosine kinase inhibitors and cancer: Seeking the golden sword of Hercules. *Cancer Research, 71*, 6576–6582.

Radeff-Huang, J., Seasholtz, T. M., Chang, J. W., Smith, J. M., Walsh, C. T., & Brown, J. H. (2007). Tumor necrosis factor-alpha-stimulated cell proliferation is mediated through sphingosine kinase-dependent Akt activation and cyclin D expression. *The Journal of Biological Chemistry, 282*, 863–870.

Ramachandran, S., Shida, D., Nagahashi, M., Fang, X., Milstien, S., Takabe, K., et al. (2010). Lysophosphatidic acid stimulates gastric cancer cell proliferation via ERK1-dependent upregulation of sphingosine kinase 1 transcription. *FEBS Letters, 584*, 4077–4082.

Raymond, M. N., Bole-Feysot, C., Banno, Y., Tanfin, Z., & Robin, P. (2006). Endothelin-1 inhibits apoptosis through a sphingosine kinase 1-dependent mechanism in uterine leiomyoma ELT3 cells. *Endocrinology, 147*, 5873–5882.

Ren, J., Dong, L., Xu, C. B., & Pan, B. R. (2002). Expression of sphingosine kinase gene in the interactions between human gastric carcinoma cell and vascular endothelial cell. *World Journal of Gastroenterology, 8*, 602–607.

Ren, S., Xin, C., Pfeilschifter, J., & Huwiler, A. (2010). A novel mode of action of the putative sphingosine kinase inhibitor 2-(p-hydroxyanilino)-4-(p-chlorophenyl) thiazole (SKI II): Induction of lysosomal sphingosine kinase 1 degradation. *Cellular Physiology and Biochemistry, 26*, 97–104.

Ricci, C., Onida, F., Servida, F., Radaelli, F., Saporiti, G., Todoerti, K., et al. (2009). In vitro anti-leukaemia activity of sphingosine kinase inhibitor. *British Journal of Haematology, 144*, 350–357.

Ruckhaberle, E., Rody, A., Engels, K., Gaetje, R., von Minckwitz, G., Schiffmann, S., et al. (2008). Microarray analysis of altered sphingolipid metabolism reveals prognostic significance of sphingosine kinase 1 in breast cancer. *Breast Cancer Research and Treatment, 112*, 41–52.

Salas, A., Ponnusamy, S., Senkal, C. E., Meyers-Needham, M., Selvam, S. P., Saddoughi, S. A., et al. (2011). Sphingosine kinase-1 and sphingosine 1-phosphate receptor 2 mediate Bcr-Abl1 stability and drug resistance by modulation of protein phosphatase 2A. *Blood, 117*, 5941–5952.

Sarkar, S., Maceyka, M., Hait, N. C., Paugh, S. W., Sankala, H., Milstien, S., et al. (2005). Sphingosine kinase 1 is required for migration, proliferation and survival of MCF-7 human breast cancer cells. *FEBS Letters, 579*, 5313–5317.

Sato, K., Ishikawa, K., Ui, M., & Okajima, F. (1999). Sphingosine 1-phosphate induces expression of early growth response-1 and fibroblast growth factor-2 through mechanism involving extracellular signal-regulated kinase in astroglial cells. *Brain Research. Molecular Brain Research, 74*, 182–189.

Sauer, L., Nunes, J., Salunkhe, V., Skalska, L., Kohama, T., Cuvillier, O., et al. (2009). Sphingosine kinase 1 inhibition sensitizes hormone-resistant prostate cancer to docetaxel. *International Journal of Cancer, 125*, 2728–2736.
Schwalm, S., Doll, F., Romer, I., Bubnova, S., Pfeilschifter, J., & Huwiler, A. (2008). Sphingosine kinase-1 is a hypoxia-regulated gene that stimulates migration of human endothelial cells. *Biochemical and Biophysical Research Communications, 368*, 1020–1025.
Sharma, A. K. (2011). Sphingo-guanidines and their use as inhibitors of sphingosine kinase (WO2010078247). *Expert Opinion on Therapeutic Patents, 21*, 807–812.
Sharma, A. K., Sk, U. H., Gimbor, M. A., Hengst, J. A., Wang, X., Yun, J., et al. (2010). Synthesis and bioactivity of sphingosine kinase inhibitors and their novel aspirinyl conjugated analogs. *European Journal of Medicinal Chemistry, 45*, 4149–4156.
Shida, D., Fang, X., Kordula, T., Takabe, K., Lepine, S., Alvarez, S. E., et al. (2008a). Cross-talk between LPA1 and epidermal growth factor receptors mediates up-regulation of sphingosine kinase 1 to promote gastric cancer cell motility and invasion. *Cancer Research, 68*, 6569–6577.
Shida, D., Takabe, K., Kapitonov, D., Milstien, S., & Spiegel, S. (2008b). Targeting SphK1 as a new strategy against cancer. *Current Drug Targets, 9*, 662–673.
Shirai, K., Kaneshiro, T., Wada, M., Furuya, H., Bielawski, J., Hannun, Y. A., et al. (2011). A role of sphingosine kinase 1 in head and neck carcinogenesis. *Cancer Prevention Research (Philadelphia, Pa.), 4*, 454–462.
Shu, X., Wu, W., Mosteller, R. D., & Broek, D. (2002). Sphingosine kinase mediates vascular endothelial growth factor-induced activation of ras and mitogen-activated protein kinases. *Molecular and Cellular Biology, 22*, 7758–7768.
Sinha, U. K., Schorn, V. J., Hochstim, C., Chinn, S. B., Zhu, S., & Masood, R. (2011). Increased radiation sensitivity of head and neck squamous cell carcinoma with sphingosine kinase 1 inhibition. *Head & Neck, 33*, 178–188.
Snider, A. J., Kawamori, T., Bradshaw, S. G., Orr, K. A., Gilkeson, G. S., Hannun, Y. A., et al. (2009). A role for sphingosine kinase 1 in dextran sulfate sodium-induced colitis. *The FASEB Journal, 23*, 143–152.
Sobue, S., Hagiwara, K., Banno, Y., Tamiya-Koizumi, K., Suzuki, M., Takagi, A., et al. (2005). Transcription factor specificity protein 1 (Sp1) is the main regulator of nerve growth factor-induced sphingosine kinase 1 gene expression of the rat pheochromocytoma cell line, PC12. *Journal of Neurochemistry, 95*, 940–949.
Sobue, S., Nemoto, S., Murakami, M., Ito, H., Kimura, A., Gao, S., et al. (2008). Implications of sphingosine kinase 1 expression level for the cellular sphingolipid rheostat: Relevance as a marker for daunorubicin sensitivity of leukemia cells. *International Journal of Hematology, 87*, 266–275.
Somenzi, G., Sala, G., Rossetti, S., Ren, M., Ghidoni, R., & Sacchi, N. (2007). Disruption of retinoic acid receptor alpha reveals the growth promoter face of retinoic acid. *PLoS One, 2*, e836.
Song, L., Xiong, H., Li, J., Liao, W., Wang, L., Wu, J., et al. (2011). Sphingosine kinase-1 enhances resistance to apoptosis through activation of PI3K/Akt/NF-kappaB pathway in human non-small cell lung cancer. *Clinical Cancer Research, 17*, 1839–1849.
Spiegel, S., & Milstien, S. (2003). Sphingosine-1-phosphate: An enigmatic signalling lipid. *Nature Reviews Molecular Cell Biology, 4*, 397–407.
Sukocheva, O. A., Wang, L., Albanese, N., Pitson, S. M., Vadas, M. A., & Xia, P. (2003). Sphingosine kinase transmits estrogen signaling in human breast cancer cells. *Molecular Endocrinology, 17*, 2002–2012.
Sukocheva, O., Wadham, C., Holmes, A., Albanese, N., Verrier, E., Feng, F., et al. (2006). Estrogen transactivates EGFR via the sphingosine 1-phosphate receptor Edg-3: The role of sphingosine kinase-1. *The Journal of Cell Biology, 173*, 301–310.

Sukocheva, O., Wang, L., Verrier, E., Vadas, M. A., & Xia, P. (2009). Restoring endocrine response in breast cancer cells by inhibition of the sphingosine kinase-1 signaling pathway. *Endocrinology, 150*, 4484–4492.

Taha, T. A., Argraves, K. M., & Obeid, L. M. (2004). Sphingosine-1-phosphate receptors: Receptor specificity versus functional redundancy. *Biochimica et Biophysica Acta, 1682*, 48–55.

Taha, T. A., Kitatani, K., El-Alwani, M., Bielawski, J., Hannun, Y. A., & Obeid, L. M. (2006a). Loss of sphingosine kinase-1 activates the intrinsic pathway of programmed cell death: Modulation of sphingolipid levels and the induction of apoptosis. *The FASEB Journal, 20*, 482–484.

Taha, T. A., Mullen, T. D., & Obeid, L. M. (2006b). A house divided: Ceramide, sphingosine, and sphingosine-1-phosphate in programmed cell death. *Biochimica et Biophysica Acta, 1758*, 2027–2036.

Tamama, K., Kon, J., Sato, K., Tomura, H., Kuwabara, A., Kimura, T., et al. (2001). Extracellular mechanism through the Edg family of receptors might be responsible for sphingosine-1-phosphate-induced regulation of DNA synthesis and migration of rat aortic smooth-muscle cells. *The Biochemical Journal, 353*, 139–146.

Tavarini, S., Colombaioni, L., & Garcia-Gil, M. (2000). Sphingomyelinase metabolites control survival and apoptotic death in SH-SY5Y neuroblastoma cells. *Neuroscience Letters, 285*, 185–188.

Theilmeier, G., Schmidt, C., Herrmann, J., Keul, P., Schafers, M., Herrgott, I., et al. (2006). High-density lipoproteins and their constituent, sphingosine-1-phosphate, directly protect the heart against ischemia/reperfusion injury in vivo via the S1P3 lysophospholipid receptor. *Circulation, 114*, 1403–1409.

Tonelli, F., Lim, K. G., Loveridge, C., Long, J., Pitson, S. M., Tigyi, G., et al. (2010). FTY720 and (S)-FTY720 vinylphosphonate inhibit sphingosine kinase 1 and promote its proteasomal degradation in human pulmonary artery smooth muscle, breast cancer and androgen-independent prostate cancer cells. *Cellular Signalling, 22*, 1536–1542.

Uddin, S., & Platanias, L. C. (2008). Sphingosine kinase is an attractive target for novel pharmacological interventions in non-Hodgkin lymphomas. *Leukemia & Lymphoma, 49*, 850–851.

Van Brocklyn, J. R., Jackson, C. A., Pearl, D. K., Kotur, M. S., Snyder, P. J., & Prior, T. W. (2005). Sphingosine kinase-1 expression correlates with poor survival of patients with glioblastoma multiforme: Roles of sphingosine kinase isoforms in growth of glioblastoma cell lines. *Journal of Neuropathology and Experimental Neurology, 64*, 695–705.

Visentin, B., Vekich, J. A., Sibbald, B. J., Cavalli, A. L., Moreno, K. M., Matteo, R. G., et al. (2006). Validation of an anti-sphingosine-1-phosphate antibody as a potential therapeutic in reducing growth, invasion, and angiogenesis in multiple tumor lineages. *Cancer Cell, 9*, 225–238.

Wang, F., Van Brocklyn, J. R., Hobson, J. P., Movafagh, S., Zukowska-Grojec, Z., Milstien, S., et al. (1999). Sphingosine 1-phosphate stimulates cell migration through a G(i)-coupled cell surface receptor. Potential involvement in angiogenesis. *The Journal of Biological Chemistry, 274*, 35343–35350.

Watson, C., Long, J. S., Orange, C., Tannahill, C. L., Mallon, E., McGlynn, L. M., et al. (2010). High expression of sphingosine 1-phosphate receptors, S1P1 and S1P3, sphingosine kinase 1, and extracellular signal-regulated kinase-1/2 is associated with development of tamoxifen resistance in estrogen receptor-positive breast cancer patients. *The American Journal of Pathology, 177*, 2205–2215.

Woodcock, J. M., Ma, Y., Coolen, C., Pham, D., Jones, C., Lopez, A. F., et al. (2010). Sphingosine and FTY720 directly bind pro-survival 14-3-3 proteins to regulate their function. *Cellular Signalling, 22*, 1291–1299.

Xia, P., Gamble, J. R., Rye, K. A., Wang, L., Hii, C. S., Cockerill, P., et al. (1998). Tumor necrosis factor-alpha induces adhesion molecule expression through the sphingosine kinase pathway. *Proceedings of the National Academy of Sciences of the United States of America, 95*, 14196–14201.

Xia, P., Wang, L., Gamble, J. R., & Vadas, M. A. (1999). Activation of sphingosine kinase by tumor necrosis factor-alpha inhibits apoptosis in human endothelial cells. *The Journal of Biological Chemistry, 274*, 34499–34505.

Xia, P., Gamble, J. R., Wang, L., Pitson, S. M., Moretti, P. A., Wattenberg, B. W., et al. (2000). An oncogenic role of sphingosine kinase. *Current Biology, 10*, 1527–1530.

Yan, G., Chen, S., You, B., & Sun, J. (2008). Activation of sphingosine kinase-1 mediates induction of endothelial cell proliferation and angiogenesis by epoxyeicosatrienoic acids. *Cardiovascular Research, 78*, 308–314.

Young, N., Pearl, D. K., & Van Brocklyn, J. R. (2009). Sphingosine-1-phosphate regulates glioblastoma cell invasiveness through the urokinase plasminogen activator system and CCN1/Cyr61. *Molecular Cancer Research, 7*, 23–32.

Zhang, H., Desai, N. N., Olivera, A., Seki, T., Brooker, G., & Spiegel, S. (1991). Sphingosine-1-phosphate, a novel lipid, involved in cellular proliferation. *The Journal of Cell Biology, 114*, 155–167.

Zhang, R., Shah, M. V., Yang, J., Nyland, S. B., Liu, X., Yun, J. K., et al. (2008). Network model of survival signaling in large granular lymphocyte leukemia. *Proceedings of the National Academy of Sciences of the United States of America, 105*, 16308–16313.

CHAPTER EIGHT

Natural Products as Platforms for the Design of Sphingolipid-Related Anticancer Agents

Antonio Delgado[*,†,‡,1], **Gemma Fabriàs**[*,†], **Josefina Casas**[*,†], **José Luis Abad**[*,†]

[*]Spanish National Research Council (Consejo Superior de Investigaciones Científicas, CSIC), Research Unit on Bioactive Molecules (RUBAM), Jordi Girona 18-26, 08034-Barcelona, Spain
[†]Departament de Química Biomèdica, Institut de Química Avançada de Catalunya (IQAC-CSIC), Jordi Girona 18-26, 08034-Barcelona, Spain
[‡]Facultat de Farmàcia, Unitat de Química Farmacèutica (Unitat Associada al CSIC), Universitat de Barcelona (UB), Avda. Joan XXIII, s/n, 08028-Barcelona, Spain
[1]Corresponding author: e-mail address: delgado@rubam.net

Contents

Abstract

Modulation of sphingolipid metabolism is a promising strategy for cancer therapy that has already opened innovative approaches for the development of pharmacological tools and rationally designed new drugs. On the other hand, natural products represent a classical and well-established source of chemical diversity that has guided medicinal chemists on the development of new chemical entities with potential therapeutic use. Based on these premises, the aim of this chapter is to provide the reader with a general overview of some of the most representative families of sphingolipid-related natural products that have been described in the recent literature as lead compounds for the design of new modulators of sphingolipid metabolism. Special emphasis is placed on the structural aspects of natural sphingoids and synthetic analogs that have found application as anticancer agents. In addition, their cellular targets and/or their mode of action are also considered.

Advances in Cancer Research, Volume 117
ISSN 0065-230X
http://dx.doi.org/10.1016/B978-0-12-394274-6.00008-X

ABBREVIATIONS

CDase ceramidase
Cer ceramide
CerK ceramide kinase
CerPP ceramide-1-phosphate phosphatase
CerS ceramide synthases
Des1 dihydroceramide desaturase
dhCer dihydroceramide
dhSo dihydrosphingosine
FB1 fumonisin B1
GalCer galactosylceramide
GCS glucosylceramide synthase
GSL glycosphingolipids
S1P sphingosine-1-phosphate
SK sphingosine kinase
SL sphingolipids
SM sphingomyelin
SMase sphingomyelinase
SMS sphingomyelin synthase
So sphingosine
SPT serine palmitoyl transferase

1. THE ROLE OF SPHINGOLIPIDS IN CANCER

Sphingolipids (SLs) are a group of lipids that derive from the common base sphingosine (So). Besides playing a structural role and regulating the physical properties of cell membranes, SL metabolites participate in cell signaling through activation or inhibition of various macromolecules to influence numerous cellular functions. SL metabolites are interconverted through a finely regulated dynamic balance that contributes to determining cell fate. This balance is synchronized by intricate metabolic pathways where specific enzymes play a capital role (Fig. 8.1). However, in addition to this inherent complexity, SL cellular location (topology) and/or compartmentalization need to be considered in order to have a full understanding of the real scope of SL actions and roles at cellular level (Futerman & Hannun, 2004; Fyrst & Saba, 2010; Hannun & Obeid, 2008). The recent advances in SL research over the past two decades have transformed our perception of SL and, hence, a fast evolution from "passive spectators" to "key actors" has been witnessed. Nowadays, many pathologies have been linked, in whole or in part, to the occurrence of dysregulated SL metabolism and the altered

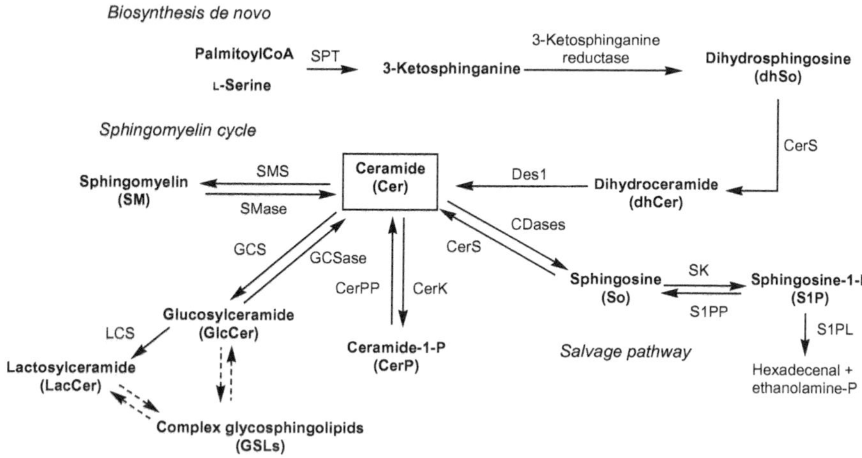

Figure 8.1 Biosynthesis of sphingolipids showing the pivotal role of ceramide.

production of specific SL (Gangoiti et al., 2010). This is the case of human cancers, a topic that has been recently reviewed in an excellent work (Ryland, Fox, Liu, Loughran, & Kester, 2011).

Among the different SL metabolites, ceramide (Cer) is one of the best studied in connection with cancer, due to its role in cellular differentiation and induction of programmed cell death. It has been observed that cancer cells are generally able to keep Cer at reduced levels as a means to resist apoptosis. This is commonly achieved by overexpression of glucosylceramide synthase (GCS; Ballereau, Levade, Genisson, & Andrieu-Abadie, 2012) and ceramidases (CDases; Furuya, Shimizu, & Kawamori, 2011), both of which have emerged as therapeutic targets for cancer treatment. Nevertheless, although Cer is decreased in several types of tumors, as compared to normal tissues or milder phenotypes (Karahatay et al., 2007; Merchant et al., 1991; Riboni et al., 2002; Rylova, Somova, & Dyatlovitskaya, 1998; Selzner et al., 2001), lipidomic analysis has evidenced that some specific Cer's are actually elevated (Karahatay et al., 2007; Schiffmann, Sandner, Birod, et al., 2009). These findings add complexity to the distinct roles of SL and their dysregulated metabolism in cancer.

In addition to Cer, recent evidences suggest that dihydroceramide (dhCer), plays a critical role in cancer biology as well (Fabrias et al., 2011). Thus, it has been reported that inducing elevations of intracellular dhCer's with pharmacological and genetic tools leads to cell cycle arrest (Kraveka et al., 2007), reduced proliferation (Schiffmann, Sandner, Schmidt, et al., 2009) and induction of autophagy (Signorelli et al., 2009). Further, dhCer's are elevated in endometrial cancer (Knapp et al., 2010) and accumulate in the hypoxic cells of tumors (Yin, Miyazaki, Shaner, Merrill, & Kannagi, 2010), which may be linked to tumor resistance through enhanced autophagy. From these results, dihydroceramide desaturase (Des) emerges as an important enzyme in cancer biology.

Sphingosine-1-phosphate (S1P) is another extensively studied SL in relation to cancer (reviewed in Brocklyn, 2010; Furuya et al., 2011; Fyrst & Saba, 2010; Maceyka, Harikumar, Milstien, & Spiegel, 2012; Ponnusamy et al., 2010; Pyne & Pyne, 2010; Stevenson, Takabe, Nagahashi, Milstien, & Spiegel, 2011; Yester, Tizazu, Harikumar, & Kordula, 2011). Several reports support that S1P contributes to favor survival, proliferation, and migration of cancer cells (Nava, Hobson, Murthy, Milstien, & Spiegel, 2002; Thamilselvan, Li, Sumpio, & Basson, 2002; Wang, Van Brocklyn, Edsall, Nava, & Spiegel, 1999), and that elevated levels of S1P are closely associated with resistance to cancer therapy (Payne, Milstien, & Spiegel, 2002) and to a poor survival rate in cancer patients (Van Brocklyn et al., 2005). Overexpression of sphingosine kinase 1 (SK1) is one of the most important contributors to the uncontrolled growth of malignant cells by increasing S1P levels at the expense of Cer's in the catabolic pathway. Expression of SK1 is upregulated in a variety of cancers (Bayerl et al., 2008; Cuvillier et al., 1996; Duan & Nilsson, 2009; French et al., 2003; Johnson et al., 2005; Kawamori et al., 2006; Ruckhaberle, Rody, et al., 2008; Van Brocklyn et al., 2005), and it is associated to chemotherapeutic resistance (Bektas et al., 2005) and poor prognosis (Akao et al., 2006; Ruckhaberle, Rody, et al., 2008; Van Brocklyn et al., 2005). Due to the oncogenic potential of SK1, it is an attractive target for cancer therapy. In this regard, several SK1 inhibitors have been developed showing *in vivo* antitumor activity (French et al., 2006; Gangoiti et al., 2010). In contrast to SK1, the endoplasmic reticulum localized sphingosine kinase 2 (SK2) is thought to participate in a proapoptotic pathway (Maceyka et al., 2005) and its expression in a variety of cell types leads to inhibition of DNA synthesis and cell cycle arrest (Igarashi et al., 2003). Thus, not just the aberrant generation of S1P but also its topology has an impact in cancer.

Another phosphorylated SL, namely, ceramide-1-phosphate, may also contribute to cancer progression, as it has been shown to stimulate DNA synthesis and to inhibit induction of apoptosis (Levi, Meijler, Gomez-Munoz, & Zor, 2009). Additionally, an increased ceramide kinase (CerK) activity has been observed in liver cancer (Sugiura et al., 2002) and augmented CerK expression correlates with poor prognosis in estrogen receptor negative breast cancer patients (Ruckhaberle et al., 2009). Reciprocally, neuroblastoma cells undergo cell growth arrest by inhibition of CerK transcription (Murakami et al., 2010) and decreased CerK expression reduces proliferation in human lung adenocarcinoma cells (Mitra et al., 2007). These results are indicative of the role of CerK as potential target for the treatment of specific cancers.

Glycosphingolipids (GSLs) are aberrantly expressed in numerous cancers and this elevated production is linked to tumor progression. Such a dysregulation in GSL balance modulates various aspects of the tumor microenvironment, including cell mobility, adhesion, angiogenesis, and the capability of cancer cells to evade immune control (reviewed in Bleicher & Cabot, 2002; Gouaze-Andersson & Cabot, 2006). Glucosylceramide has been shown to be elevated in certain tumors and this feature is associated to the development of multidrug resistant phenotypes (Liu, Han, Giuliano, & Cabot, 1999; Morjani et al., 2001). Likewise, galactosylceramide (GalCer; Liu et al., 2010), sulfatides (Boggs et al., 2010; Gnewuch, Jaques, Havemann, & Wiegandt, 1994; Hiraiwa et al., 1990; Morichika, Hamanaka, Tai, & Ishizuka, 1996; Sakakibara et al., 1991), and more complex GSLs (Hakomori, 1985; Hakomori & Handa, 2002; Hanai, Nakamura, & Shitara, 2000) are also increased in several cancers. Specifically, lactosylsulfatide enhances cell adhesion and appears to be involved in metastasis (Zhong Wu, Honke, Long Zhang, Liang Zha, & Taniguchi, 2004), while gangliosides GD2, GD3, and GM2 contribute to chemotherapeutic resistance and the metastatic potential of the tumor (Li, Villacreses, & Ladisch, 1995). Overexpression of GCS is the main contributor to the high GSL levels present in tumors (Bleicher & Cabot, 2002; Duan & Nilsson, 2009; Gouaze et al., 2004; Liu, Han, Giuliano, & Cabot, 2001; Liu et al., 2004; Morjani et al., 2001; Ogretmen & Hannun, 2001; Ruckhaberle, Karn, et al., 2008). Besides augmenting GSL levels, such overexpression limits de buildup of Cer, which often results in the development of multidrug resistance in cancer cells (Bleicher & Cabot, 2002; Liu et al., 2001). In accordance, inhibition of GCS increases apoptosis (Bleicher & Cabot, 2002; Cakir, Saydam, Sahin, & Baran, 2011; Maurer, Metelitsa, Seeger, Cabot, & Reynolds, 1999; Uchida et al., 2002), reduces tumor growth (Weiss, Hettmer, Smith, & Ladisch, 2003), enhances the cytotoxic response to Cer (Liu et al., 2001, 2004), and restores sensitivity of cancer cells to chemotherapeutics (Gouaze et al., 2005).

Additional experimental evidence in support of the role of SL in cancer includes the activation of SL-metabolizing enzymes by a number of therapeutic strategies. Several drugs, such as daunorubicin (Bose et al., 1995), etoposide (Bonhoure et al., 2006), and THC (Gustafsson, Sander, Bielawski, Hannun, & Flygare, 2009), induce cell death in cancer cells by activation of serine palmitoyl transferase (SPT; reviewed in Perry, 2002), the first enzyme in SL biosynthesis *de novo*. Moreover, inhibition of SPT with myriocin (see Section 2.1) decreased Cer production and apoptosis in prostate cancer cells (Sauane et al., 2010). Further, the occurrence of reduced SPT levels and decreased amounts of Cer has also been reported in colon cancer cells (Duan & Nilsson, 2009).

Ceramide synthases (CerS) are activated in some instances after treatment with chemotherapy and radiotherapy (reviewed in Henry, Moller, Dimanche-Boitrel, Gulbins, & Becker, 2011; Mullen, Hannun, & Obeid, 2012; Ponnusamy et al., 2010). Thus, treatment of myeloid leukemia cells with daunorubicin results in increased Cer production and cell death by activation of CerS, while pretreatment with fumonisin B1 (FB1; Section 2.3) blocked daunorubicin-induced apoptosis (Bose et al., 1995). Further, cannabinoid treatment led to transcriptional induction of CerS3 and CerS6 in mantle cell lymphoma cells (Gustafsson et al., 2009). Thus, the regulation of specific CerS may be an underlying mechanism by which the concentrations of specific Cer or dhCer species are dysregulated in cancer.

Several anticancer drugs and radiotherapy strategies are known to involve activation of sphingomyelinases (SMases; reviewed in Canals, Perry, Jenkins, & Hannun, 2011; Carpinteiro, Dumitru, Schenck, & Gulbins, 2008; Henry et al., 2011; Kolesnick & Fuks, 2003; Kolesnick, 2002; Smith & Schuchman, 2008b; Wu, Clarke, & Hannun, 2010), which produce Cer from sphingomyelin (SM). In fact, administration of exogenous dietary SM after gemcitabine treatment increased cell death in pancreatic cancer cells (Modrak, Cardillo, Newsome, Goldenberg, & Gold, 2004) and inhibited malignant cancer progression in mice (Dillehay, Webb, Schmelz, & Merrill, 1994). Therefore, the chemotherapeutic and chemopreventive potential of SM have been investigated (Lemonnier et al., 2003; Modrak, Lew, Goldenberg, & Blumenthal, 2000). Among the several SMases, acid SMase seems to be particularly important in cancer therapy. Thus, overexpression of acid SMase, along with irradiation treatment (Smith & Schuchman, 2008a) or chemotherapy (Grammatikos et al., 2007), has been reported to improve tumor reduction in some cancer cells, as compared to the treatments alone. Reciprocally, acid SMase-deficient mice are unable to increase intracellular Cer levels and develop resistance to cell death induced by radiation (Santana et al., 1996).

Increased generation of dihydrosphingosine (dhSo) by induced expression of alkaline ceramidase 2 has been implicated in the cytotoxicity of fenretinide to tumor cells (Mao et al., 2010). Elevated dhSo levels by inhibition of CerS contribute to the cytotoxicity of fumonisins (Section 2.3; Soriano, Gonzalez, & Catala, 2005), treatment with exogenous dhSo induces apoptosis in some leukemic cells (Jarvis et al., 1996; Ohta et al., 1995), and both So- and dhSo-induced apoptotic cell death in human colon cancer cells (Ahn & Schroeder, 2010). These findings support a role of sphingoid bases in cancer. In fact, alterations in long-chain bases have been reported to occur in certain types of cancer (Kandyba et al., 2002, 2000).

The occurrence of dysregulated metabolic cascades is the source of new therapeutic strategies, most of them aimed at the use of SL analogs, synthetic inhibitors of specific SL-metabolizing enzymes or agonists and antagonists of specific SL receptors and effectors. In this context, natural products constitute a well-established source of bioactive agents and privileged scaffolds evolved to interact with specific biomolecules. For this reason, it is accepted that the screening of natural products has yielded even higher hit rates than those obtained from small molecule libraries (Koch, Schuffenhauer, et al., 2005; Kombarov et al., 2010). More recently, the notion of natural products as chemical probes, specially suited for the exploration of biological phenomena and/or biochemical pathways, has emerged (Carlson, 2010). With all these considerations in mind, this chapter is intended to provide the reader with a general overview of some of the most representative families of SL-related natural products that have been reported as lead compounds for the design of new modulators of SL metabolism. Special attention will be paid to the structural aspects of natural sphingoids and synthetic analogs with well-established anticancer properties, as well as to their cellular targets and/or their mode of action.

2. FUNGAL METABOLITES AS INHIBITORS OF SL BIOSYNTHESIS

SLs are present not only in mammals but also in other organisms, both from the prokaryote and the eukaryote kingdoms. The multiplicity of biosynthetic pathways implicated in these species is responsible for the structural diversity found among the natural sphingoid bases reported so far (Pruett et al., 2008). One of such types of organisms is fungi, which represent an invaluable source of mycotoxins, low-molecular weight secondary metabolites responsible for a variety of toxic effects in vertebrates (Brase, Encinas, Keck, & Nising, 2009). In this section, we want to collect the most relevant information about some of the different fungal metabolites that have been described in the literature as disruptors of SL metabolism. Interestingly, some of them have also become representative lead compounds for the design of synthetic analogs with relevant therapeutic applications or, alternatively, they have evolved as pharmacological tools for basic biomedical research.

2.1. Myriocin: From an SPT inhibitor to the development of fingolimod (FTY720)

SPT is a pyridoxal phosphate-dependent enzyme (Amadasi et al., 2007) that catalyzes the condensation of serine with palmitoyl-CoA in the first step of the biosynthesis of SL (Merrill, 2002). Several potent fungal metabolites have been characterized as potent inhibitors of this enzyme (see Fig. 8.2). According to the postulated mechanism for the SPT catalyzed condensation reaction, these metabolites are assumed to be substrates of the target enzyme that give rise to aberrant forms of the reaction intermediate A (Fig. 8.2) that is postulated in this enzymatic process. This fact accounts for their strong inhibitory activity (Hanada, 2003).

Many of the above fungal metabolites have attracted the attention of organic chemists as platforms for the design or application of new or well-established synthetic methodologies in the context of the total synthesis of natural products. This is the case of mycestericin F and G (Berhal,

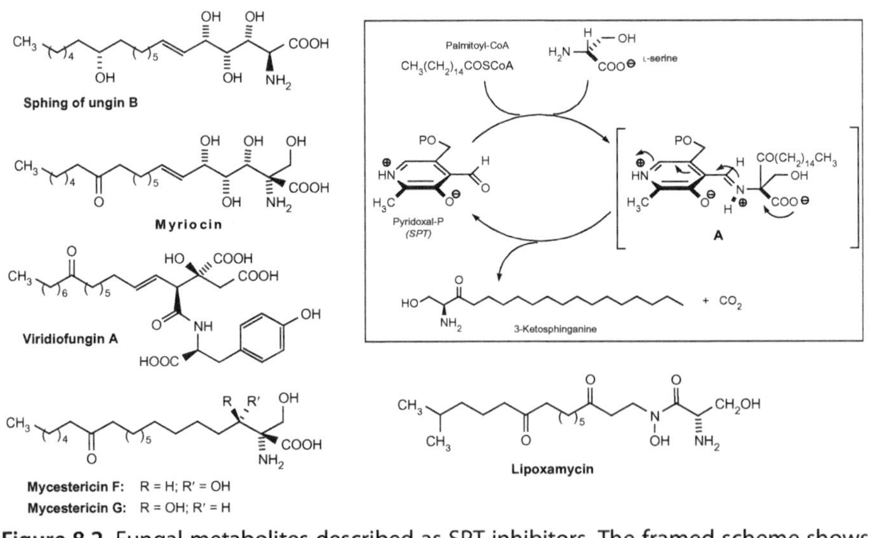

Figure 8.2 Fungal metabolites described as SPT inhibitors. The framed scheme shows the postulated mechanism for the SPT catalyzed reaction leading to 3-ketosphinganine from palmitoyl-CoA and L-serine.

Takechi, Kumagai, & Shibasaki, 2011), sphingofungins (Ohfune & Shinada, 2005), and viridofungins (Mandala, Thornton, Frommer, Dreikorn, & Kurtz, 1997; Morokuma, Takahashi, Ishihara, & Hatakeyama, 2005). From a structural standpoint, the above fungal metabolites are characterized by their amphiphilic nature resulting from the combination of a lipidic chain and a densely functionalized polar head. For this reason, these compounds have attracted the interest of both synthetic and medicinal chemists interested in the development of new synthetic protocols as well as for the development of simplified analogs with improved biological properties.

Among the different fungal metabolites reported as SPT inhibitors, structural modifications of myriocin (ISP-I or thermozymocidin) have been very fruitful from a clinical standpoint. Thus, fingolimod (FTY720, Fig. 8.3) has been recently approved (under the trademark Gilenya®) for the treatment of multiple sclerosis (Brinkmann et al., 2010; Strader, Pearce, & Oberlies, 2011; Yeung, 2011) and it is also being studied as anticancer agent for its properties as SphK1 inhibitor (see below; Pyne, Bittman, & Pyne, 2011). As in many case studies of drug discovery, the story of fingolimod started from the initial observation of the immunosuppressant activity of myriocin, after its isolation from the culture broth of the fungus *Isaria sinclairii* by Fujita et al. (1994). Interestingly, myriocin turned out to be identical to another metabolite already known from other fungal species. However, Fujita was the first to show this metabolite to be 10- to 100-fold more potent than cyclosporin A as an immunosuppressive agent of the immune response *in vitro* and *in vivo*. However, the poor solubility and high toxicity of myriocin prompted researchers to develop a program aimed at the optimization of this interesting hit (Yoshikawa, Yokokawa, Okuno, Yagi, & Murakami, 1995). Thus, the triol ISP-I-28 represents a simplified analog at the polar head, which

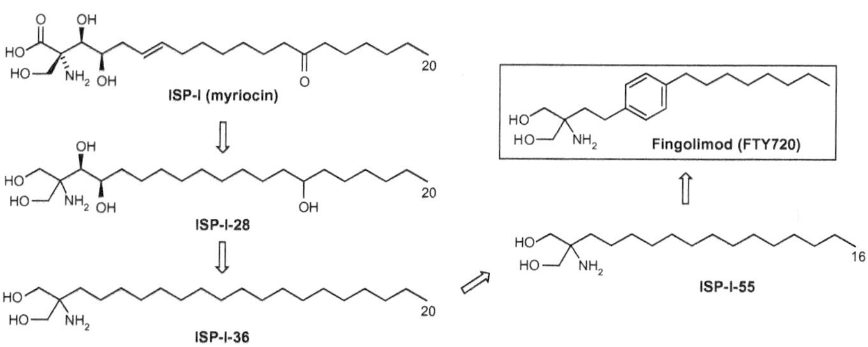

Figure 8.3 Structural modifications leading from myriocin (ISP-I) to fingolimod.

further evolved to the nonchiral analogs ISP-I-36 and ISP-I-55, differing at the length of the aliphatic long chain. The final modification consisted on the introduction of a phenyl moiety as part of the aliphatic chain in order to restrict the conformational flexibility of the molecule (Kiuchi et al., 2000). Unlike other immunosuppressant agents, such as cyclosporin A or FK506, fingolimod is not active as SPT inhibitor, suggesting a different mechanism of action for this compound (Strader et al., 2011).

As stated above, fingolimod has also been evaluated as potential anticancer agent for its ability to inhibit SK1, whose association in proliferation and metastasis in several types of human cancer has been shown (Pyne et al., 2012). This enzyme is implicated in the biosynthesis of S1P, a SL metabolite involved in the regulation of angiogenesis (Pyne et al., 2011). Fingolimod has also been reported as a S1P lyase inhibitor *in vitro* and also in mice (Bandhuvula, Tam, Oskouian, & Saba, 2005; Reina, Camacho, Casas, Van Veldhoven, & Fabrias, 2012; Serra & Saba, 2010).

Fingolimod is a pro-chiral compound that gives rise to a chiral (*S*)-phosphate (fingolimod-P, Fig. 8.4) after metabolization by the SK2 isoform of SK. Fingolimod-P is the active metabolite responsible for the therapeutic usefulness of this compound in the treatment of multiple sclerosis. Although it is not a S1P lyase inhibitor (Bandhuvula et al., 2005), fingolimod-P behaves as a structural analog of S1P that shows affinity for S1P receptors, mainly $S1P_1$ (Brinkmann, 2004), which complicates somehow the pharmacological profile of the drug. As anticancer agent, the (*S*)-phosphate is not as efficient as the parent fingolimod as cell growth inhibitor in several cancer cell lines. This is the case of MCF-7, MDA-MC-231, and Sk-Br-3 breast cancer lines (Nagaoka, Otsuki, Fujita, & Uesato, 2008; Tonelli et al., 2010) and the colorectal cancer lines HCT116 and SW620 (Nagaoka et al., 2008). Other studies of fingolimod as anticancer agent are related to its use against MGC803 gastric adenocarcinoma cell line with promising results (Zheng et al., 2010).

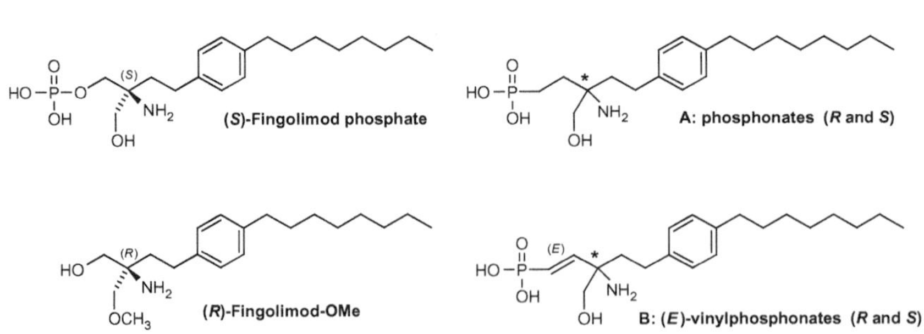

Figure 8.4 (*S*)-Fingolimod and structural analogs.

Structural modifications of fingolimod have been extensively studied both for basic structure–activity relationships and also to improve the pharmacological properties of the parent compound. In this context, modifications at the polar head have been extensively considered. Thus, in addition to the already mentioned (*S*)-fingolimod-P, a series of C-analogs have been obtained by replacement of one of the polar hydroxyl groups with a phosphonate or a (*E*)-vinylphosphonate moiety (Fig. 8.4). Thus, phosphonates A and vinylphosphonates B were obtained in both configurations (Lu et al., 2009). Interestingly, the biochemical profile elicited by these analogs was closely related to their configuration, since the (*S*)-isomers showed antiapoptotic activity in intestinal epithelial cells and the (*R*)-isomers were devoid of antiapoptotic activity but were able to activate the $S1P_1$ receptors (Lu et al., 2009). Additional studies on these analogs as regulators of the pulmonary vascular permeability *in vivo* and *in vitro* (Camp et al., 2009) and as potential anticancer agents in MCF-7 breast cancer cells have also been described (Lim, Tonelli, et al., 2011). Very recently, the methyl ether (*R*)-FTY720-OMe has been reported as a new fingolimod analog modified at the polar head. This compound has been shown to be a selective SK2 competitive inhibitor in HEK 293 cells and also able to prevent the S1P-stimulated rearrangement of actin in MCF-7 breast cancer cells, this showing the role of SK2 as a prosurvival protein (Lim, Sun, Bittman, Pyne, & Pyne, 2011).

Some formal deoxy analogs of fingolimod have also been reported. These compounds arise from removal of one of the terminal hydroxyl groups and its replacement with a hydrogen atom. To simplify the synthesis, an oxygen atom linked to the aromatic ring is used in the common scaffold. The simpler members of this series are represented by both enantiomers of AAL (Fig. 8.5). Recombinant mouse SK1a catalyzed the phosphorylation of (*R*)-AAL, but not that of the (*S*) enantiomer. As expected by its stereochemical correlation with (*S*)-fingolimod-P, phosphorylated (*R*)-AAL ((*R*)-AAL-P) behaved as an S1P agonist on four out of the five S1P receptors, unlike the corresponding alcohol (Brinkmann et al., 2002). Both (*R*)-AAL and (*R*)-AAL-P have been used as pharmacological tools in several studies on S1P receptors (Butler, Lana, Round, & LaMontagne, 2004; Sensken, Bode, & Graler, 2009). More recently, a series of (*R*)-AAL-P analogs containing a phosphate surrogate (compounds 3, 4, 6, 7, and 11, Fig. 8.5) have been prepared and evaluated as S1P agonists. In particular, compound (*R*)-11 was shown to reversibly reduce blood lymphocyte counts in rats after oral administration (Hogenauer, Hinterding, & Nussbaumer, 2010).

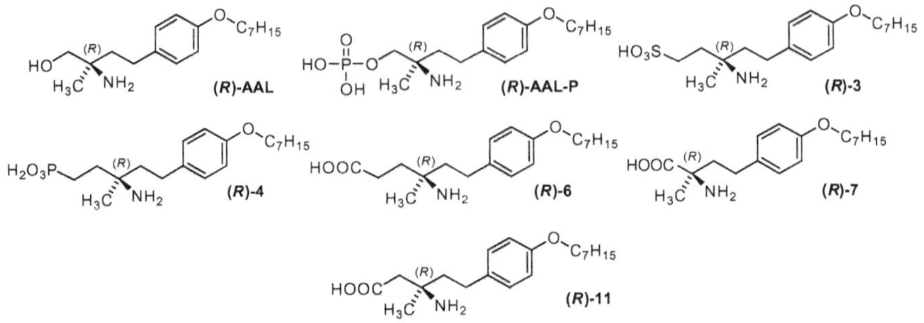

Figure 8.5 Deoxy analogs of fingolimod.

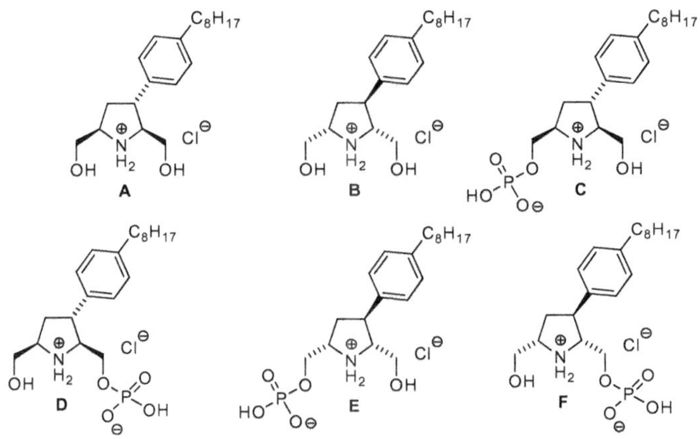

Figure 8.6 Conformationally constrained pyrrolidine analogs of fingolimod.

Conformationally, constrained analogs of fingolimod have also been reported as molecular probes for S1P receptors. One type of analogs is based on pyrrolidine derivatives in which the amino group has been incorporated as part of the heterocyclic system (compounds A–F, Fig. 8.6; Hanessian, Charron, Billich, & Guerini, 2007). In particular, analog A was shown to be a substrate of SK2, being phosphorylated four times more efficiently than fingolimod. The corresponding phosphates were also prepared in this work and used to explore the biology and binding of S1P receptors.

In a related work, a set of cyclopentane derivatives were designed as another family of conformationally constrained analogs of fingolimod (Fig. 8.7; Zhu et al., 2007). Among the four stereoisomers described in this study, compounds A and C behaved as SK2 substrates and were phosphorylated to afford potent $S1P_1$ agonists. Interestingly, phosphorylated compound

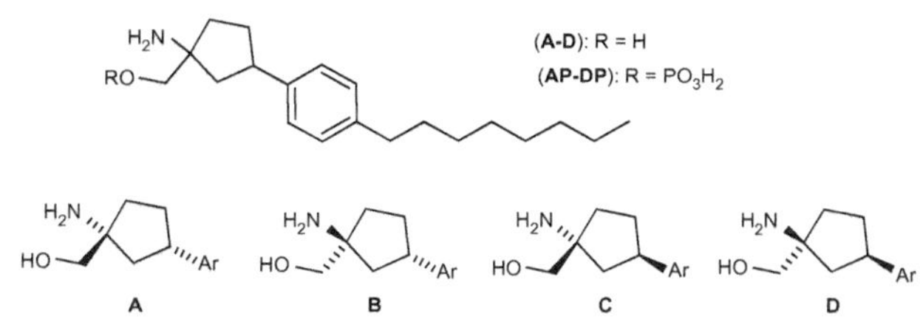

Figure 8.7 Cyclopentane analogs of fingolimod.

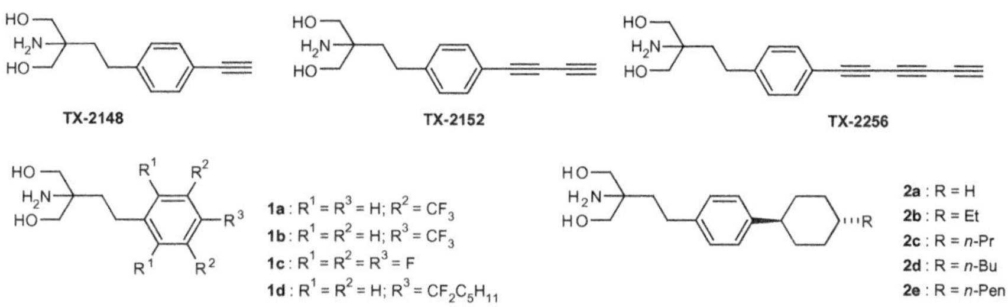

Figure 8.8 Ring substituted and side chain conformationally constrained analogs.

AP turned out to be a potent $S1P_3$ antagonist with interesting applications as pharmacological tool.

Conformational restriction at the aliphatic chain of fingolimod is found in the electron-rich acetylenic analogs TX (Fig. 8.8). Among the different members of this series, the diyne TX-2152 was a potent *in vivo* anti-angiogenic compound and was considered a promising candidate for further development as antineoplastic agent (Nakayama et al., 2008).

Other modifications at the aliphatic chain level include the introduction of fluorine atoms (compound 1d, Fig. 8.8) or its replacement with fluorinated (compounds 1a–c, Fig. 8.8; Ko, Chu, & Chiu, 2011) or cycloaliphatic groups (compounds 2a–e, Fig. 8.8; Feng et al., 2012). At the time this chapter was written, no biological information on the fluorinated analogs 1a-d was still available. Concerning the cycloaliphatic analogs 2a-e, a modest *in vivo* immunomodulatory activity has been reported (Feng et al., 2012).

In summary, the development of fingolimod (Gilenya®) for the treatment of multiple sclerosis represents an interesting example of rational drug design arising from the structural modification of a natural product. This drug has already received the U.S. FDA approval and is the first orally available drug for this disease. Despite this success, active research in the field seeking second- and third-generation analogs with reduced side effects is underway and will surely afford improved drugs with a safer pharmacological profile.

2.2. Scyphostatin and related fungal metabolites as neutral SMase inhibitors

Scyphostatin (Fig. 8.9) is a fungal metabolite isolated from the fermentation broth of *Dasyscyphus mollisimus* SANK-13892 (later renamed as *Trichopeziza mollissima*) as part of a program addressed at the search of neutral SMase inhibitors. Scyphostatin exhibited a potent, reversible (Wascholowski, Giannis, & Pitsinos, 2006), and selective inhibitory activity against N-SMase in a rat brain microsome assay (Tanaka, Nara, Suzuki Konagai, Hosoya, & Ogita, 1997). The breakdown of SM by membrane-bound neutral SMase is one of the effects that appear downstream of several signaling events mediated by cytokines, as well as radiation, heat, and oxidative agents, *inter alia*. Since Cer, the product of hydrolysis of SM, is one of the lipid second messengers that play a key role in the cellular fate, neutral SMase has become a target for drug research (Wascholowski & Giannis, 2001). In addition to its therapeutic potential, neutral SMase inhibitors represent also valuable pharmacological tools to better understand the molecular basis of several cellular alterations leading ultimately to the development of certain diseases. Along this line, the implication of neutral SMase and cancer, via the generation of the proapoptotic Cer, has been well established.

The close structural analogy between scyphostatin and the antibiotic manumycin A (Fig. 8.9; Thiericke, Stellwaag, Zeeck, & Snatzke, 1987; Zeeck, Frobel, Heusel, Schroder, & Thiericke, 1987; Zeeck, Schroder, Frobel, Grote, & Thiericke, 1987) prompted researchers to postulate a similar biological profile for both natural products, as it turned out to be (Arenz et al., 2001). In fact, manumycin A is an irreversible inhibitor of neutral SMase that shares with scyphostatin an electrophilic epoxy cyclohexenone moiety as a reactive polar head. In analogy with other phosphodiesterases, neutral SMase may contain nucleophilic groups in its active

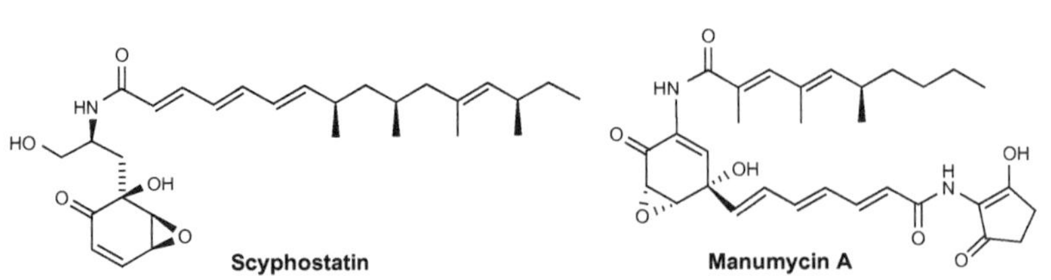

Figure 8.9 Fungal metabolites reported as N-SMase inhibitors.

site suitably placed to react with the reactive electrophilic groups present in these natural products. With these considerations in mind, it is easy to understand that the structural analogs of scyphostatin and manumycin reported so far contain an electrophilic moiety that resembles that found in the natural products. This is the case of epoxides 2 (Arenz et al., 2001), 3, and 4 (Block, Klein, Altenbach, & Brauer, 2000; Fig. 8.10), described as irreversible N-SMase inhibitors and designed as a simplified structural analogs of the above natural products.

In another series of analogs, cyclohexadienones 5 and 6 (Fig. 8.10; Pitsinos et al., 2003), lacking the epoxide moiety, also behaved as irreversible inhibitors, similarly to spiroepoxide 2. This finding seems to indicate that the epoxy moiety is not essential for enzyme inhibition. In agreement with this hypothesis, cyclohexenone 7 has been reported as the first reversible neutral SMase inhibitor arising from structural simplification of the above natural products (Claus et al., 2005). Concerning the fatty acyl chain, several analogs have also been reported. Thus, in a series of *N*-acyl scyphostatin analogs, compounds 8a–d behaved as irreversible inhibitors of N-SMase (Wascholowski et al., 2006). This was a quite surprising finding in light of the claimed reversible inhibition of neutral SMase by scyphostatin. Nevertheless, taking into account the expected reactivity of the epoxy cyclohexene moiety toward nucleophiles, a reinvestigation of the mechanism of inhibition elicited by Scyphostatin should not be disregarded. In another report, a series of scyphostatin analogs possessing a polyunsaturated acyl chain (compound A, Fig. 8.11; Cha, Burnett, Huang, Davidson, & Pettus, 2011) or a series of saturated acyl side chains (compound B, Fig. 8.11; Watanabe et al., 2007) were synthesized in order to improve the stability of the natural product. However, no biological data have been reported for these analogs.

A further structural simplification of the cyclic skeleton present in scyphostatin and manumycin led to the design of the sphingolactones (Fig. 8.12; Wascholowski & Giannis, 2006). These analogs present the minimal structural requirements for N-SMase inhibition, as deduced from the

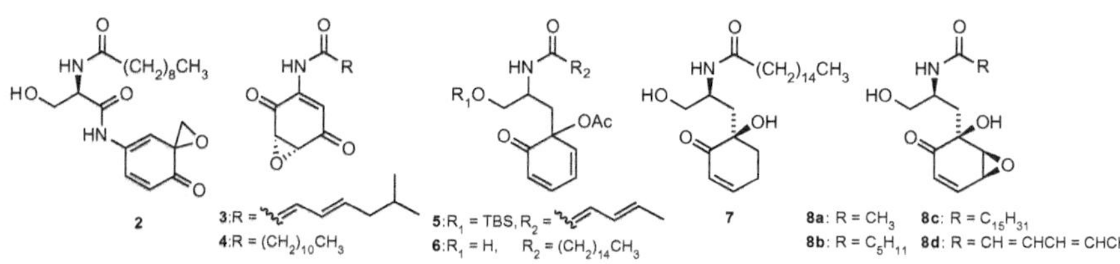

Figure 8.10 Simplified analogs of scyphostatin and manumycin.

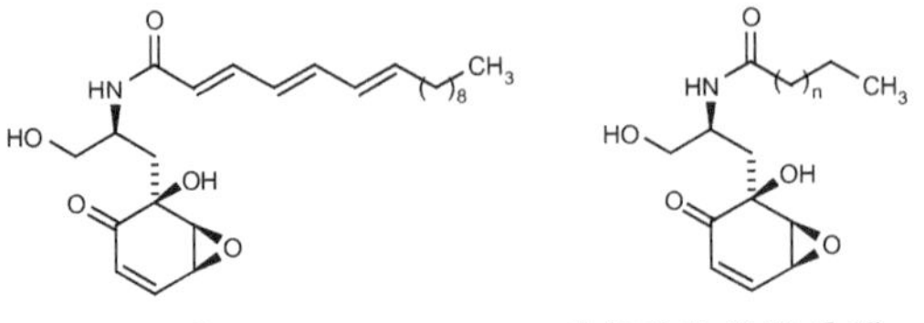

Figure 8.11 Scyphostatin analogs modified at the acyl chain.

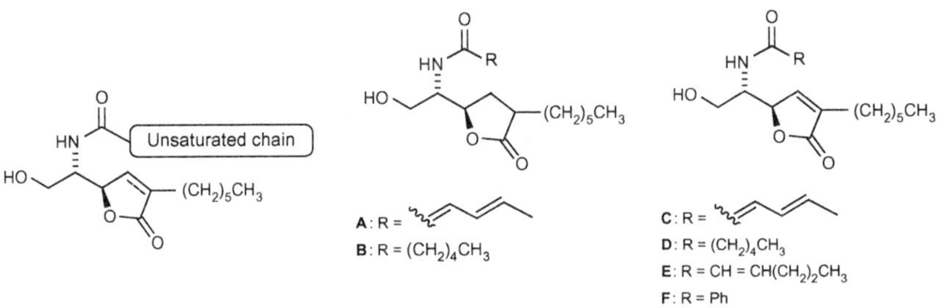

Figure 8.12 Sphingolactones designed as neutral SMase inhibitors.

structure–activity relationships observed in the above natural products. Thus, an unsaturated acyl chain, a primary alcohol, and a reactive γ-buyrolactone scaffold are the minimum structural requirements common to these types of analogs. Sphingolactones have been reported as irreversible N-SMase inhibitors with very slight activity on acid SMase. Among the sphingolactones tested (compounds A-F, Fig. 8.12), the saturated lactones were the best inhibitors, in particular, compound A (Wascholowski & Giannis, 2006).

2.3. The CerS inhibitor fumonisin and related 1-deoxysphingolipids

Fumonisins (Fig. 8.13) are a family of toxins of fungal origin (*Fusarium verticilloides*) that are found as food contaminants of corn and other grains and cereals. They are responsible for a large variety of diseases associated to liver and kidney toxicity, neurotoxicity, and immunological disorders, among others (Thiel, Marasas, Sydenham, Shephard, & Gelderblom, 1992). Many of the above adverse effects are associated to their ability to disrupt SL metabolism, in particular, inhibition of CerS to raise the concentration of the enzyme substrates So (from salvage pathway) and dhSo (from *de novo* synthesis) (Merrill, van Echten, Wang, & Sandhoff, 1993). These metabolic intermediates can be diverted to the corresponding phosphates (S1P and

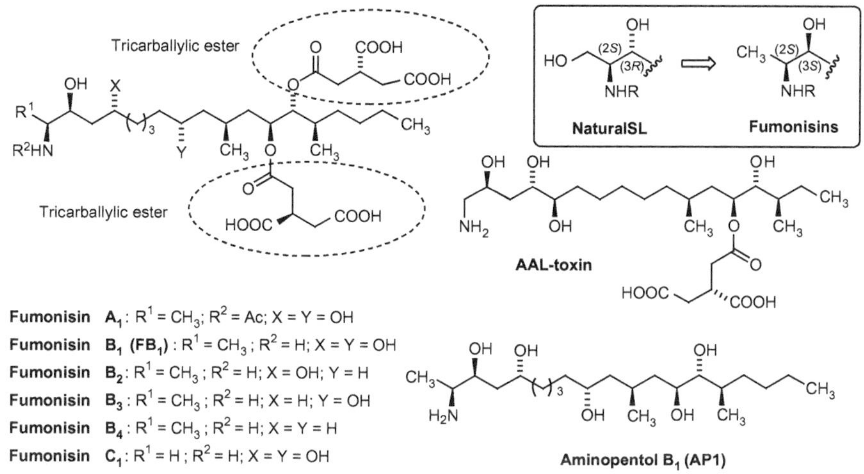

Figure 8.13 Fumonisins and related natural products. The framed scheme shows the opposite stereochemical relationship between natural SL and fumonisins at the terminal amino alcohol end.

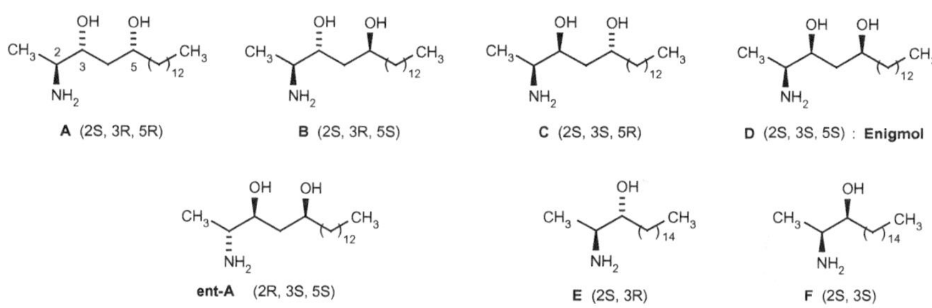

Figure 8.14 Some of the most relevant 1-deoxySL reported in the literature.

dihydrosphingosine-1-phosphate, respectively), which can also account for growth stimulation and the inhibition of apoptosis also associated to fumonisins. The short-term toxicity effects of fumonisins have been related to the elevated levels of dhSo, since inhibition of the upstream enzyme SPT suppresses these effects (Desai et al., 2002). More recently, in addition to the elevation of dhSo and So, 1-deoxydhSo (compound E, Fig. 8.14) has been identified as a new metabolite resulting from inhibition of CerS by fumonisins (Zitomer et al., 2009). The origin of this novel sphingoid base lies in the utilization of alanine instead of serine by SPT and its accumulation may also contribute to the toxic effects associated to fumonisins and related mycotoxins. Despite 1-deoxysphingoid bases had already been identified in other organisms (Pruett et al., 2008) and the first evidences in mammalian cells came from studies of a mutation in the *SPTLC1* gene that is found in human sensory neuropathy type 1 (HSN1; Zitomer et al., 2009), it is now accepted that 1-deoxydhSo and their corresponding *N*-acyl metabolites are also formed in mammalian cells. Interestingly, in a recent study, the presence of these deoxySL in plasma has been postulated as a novel class of biomarkers for the metabolic syndrome (Othman et al., 2012).

Among the different fumonisins, FB1 (Fig. 8.13) is largely used in basic research as pharmacological tool to block SL metabolism at the CerS level. Structurally, FB1 is composed of a linear amino pentahydroxyeicosane chain in which two of the hydroxyl groups are ester bound to a tricarballylic acid molecule each. The postulated mode of action of FB1 to account for CerS inhibition implies the competition of the amino poliol end with the enzyme substrate (So or dhSo), whereas the tricarballylic acid moiety interferes with the fatty acid acyl-CoA binding site (Merrill, Sullards, Wang, Voss, & Riley, 2001). According to this interpretation, CerS would exhibit a low stereoselectivity toward the terminal amino alcohol end of FB1, taking into the account the opposite configuration between the C3-hydroxy group of natural SL and the equivalent hydroxyl group in FB1 (Fig. 8.13).

The terminal methyl group is absent in a series of related toxins, such as fumonisin C1 and the *Alternaria alternata* toxin (AAL-toxin; Fig. 8.13), whose toxic effects are similar to those exhibited by FB1. Altogether, these observations are consistent with the apparent low structural demanding effects of CerS towards the terminal amino portion in this kind of compounds. A different scenario, however, is observed in aminopentol B1 (AP1), the dideacylated form of FB1, which behaves as a weak CerS inhibitor. However, AP1 is also a CerS substrate that leads to highly cytotoxic CerS inhibitors upon acylation, such as the corresponding *N*-palmitoyl-AP1 (PAP_1), which was around 10 times more toxic than FB1 or AP1 in HT29 human cancer cell line (Humpf et al., 1998). Acylation with other fatty acids has also been reported and the corresponding *N*-acylated derivatives have been related to the organ lesions and carcinogenicity associated to these metabolites (Desai et al., 2002).

In an attempt to uncover the structure–activity relationships in this family of compounds, a series of FB1 simplified synthetic analogs bearing a C_{18} alkyl chain with two stereogenic centers (at C2 and C3, compounds E and F, Fig. 8.14) or three stereogenic centers (at C2, C3, and C5, compounds A–D, Fig. 8.14) bearing all possible configurations were studied. All these analogs behaved as moderate to good CerS inhibitors and also as enzyme substrates (Humpf et al., 1998). Interestingly, the (2*S*,3*R*,5*R*) isomer (analog A,

Fig. 8.14) was the most potent inhibitor, only slightly less potent than FB1 using [^{3}H]sphingosine as substrate. This same analog was also the most efficient as CerS substrate, with an apparent $V_{max}/K_m = 125$ ($V_{max}/K_m = 248$ for D-*erythro*-dhSo). Among the analogs lacking the stereogenic center at C5, compound E (2*S*,3*R*) (1-deoxydhSo), identical to the SL metabolite that accumulates in the presence of FB1 and in HSN1 (see above and also Section 3) was also found to be the most potent CerS inhibitor and the best enzyme substrate, although at a lesser extent than analog A in both assays. These results stress the importance of the "natural" (2*S*,3*R*) stereochemistry for the interaction with CerS in this series of analogs, as well as the modulating role of the additional C5(*R*) stereocenter. Interestingly, the (2*S*,3*R*) configuration does not match that present in FB1 at this level (see Fig. 8.13).

Some of the above synthetic analogs A–D have received special attention in the past years. They all bear a 5-hydroxy dihydrosphingosine backbone and some of them have been studied in depth as SL modulators. The (2*S*,3*S*,5*S*) isomer (named enigmol in the recent literature) is particularly interesting. In addition of not being phosphorylated by SK's, as found in all 1-deoxySL, it is slowly *N*-acylated by CerS and therefore is more persistent than the natural sphingoid bases (Humpf et al., 1998). Enigmol has recently been identified as a potent antitumor agent from a NCI-60 human tumor cell line screen and its cytotoxicity has been confirmed in the HT29 colon cancer cell line (Symolon et al., 2011). In addition, enigmol has also been reported as a promising therapeutic agent to treat prostate cancer, as evidenced by a recent study using PC-3 and LNCaP cancer cell lines (Garnier-Amblard et al., 2011). In this study, the cytotoxicities of the four enigmol stereoisomers at C3 and C5 positions, together with *N*-methylenigmol, were compared in the above two cell lines. Only the (2*S*,3*R*,5*R*) isomer (compound A, Fig. 8.14) was not cytotoxic in the assay used, whereas isomers B (2*S*,3*R*,5*S*) and C (2*S*,3*S*,5*R*) (Fig. 8.14) caused statistically significant inhibition of tumor growth. Also recently, enigmol and a new stereoisomer (compound *ent*-A: 2*R*,3*S*,5*S*), have been evaluated in a panel of six human cancer cell lines (Esteve et al., 2011). The observed 50% growth inhibition (GI_{50}) concentration values ranged between 3 and 17 μM, enigmol being more active than *ent*-A in ovarian, lung, and colon cell lines. Interestingly, although not related to the topic of this chapter, it is worth mentioning that a small collection of *N*-acyl and *N*-alkyl enigmol stereoisomers have been synthesized and tested for their activity against two strains of the predominant human malaria species *Plasmodium falciparum* and *Plasmodium knowlesi*, with *in vitro* antimalarial activity (Meyer et al., 2012). Since plasmodium-infected erythrocytes use SL from both endogenous metabolism and host pools, this represents a promising strategy in the search of new antimalarial agents.

3. DEOXYSPHINGOLIPIDS OF MARINE ORIGIN

In addition to the 1-deoxy-3,5-dihydroxysphingolipids considered in the previous section, a variety of "simpler" natural 1-deoxy-3-hydroxysphingolipids have also been reported in the literature. This is the case of spisulosine (1-deoxydhSo) (Fig. 8.15), an antiproliferative compound of marine origin isolated from the clam *Spisula polynyma* (syn. *Mactromeris polynima*) (Cuadros et al., 2000). This compound resembles the natural SL in the C18 sphingoid backbone and in the (2*S*,3*R*) configuration of the amino and hydroxyl groups, respectively. It is interesting to note that, prior to its isolation and characterization as a marine natural product, this compound had already been synthesized and studied as a synthetic analog of enigmol (Humpf et al., 1998; Fig. 8.14, compound E) and also identified as a SL metabolite arising from inhibition of CerS by fumonisins in mammals (see Section 2.3). Spisulosine, also known as ES-285, was initially developed as a promising anticancer agent due to its ability to inhibit the proliferation in the prostate tumor PC-3 and LNCaP cell lines via stimulation of *de novo* synthesis of Cer (Sanchez et al., 2008), although additional data suggest that ES-285 triggers an atypical cell death program when compared to other So-dependent apoptosis pathways (Salcedo et al., 2007). Since in the absence of FB1, substantial amounts of 1-deoxydhSo are biosynthesized and transformed into *N*-acyl-1-deoxydhSo (Humpf et al., 1998), these compounds constituted a relatively unexplored and promising type of SL that might play important roles in cell regulation. Nevertheless, in year 2008, spisulosine was

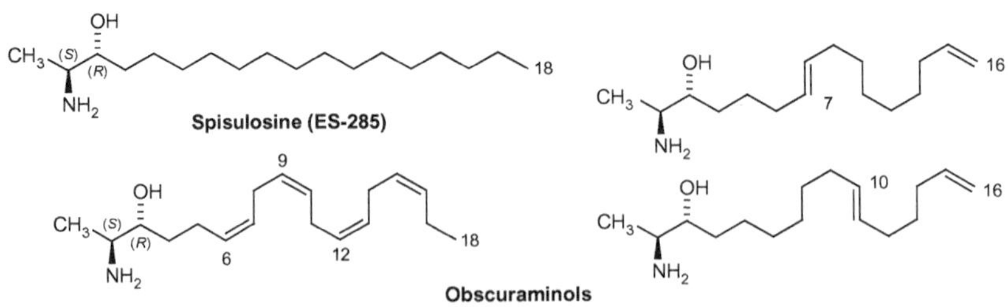

Figure 8.15 The marine SL spisulosine (ES-285) and obscuraminols.

discontinued from Phase I development (Baird et al., 2009), following a review of its risk–benefit ratio (Williams, 2009).

Obscuraminols (Fig. 8.15) are a group of natural sphingoids isolated from marine ascidians from the southern coast of Spain. In particular, they were isolated from specimens of the tunicate *Pseudodistoma obscurum* whose chloroform extracts showed a remarkable cytotoxicity against mouse lymphoma P388, human lung carcinoma A549, and human colon carcinoma HT29 cell lines (Garrido, Zubia, Ortega, Naranjo, & Salva, 2001). Obscuraminols resemble natural SL in the (2*S*,3*R*) configuration of the amino alcohol moiety, as determined by Mosher ester analysis of the corresponding *N*-acetyl derivatives, and present C_{16} and C_{18} aliphatic chains with different degrees of unsaturations. Despite not being of marine origin, another structurally related type of natural sphingoids is crucigasterins (Fig. 8.16), isolated from *Pseudomonas crucigaster*. As obscuraminols, they present one or several unsaturations as part of the sphingoid backbone but with an opposite (2*S*,3*R*) configuration at the amino alcohol moiety. Crucigasterins have been reported as antimicrobial agents and cytotoxic against mouse lymphocytic leukemia L1210 cells (Jareserijman, Bapat, Lithgowbertelloni, Rinehart, & Sakai, 1993).

Clavaminols (Fig. 8.17) are another group of 2-amino-3-alkanol derivatives isolated from a Mediterranean ascidian, namely, *Clavelina phlegraea* (Aiello et al., 2007). Their structure was established by spectroscopic methods and chemical correlation studies. Unlike the most common natural SL, clavaminols present a (2*R*,3*S*) configuration and, most of them, a C_{12} sphingoid chain. Thus, clavaminols A, C, D, F, I, J, and N are saturated

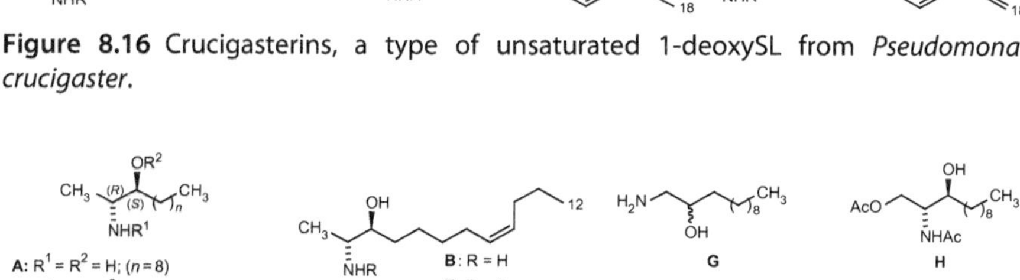

Figure 8.16 Crucigasterins, a type of unsaturated 1-deoxySL from *Pseudomonas crucigaster.*

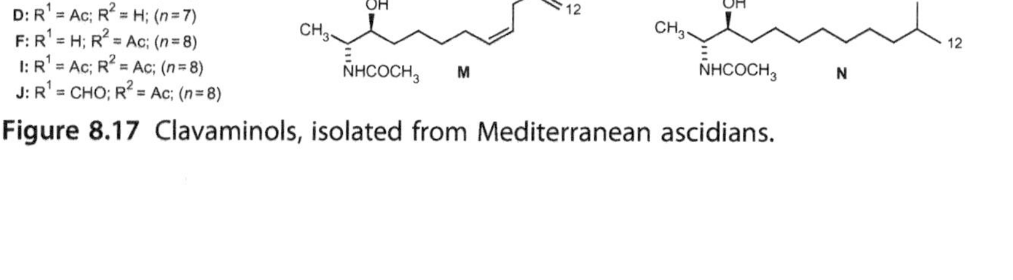

Figure 8.17 Clavaminols, isolated from Mediterranean ascidians.

amino alcohols, whereas clavaminols B, E, and M present a mono- or a diunsaturated sphingoid backbone. On the other hand, clavaminol G presents an isomeric 1-amino-2-hydroxy moiety and clavaminols H is, in fact, a diacylated amino diol sphingoid analog. Curiously, none of the reported clavaminols have been named as K and L. Some clavaminols present interesting biological properties. For example, clavaminol A was shown to be the most cytotoxic member of the series in AGS (càncer carcinoma) cells. Interestingly, the toxicity of *N*-acetylated clavaminols increased by chemical deacylation (Aiello et al., 2009).

Other structurally related 1-deoxySL have been isolated from marine sources. Despite they have not been reported as potential anticancer agents, they contain a sphingoid backbone and, hence, their general structures are collected in this section. This is the case of xestoaminols (Fig. 8.18), a group of 1-deoxysphingoids with a shorter (C_{14}) saturated or polyunsaturated backbone and (2*S*,3*R*) or (2*S*,3*S*) configurations. Xestoaminols have been isolated from the Fiji sponge *Xestospongia* sp. and were active against parasites, microbes, and reverse transcriptase (Jimenez & Crews, 1990). Along the same line, the related (2*S*,3*R*)-2-aminododecan-3-ol (Fig. 8.18), isolated from the ascidian *Clavelina oblonga* has been reported as an antifungal agent (Kossuga et al., 2004).

A more complex structural pattern is found in the so-called two-headed sphingoid bases. They constitute a relatively wide group of marine natural products characterized by the formal "tail-to-tail" connection of two sphingoid molecules. Since most of them show antibacterial or antifungal activities, they will not be considered in detail in this review and only the most relevant structural features concerning their potential as anticancer agents will be highlighted. The two-headed pattern characteristic of this type of compounds can arise either from combination of two units of a 1-deoxysphingoid base or hybrid compounds arising from combination

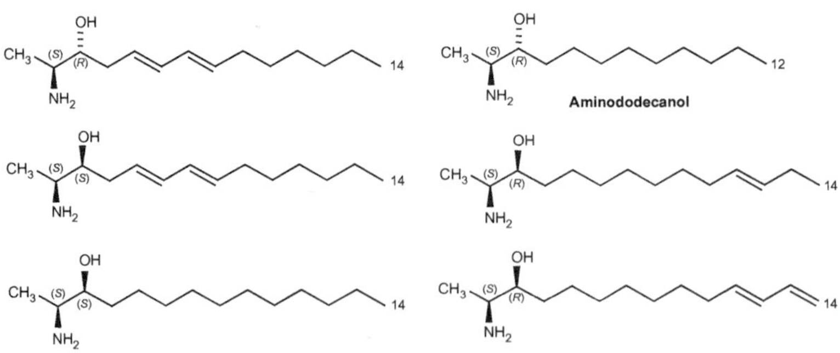

Figure 8.18 Xestoaminols and the saturated aminododecanol.

of one unit of a 1-deoxysphingoid base and one unit of a 1-hydroxy sphingoid base.

Rhizochalins (Fig. 8.19) are representative members of the first group that were isolated from the sponge *Rhizochalina incrustata* (Fedorov et al., 2009; Makarieva, Guzii, Denisenko, Dmitrenok, & Stonik, 2008). They are bis-β-galactosides in which the two-headed (2*R*,3*R*,26*R*,27*R*)-1-deoxysphingoid base is the aglycon part of the glycoside. Rhizochalins have been reported to possess antibacterial, antifungal, and cytotoxic activities. Recent studies have shown that the aglycon moiety is responsible for the cytotoxic effects of rhizochalin in HT29 colon cancer cells by DNA cleavage (Khanal et al., 2011; Sibirtsev, Shastina, Menzorova, Makarieva, & Rasskazov, 2011).

Another family of related bis-1-deoxysphingoids is represented by leucettamol A and B (Fig. 8.19). They were first identified as antimicrobial lipids from the calcareous sponge *Leucetta microraphis* (Kong & Faulkner, 1993). Further studies stressed the potential of these compounds as anticancer agents by their ability to upregulate the activity of the tumor suppressor p53 protein (Tsukamoto et al., 2008). In the same study, the saturated derivatives arising from hydrogenation of the double bonds showed higher activity in this assay, while its tetraacetate derivative was inactive, indicating that the hydroxyl and/or amino groups in leucettamols are required for the inhibition. Structural studies revealed that the configuration of each end of the chain has *erythro* stereochemistry with an absolute configuration of (2*R*,3*S*,28*S*,29*R*) (Dalisay, Tsukamoto, & Molinski, 2009).

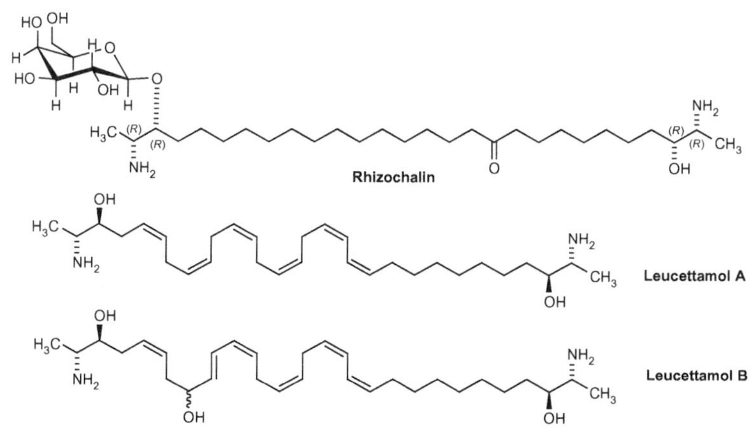

Figure 8.19 Rhizochalin and leucettamols, two representative members of "two-headed" sphingoid bases containing two units of a 1-deoxysphingoid base.

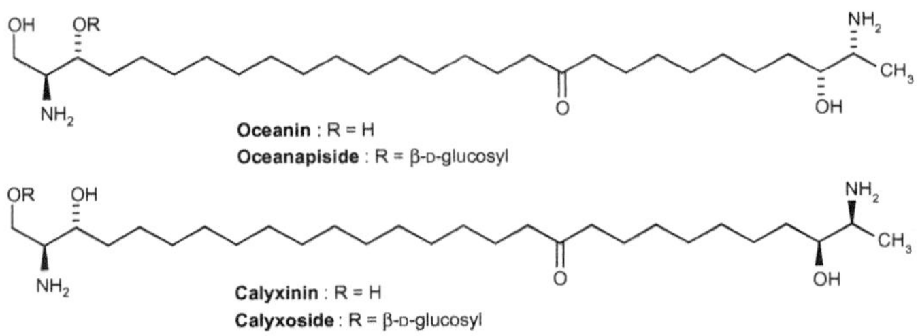

Figure 8.20 Oceanin and calyxinin, examples of hybrid structures containing one unit of a 1-deoxysphingoid base and one unit of a 1-hydroxy sphingoid base.

Among hybrid structures that combine two different types of terminal sphingoid heads, oceanin (Fig. 8.20), the aglycone moiety of oceanapiside, isolated from the sponge *Oceanapia phillipensis*, has been reported as antifungal agent (Nicholas et al., 2003, 1999; Nicholas, Li, MacMillan, & Molinski, 2002). A related marine compound is calyxoside (Fig. 8.20), a glucoside whose aglycon (calyxinin) is composed of a (2*S*,3*R*)-2-amino-1,3-diol moiety and a (2*S*,3*S*)-2-amino-3-hydroxy terminal end (Zhou, Mattern, Johnson, & Kingston, 2001). This glucoside induces DNA damage in cells and inhibits protein kinase C (Dalisay et al., 2009).

4. JASPINE B, A MARINE HETEROCYCLIC SPHINGOID

In some families of natural SLs, the sphingoid base can be recognized as part of a heterocyclic system. This is the case of Jaspine B (pachastrissamine) (Fig. 8.21), isolated as a cytotoxic component of the sponge *Pachastrissa* sp. by Kuroda et al. (2002) and later also isolated from the sponge *Jaspis* sp. From a structural standpoint, Jaspine B can be considered as the result of the formal cyclodehydration between the C1 and C4 hydroxyl groups of L-*lyxo*-phytosphingosine. For this reason, the term "anhydro phytosphingosine" is

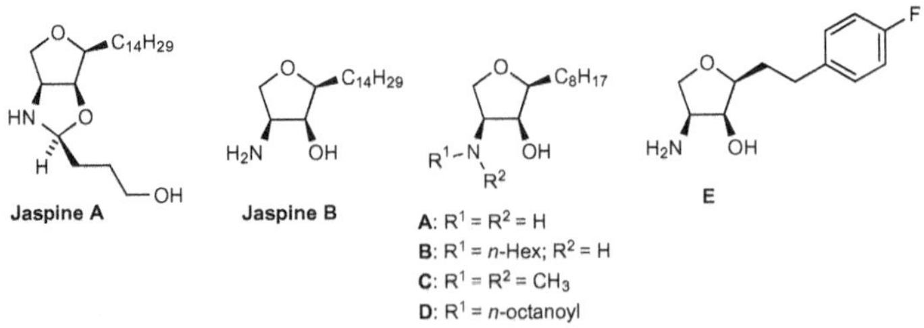

Figure 8.21 Natural jaspines and structural analogs.

sometimes used to refer to this type of compounds. Other structurally related analogs have also been isolated from the same species, such as Jaspine A (Fig. 8.21), a hemiaminal derivative of Jaspine B (Ledroit, Debitus, Lavaud, & Massiot, 2003). Since its isolation, several total synthesis for Jaspine B and structural analogs thereof have been developed, as it has been compiled in a very recent review on this topic (Ballereau, Baltas, & Genisson, 2011).

In the original work of Kuroda, a marked cytotoxic activity of Jaspine B ($IC_{50} = 0.01$ μg/mL) against P388, A549, HT29, and MEL28 cell lines was reported (Kuroda et al., 2002). Some of these results were also confirmed shortly after on A549 lung tumor cell line ($IC_{50} = 3.4 \times 10^{-7}$ M) (Ledroit et al., 2003). Some years later, a more thorough biological study revealed that the cytotoxicity in murine B16 and human SK-MEL28 melanoma cells was related to interference with SL metabolism via typical apoptosis, release of cytochrome c and caspase processing associated with increased Cer levels. In particular, sphingomyelin synthase (SMS) was proposed as the putative target for Jaspine B to account for the observed effects in B16 melanoma cells (Salma et al., 2009). In a related work (Salma et al., 2011), analogs A-E (Fig. 8.21) were also evaluated for their cytotoxicity in murine B16 melanoma cells. None of them were more cytotoxic than Jaspine B, but the short chain analog A and its *N*-substituted analogs B and D showed a comparable cytotoxicity profile. However, the dialkylated analog C and the aryl substituted analog E were much less cytotoxic.

In a research on diastereomeric Jaspine B at C2 and C3 positions (Fig. 8.22), the cytotoxicity of jaspines in A549 human alveolar cancer cells was associated to a strong increase of dhCer levels, but not Cer (Canals et al., 2009). Jaspine B was the most cytotoxic compound of the series and this effect decreased in the presence of the autophagy inhibitor 3-metyladenine (3MA). Collectively, these results suggest that dhCer-mediated autophagy might be involved in the cytotoxicity of Jaspine B in

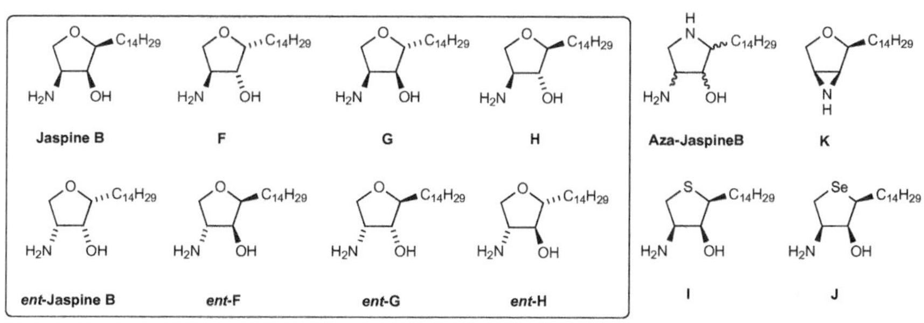

Figure 8.22 The eight stereoisomeric jaspines and some heteroatom analogs.

this cell model. The above four isomers, together with their enantiomeric analogs (Fig. 8.22), have been recently evaluated as SK1 and SK2 inhibitors (Yoshimitsu et al., 2011). All eight stereoisomers exhibited moderate to potent inhibitory activity against these targets. Interestingly, isomers *ent*-F and *ent*-H (Fig. 8.22) were around three to five times more potent, although less selective, than the reference SK1 inhibitor dimethylsphingosine. In a recent work, bioisosteric Jaspine B analogs containing sulfur and selenium atoms replacing the oxygen in the ring system have been described (Fig. 8.22, compounds I and J; Jeon et al., 2011). These new analogs displayed cytotoxicities comparable to those of Jaspine B in HCT116, A549, and PC-3 cancer cell lines. A Jaspine B analog in which the amino group is part of an aziridine ring has also been described (Fig. 8.22, compound K), although it showed no effect on cell viability in murine B16 melanoma cells (Ballereau, Andrieu-Abadie, Saffon, & Genisson, 2011). Along this line, a series of "aza analogs" of Jaspine B ("aza-Jaspine B") have also been reported (Fig. 8.22; Rives, Ladeira, Levade, Andrieu-Abadie, & Génisson, 2010). The design of these new analogs was inspired in the combination of the structural frameworks present in Jaspine B and in a series of GCS pyrrolidine-derived sphingosine mimetics with marked proapoptotic behavior (Faugeroux et al., 2006). Five stereoisomers of the above aza analogs were prepared, including the all-*cis* isomer matching the stereochemistry of natural Jaspine B. All compounds displayed cytotoxicity against various melanoma cells with potencies comparable to that of Jaspine B. Some of them impaired the conversion of Cer into SM in B16 cells, as elicited by the natural compound in the same assay.

5. OTHER HETEROCYCLIC NATURAL SPHINGOIDS

A large number of natural heterocyclic sphingoid compounds are characterized by the presence of the amino group as part of the heterocyclic ring. The more strained sphingoid derivatives are found among several families of natural azetidines isolated from sponges of the genus *Penares*. Penaresidins A and B (from *Penares* sp.) (Kobayashi et al., 1991) and Penazetidine A (from *Penares sollasi*) are two representative members described as PKC inhibitors (Fig. 8.23; Alvi, Jaspars, Crews, Strulovici, & Oto, 1994). The modest cytotoxic activity against L1210 (murine lymphoma) cells found for Penaresidin B stimulated the development of a SAR study of stereoisomeric simplified analogs. Two of them (compounds

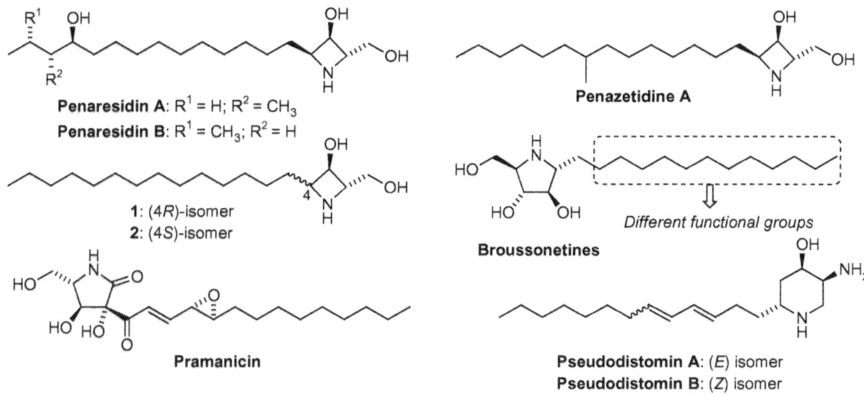

Figure 8.23 Natural heterocyclic sphingoid compounds.

1 and 2, Fig. 8.23) showed cytotoxic activity against A549 and HT29 cancer cell lines at IC_{50} around 0.2-0.6 μg/mL (Ohshita et al., 2007).

Heterocyclic SLs embedded into a pyrrolidine system have also been reported. This is the case of the broussonetines (Fig. 8.23), a large family of plant alkaloids initially reported in 1997 as glycosidase inhibitors (Shibano, Kitagawa, & Kusano, 1997). Since then, around 30 different broussonetines have been described (Shibano, Tsukamoto, & Kusano, 2002) and several synthetic strategies have been devised for these types of compounds (Ribes, Falomir, Murga, Carda, & Marco, 2009a,b). Other heterocyclic SL-related systems are found in the nature and several excellent reviews on this topic have been recently published (Ballereau, Baltas, et al., 2011; Pruett et al., 2008). However, as far as their potential as anticancer agents, only scattered information can be found, since they show a complex biological profile. In this context, the ability of pramanicin (Fig. 8.23) to induce apoptosis in Jurkat T leukemia cells in a dose- and time-dependent manner is worthy of mention (Kutuk, Pedrech, Harrison, & Basaga, 2005). Another interesting system is that represented by Pseudodistomins A and B (Fig. 8.23), two cytotoxic piperidine alkaloids isolated from the Okinawan tunicate *Pseudodistoma kanoko* and representative of the pseudodistomins (Kobayashi, Naitoh, Doi, Deki, & Ishibashi, 1995). They both were tested for antitumor activity against L1210 and L5178Y murine leukemia cells *in vitro* showing a remarkable cytotoxicity. In a later study, new pseudodistomins (Pseudodistomins D-F) were isolated from the ascidian *Pseudodistoma megalarva*. Together with Pseudodistomin B, they were found to be active in a cell-based assay for DNA damage induction (Freyer et al., 1997).

6. α-GALACTOSYLCERAMIDES AND STRUCTURAL MODIFICATIONS THEREOF

Agelasphines are a group of GSLs isolated from the marine sponge *Agelas mauritanus*. Among the different agelasphines reported, agelasphine 9b (Fig. 8.24) emerged as the most interesting member of this family in light of the antitumor activity against *in vivo* models of murine B16 melanoma (Natori, Koezuka, & Higa, 1993; Natori, Morita, Akimoto, & Koezuka, 1994). Synthetic efforts were soon addressed at the total synthesis of agelasphin 9b (Akimoto, Natori, & Morita, 1993), to confirm its chemical and stereochemical identity, and also at the scale-up of the synthesis in order to obtain enough quantities of material for further biological studies. Simultaneously, initial structure–activity studies led to the design of several synthetic analogs, from which KRN 7000 (Fig. 8.24) emerged as a suitable candidate for further development (Morita et al., 1995).

α-Galactosylceramides (α-GalCer) play an essential role in the triggering of the immune system. Like some other natural glycolipids (Brutkiewicz, 2006), α-GalCer can play the role of lipid antigens by binding to the CD1d protein present in the surface of the so-called antigen-presenting cells (Yu & Porcelli, 2005). This CD1d/lipid antigen "binary complex" (Koch, Stronge, et al., 2005; Zajonc et al., 2005) is presented to invariant natural killer (iNKT) cells by association with a semi-invariant T-cell receptor (TCR) expressed at the iNKT cell surface that specifically recognizes the above complex. The CD1d/glycolipid/TCR "ternary complex" thus formed (Borg et al., 2007) is responsible for the activation of the immune system by allowing the production of a vast variety of cytokines with different cellular effects (Fig. 8.25).

Cytokine production is a complex process that gives rise to a cascade of intracellular metabolic events. Several families of cytokines (interleukins, IL;

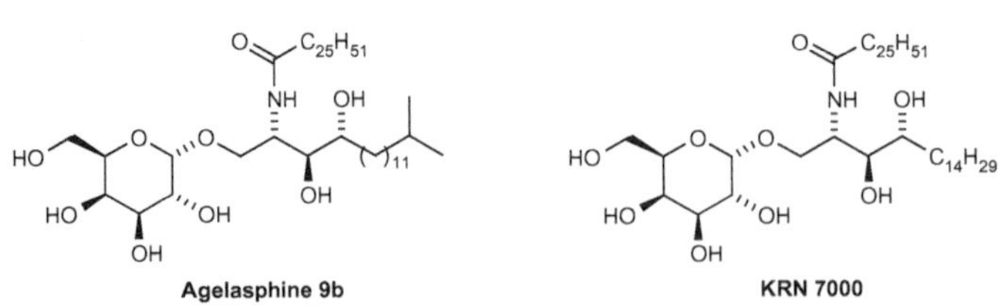

Figure 8.24 Agelasphine 9b and the synthetic analog KRN 7000.

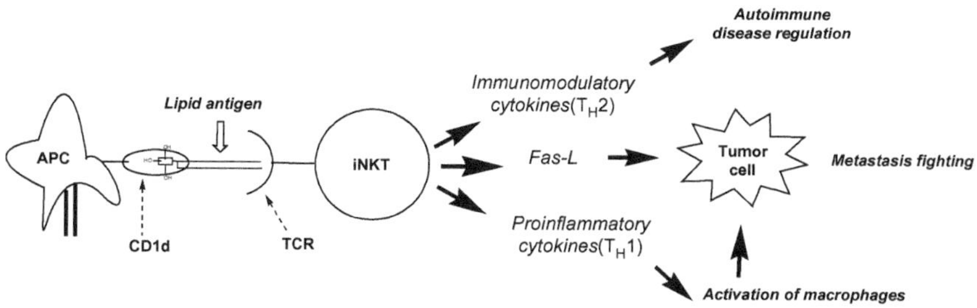

Figure 8.25 Schematic representation of the events triggered by the CD1d/glycolipid/TCR "ternary complex."

interferons, IFN; chemokines; tumor necrosis factor, TNF; colony-stimulating factors, etc.) with a vast array of biochemical implications have been identified. According to their biochemical profile, cytokines can be grouped into proinflammatory cytokines (IFN-γ, IL-2, IL-3, TNF-α, which define a "T-helper 1" (T_H1) type response or cell-mediated immunity), immunomodulatory cytokines (IL-4, IL-5, IL-6, IL-10, which define a T_H2 type response or humoral immunity) and effector cytokines (Fas-L), which provide direct defense against infections and certain types of tumor cells. As far as the topic of this review is concerned, NKT cells are recognized actors of the regulation of tumor immunity by modulation of the so-called tumor immunosurveillance, a part of the dynamic process of interaction between the abnormal cells and the host immune system (also known as "immunoediting") (Ambrosino, Berzofsky, & Terabe, 2008). NKT cells are a heterogeneous cell population with different functions among its multiple subsets. Interestingly, there is a correlation between the NKT cell cytokine profile (T_H1 vs. T_H2), the immune response arising after NKT cell activation and the final disease outcome. Thus, the "quality" of the TCR signal received by NKT cells seems to be a key factor in the regulation of the T_H1/T_H2 balance. In fact, different antigens, with different affinity for TCR or CD1d have been reported to alter the above balance (Oki, Chiba, Yamamura, & Miyake, 2004). Despite the complexity of these intricate processes, NKT cells have been shown to be phenotypically and functionally diverse, depending on their tissue distribution or their surface expression of CD4/CD8 glycoproteins.

Returning to α-GalCer, the synthetic ligand KRN 7000 is one of the most potent stimulators of type I NKT cells and protects mice from tumor growth in several tumor models, via a mild T_H1 skewed response (Crowe et al., 2003). However, its therapeutic use has been hampered by the lack of selectivity towards a selective T_H-type response. This fact, together with the inherent complexity of the immune response, has boosted the search of improved analogs with a more selective cytokine production profile. Since the design of analogs of KRN 7000 has been thoroughly reviewed in a recent and excellent work (Banchet-Cadeddu et al., 2011), in this chapter, we just want to stress on the key structural aspects that have guided the development of this interesting family of compounds.

Taking the structure of KRN 7000 as template, three main points of structural variation have been considered: (a) lipidic chains, (b) aminodiol core, and (c) sugar moiety (Fig. 8.26).

From the large amount of compounds synthesized and tested, some of them have emerged as promising candidates for further development. This is the case of the close KRN 7000 structural analogs OCH, A, and B (Fig. 8.27), showing modifications at the sphingoid chain, the carba-analogs C-KRN and RCAI-56, and compounds C–E, modified at the amide lipidic chain (Fig. 8.27; Banchet-Cadeddu et al., 2011).

By a combination of structure–activity studies and structural data from crystallographic crystallographic data of the human-CD1d/a-GalCer complex (Koch, Stronge, et al., 2005; Zajonc et al., 2005) and molecular modeling considerations (Banchet-Cadeddu et al., 2011; Henon, Dauchez, Haudrechy, & Banchet, 2008), some trends for the T_H1 response can be envisioned. Thus, a glycoside bond, a rigid spacer between the D-galactose and the sphingoid moieties and the presence of aromatic rings as part of the lipidic chains are part of the most prominent structural requirements for T_H1 response. Conversely, the T_H2 response is predominant, for example, in analogs with truncated and/or unsaturated lipid chains or in amide modified analogs (compound C, Fig. 8.27). Needless to say that the precise nature of the cytokine profile arising from a structural modification is still quite unpredictable, since even subtle changes in the glycolipid structure can

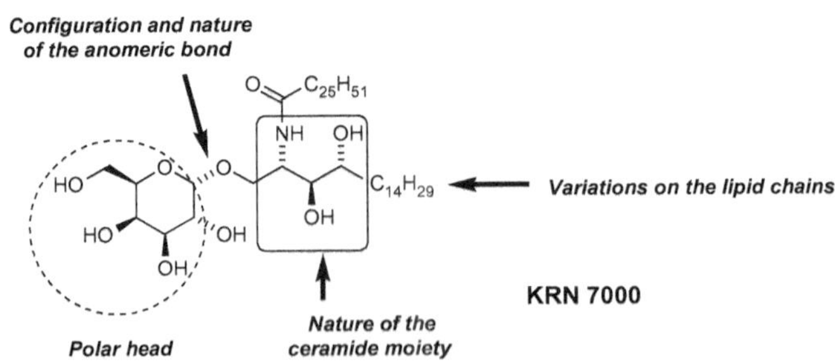

Figure 8.26 Structural modifications on KRN 7000.

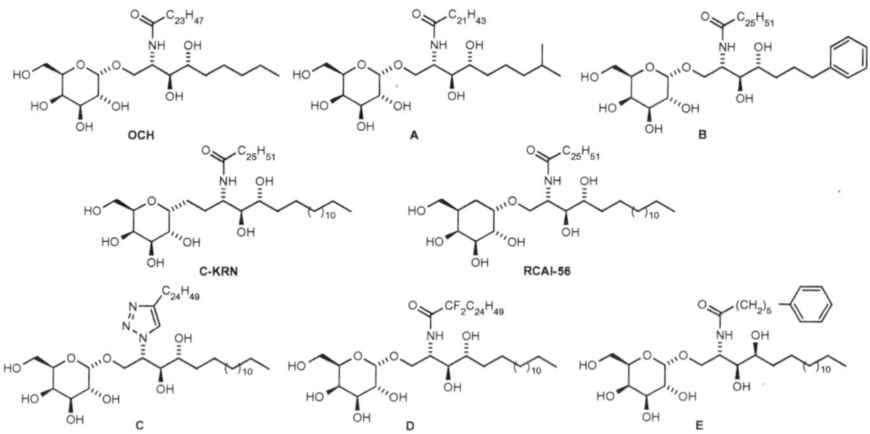

Figure 8.27 Some of the most promising analogs of KRN 7000.

dramatically modify its affinity towards the formation of the binary (CD1d/glycolipid) or ternary (CD1d/glycolipid/iNKT) complexes required for cytokine production. Last, but not least, the biochemical tests that should guide the structural evolution of this type of glycolipids should be performed in a unified model system, preferably on human iNKT cells, due to the observed differences with the mouse cells that are commonly used in preliminary screenings.

7. CONCLUDING REMARKS

Natural products have evolved as the major source of chemical diversity over the past century. As far as medicinal chemistry is concerned, the origin of many of the currently used drugs can be traced from the development of lead compounds arising ultimately from natural products. Despite the recent advances of proteomics and high throughput screening, and the postulated gap between lead finding strategies based on these techniques and those arising from classical approaches (Harvey, 1999), drug discovery is still benefiting from the classical research based on natural products, as evidenced by the nonnegligible number of recently approved drugs in different therapeutic areas (Mishra & Tiwari, 2011).

Although relatively less studied than other families of natural products, SL are important biomolecules with crucial roles in regulating different cellular processes. In addition, alterations in SL metabolism result in the outcome and progression of diseases and also in drug resistance (Gangoiti et al., 2010). From the assumption of the clinical potential of SL-based therapeutics (Fox, Finnegan, Blumenthal, & Kester, 2006), the search of natural sources of SL as scaffolds for the design of new chemical entities with potential therapeutic applications has increased along the past years. However, one inherent limitation of natural SL is their relatively low druggable properties as a result of their low water solubility and the scarce bioavailability associated to their lipidic nature. In this context, several structural modifications (Delgado, Casas, Llebaria, Abad, & Fabrias, 2007; Delgado, Fabrias, Bedia, Casas, & Abad, 2012) or formulations (Kaur et al., 2011) can be devised for the design of more suitable analogs with a better therapeutic potential. In the years to come, the use of nanoscale delivery devices for a more efficient targeting of Cer and related analogs (Barth, Cabot, & Kester, 2011), coupling with the structural modifications of SL-derived natural products will open new perspectives in this fascinating field of research.

ACKNOWLEDGMENTS

Partial financial support from the "Ministerio de Ciencia e Innovación," Spain (Project SAF-2011-2244) and "Generalitat de Catalunya" (Grant 2009SGR-1072) is acknowledged.

REFERENCES

Ahn, E. H., & Schroeder, J. J. (2010). Induction of apoptosis by sphingosine, sphinganine, and C(2)-ceramide in human colon cancer cells, but not by C(2)-dihydroceramide. *Anticancer Research, 30*, 2881–2884.

Aiello, A., Fattorusso, E., Giordano, A., Menna, M., Navarrete, C., & Munoz, E. (2007). Clavaminols A-F, novel cytotoxic 2-amino-3-alkanols from the ascidian Clavelina phlegraea. *Bioorganic & Medicinal Chemistry, 15*, 2920–2926.

Aiello, A., Fattorusso, E., Giordano, A., Menna, M., Navarrete, C., & Munoz, E. (2009). Clavaminols G-N, six new marine sphingoids from the Mediterranean ascidian Clavelina phlegraea. *Tetrahedron, 65*, 4384–4388.

Akao, Y., Banno, Y., Nakagawa, Y., Hasegawa, N., Kim, T. J., Murate, T., et al. (2006). High expression of sphingosine kinase 1 and S1P receptors in chemotherapy-resistant prostate cancer PC3 cells and their camptothecin-induced up-regulation. *Biochemical and Biophysical Research Communications, 342*, 1284–1290.

Akimoto, K., Natori, T., & Morita, M. (1993). Synthesis and stereochemistry of Agelasphin-9b. *Tetrahedron Letters, 34*, 5593–5596.

Alvi, K. A., Jaspars, M., Crews, P., Strulovici, B., & Oto, E. (1994). Penazetidine-A, an alkaloid inhibitor of protein-kinase-C. *Bioorganic & Medicinal Chemistry Letters, 4*, 2447–2450.

Amadasi, A., Bertoldi, M., Contestabile, R., Bettati, S., Cellini, B., di Salvo, M. L., et al. (2007). Pyridoxal 5′-phosphate enzymes as targets for therapeutic agents. *Current Medicinal Chemistry, 14*, 1291–1324.

Ambrosino, E., Berzofsky, J. A., & Terabe, M. (2008). Regulation of tumor immunity: The role of NKT cells. *Expert Opinion on Biological Therapy, 8*, 725–734.

Arenz, C., Thutewohl, M., Block, O., Waldmann, H., Altenbach, H. J., & Giannis, A. (2001). Manumycin A and its analogues are irreversible inhibitors of neutral sphingomyelinase. *Chembiochem, 2*, 141–143.

Baird, R. D., Kitzen, J., Clarke, P. A., Planting, A., Reade, S., Reid, A., et al. (2009). Phase I safety, pharmacokinetic, and pharmacogenomic trail of ES-285, a novel marine cytotoxic agent, administered to adult patients with advanced solid tumors. *Molecular Cancer Therapeutics*, *8*, 1430–1437.

Ballereau, S., Andrieu-Abadie, N., Saffon, N., & Genisson, Y. (2011). Synthesis and biological evaluation of aziridine-containing analogs of phytosphingosine. *Tetrahedron*, *67*, 2570–2578.

Ballereau, S., Baltas, M., & Genisson, Y. (2011). Sphingolipids cyclic derivatives: Occurrence, biological relevance and synthetic approaches. *Current Organic Chemistry*, *15*, 953–986.

Ballereau, S., Levade, T., Genisson, Y., & Andrieu-Abadie, N. (2012). Alteration of ceramide 1-O-functionalization as a promising approach for cancer therapy. *Anti-Cancer Agents in Medicinal Chemistry*, *12*, 316–328.

Banchet-Cadeddu, A., Henon, E., Dauchez, M., Renault, J.-H., Monneaux, F., & Haudrechy, A. (2011). The stimulating adventure of KRN 7000. *Organic and Biomolecular Chemistry*, *9*, 3080–3104.

Bandhuvula, P., Tam, Y. Y., Oskouian, B., & Saba, J. D. (2005). The immune modulator FTY720 inhibits sphingosine-1-phosphate lyase activity. *The Journal of Biological Chemistry*, *280*, 33697–33700.

Barth, B. M., Cabot, M. C., & Kester, M. (2011). Ceramide-based therapeutics for the treatment of cancer. *Anti-Cancer Agents in Medicinal Chemistry*, *11*, 911–919.

Bayerl, M. G., Bruggeman, R. D., Conroy, E. J., Hengst, J. A., King, T. S., Jimenez, M., et al. (2008). Sphingosine kinase 1 protein and mRNA are overexpressed in non-Hodgkin lymphomas and are attractive targets for novel pharmacological interventions. *Leukemia & Lymphoma*, *49*, 948–954.

Bektas, M., Jolly, P. S., Muller, C., Eberle, J., Spiegel, S., & Geilen, C. C. (2005). Sphingosine kinase activity counteracts ceramide-mediated cell death in human melanoma cells: Role of Bcl-2 expression. *Oncogene*, *24*, 178–187.

Berhal, F., Takechi, S., Kumagai, N., & Shibasaki, M. (2011). Catalytic asymmetric amination of N-nonsubstituted alpha-alkoxycarbonyl amides: Concise enantioselective synthesis of mycestericin F and G. *Chemistry European Journal*, *17*, 1915–1921.

Bleicher, R. J., & Cabot, M. C. (2002). Glucosylceramide synthase and apoptosis. *Biochimica et Biophysica Acta*, *1585*, 172–178.

Block, O., Klein, G., Altenbach, H. J., & Brauer, D. J. (2000). New stereoselective route to the epoxyquinol core of manumycin-type natural products. Synthesis of enantiopure (+)-bromoxone, (−)-LL-C10037 alpha, and (+)-KT 8110. *The Journal of Organic Chemistry*, *65*, 716–721.

Boggs, J. M., Gao, W., Zhao, J., Park, H. J., Liu, Y., & Basu, A. (2010). Participation of galactosylceramide and sulfatide in glycosynapses between oligodendrocyte or myelin membranes. *FEBS Letters*, *584*, 1771–1778.

Bonhoure, E., Pchejetski, D., Aouali, N., Morjani, H., Levade, T., Kohama, T., et al. (2006). Overcoming MDR-associated chemoresistance in HL-60 acute myeloid leukemia cells by targeting sphingosine kinase-1. *Leukemia*, *20*, 95–102.

Borg, N. A., Wun, K. S., Kjer-Nielsen, L., Wilce, M. C., Pellicci, D. G., Koh, R., et al. (2007). CD1d-lipid-antigen recognition by the semi-invariant NKT T-cell receptor. *Nature*, *448*, 44–49.

Bose, R., Verheij, M., Haimovitz-Friedman, A., Scotto, K., Fuks, Z., & Kolesnick, R. (1995). Ceramide synthase mediates daunorubicin-induced apoptosis: An alternative mechanism for generating death signals. *Cell*, *82*, 405–414.

Brase, S., Encinas, A., Keck, J., & Nising, C. F. (2009). Chemistry and biology of mycotoxins and related fungal metabolites. *Chemical Reviews*, *109*, 3903–3990.

Brinkmann, V. (2004). FTY720: Mechanism of action and potential benefit in organ transplantation. *Yonsei Medical Journal*, *45*, 991–997.

Brinkmann, V., Billich, A., Baumruker, T., Heining, P., Schmouder, R., Francis, G., et al. (2010). Fingolimod (FTY720): Discovery and development of an oral drug to treat multiple sclerosis. *Nature Reviews. Drug Discovery*, *9*, 883–897.

Brinkmann, V., Davis, M. D., Heise, C. E., Albert, R., Cottens, S., Hof, R., et al. (2002). The immune modulator FTY720 targets sphingosine 1-phosphate receptors. *The Journal of Biological Chemistry*, *277*, 21453–21457.

Brocklyn, J. R. C. (2010). Regulation of cancer cell migration and invasion by sphingosine-1-phosphate. *World Journal of Biological Chemistry*, *1*, 307–312.

Brutkiewicz, R. R. (2006). CD1d ligands: The good, the bad and the ugly. *Journal of Immunology*, *177*, 769–775.

Butler, J., Lana, D., Round, O., & LaMontagne, K. (2004). Functional characterization of sphingosine 1-phosphate receptor agonist in human endothelial cells. *Prostaglandins & Other Lipid Mediators*, *73*, 29–45.

Cakir, Z., Saydam, G., Sahin, F., & Baran, Y. (2011). The roles of bioactive sphingolipids in resveratrol-induced apoptosis in HL60: Acute myeloid leukemia cells. *Journal of Cancer Research and Clinical Oncology*, *137*, 279–286.

Camp, S. M., Bittman, R., Chiang, E. T., Moreno-Vinasco, L., Mirzapoiazova, T., Sammani, S., et al. (2009). Synthetic analogues of FTY720 differentially regulate pulmonary vascular permeability in vivo and in vitro. *The Journal of Pharmacology and Experimental Therapeutics*, *331*, 54–64.

Canals, D., Mormeneo, D., Fabrias, G., Llebaria, A., Casas, J., & Delgado, A. (2009). Synthesis and biological properties of Pachastrissamine (jaspine B) and diastereoisomeric jaspines. *Bioorganic & Medicinal Chemistry*, *17*, 235–241.

Canals, D., Perry, D. M., Jenkins, R. W., & Hannun, Y. A. (2011). Drug targeting of sphingolipid metabolism: Sphingomyelinases and ceramidases. *British Journal of Pharmacology*, *163*, 694–712.

Carlson, E. E. (2010). Natural products as chemical probes. *ACS Chemical Biology*, *5*, 639–653.

Carpinteiro, A., Dumitru, C., Schenck, M., & Gulbins, E. (2008). Ceramide-induced cell death in malignant cells. *Cancer Letters*, *264*, 1–10.

Cha, J. Y., Burnett, G. L., Huang, Y., Davidson, J. B., & Pettus, T. R. R. (2011). A strategy for the late-stage divergent syntheses of scyphostatin analogues. *The Journal of Organic Chemistry*, *76*, 1361–1371.

Claus, R. A., Wustholz, A., Muller, S., Bockmeyer, C. L., Riedel, N. H., Kinscherf, R., et al. (2005). Synthesis and antiapoptotic activity of a novel analogue of the neutral sphingomyelinase inhibitor scyphostatin. *Chembiochem*, *6*, 726–737.

Crowe, N. Y., Uldrich, A. P., Kyparissoudis, K., Hammond, K. J., Hayakawa, Y., Sidobre, S., et al. (2003). Glycolipid antigen drives rapid expansion and sustained cytokine production by NK T cells. *Journal of Immunology*, *171*, 4020–4027.

Cuadros, R., Montejo de Garcini, E., Wandosell, F., Faircloth, G., Fernandez-Sousa, J. M., & Avila, J. (2000). The marine compound spisulosine, an inhibitor of cell proliferation, promotes the disassembly of actin stress fibers. *Cancer Letters*, *152*, 23–29.

Cuvillier, O., Pirianov, G., Kleuser, B., Vanek, P. G., Coso, O. A., Gutkind, S., et al. (1996). Suppression of ceramide-mediated programmed cell death by sphingosine-1-phosphate. *Nature*, *381*, 800–803.

Dalisay, D. S., Tsukamoto, S., & Molinski, T. F. (2009). Absolute configuration of the alpha, omega-bifunctionalized sphingolipid leucettamol A from Leucetta microrhaphis by deconvoluted exciton coupled CD. *Journal of Natural Products*, *72*, 353–359.

Delgado, A., Casas, J., Llebaria, A., Abad, J. L., & Fabrias, G. (2007). Chemical tools to investigate sphingolipid metabolism and functions. *ChemMedChem*, *2*, 580–606.

Delgado, A., Fabrias, G., Bedia, C., Casas, J., & Abad, J. L. (2012). Sphingolipid modulation: A strategy for cancer therapy. *Anti-Cancer Agents in Medicinal Chemistry*, *12*, 285–302.

Desai, K., Sullards, M. C., Allegood, J., Wang, E., Schmelz, E. M., Hartl, M., et al. (2002). Fumonisins and fumonisin analogs as inhibitors of ceramide synthase and inducers of apoptosis. *Biochimica et Biophysica Acta*, *1585*, 188–192.

Dillehay, D. L., Webb, S. K., Schmelz, E. M., & Merrill, A. H., Jr. (1994). Dietary sphingomyelin inhibits 1,2-dimethylhydrazine-induced colon cancer in CF1 mice. *The Journal of Nutrition*, *124*, 615–620.

Duan, R. D., & Nilsson, A. (2009). Metabolism of sphingolipids in the gut and its relation to inflammation and cancer development. *Progress in Lipid Research*, *48*, 62–72.

Esteve, J., Lorente, A., Romea, P., Urpi, F., Rios-Luci, C., & Padron, J. M. (2011). Synthesis and biological evaluation of 1-deoxy-5-hydroxysphingosine derivatives. *European Journal of Organic Chemistry*, *2011*, 960–967.

Fabrias, G., Munoz-Olaya, J., Cingolani, F., Signorelli, P., Casas, J., Gagliostro, V., et al. (2011). Dihydroceramide desaturase and dihydrosphingolipids: Debutant players in the sphingolipid arena. *Progress in Lipid Research*, *51*, 82–94.

Faugeroux, V., Genisson, Y., Andrieu-Abadie, N., Colie, S., Levade, T., & Baltas, M. (2006). C-alkyl 5-membered ring imino sugars as new potent cytotoxic glucosylceramide synthase inhibitors. *Organic and Biomolecular Chemistry*, *4*, 4437–4439.

Fedorov, S. N., Makarieva, T. N., Guzii, A. G., Shubina, L. K., Kwak, J. Y., & Stonik, V. A. (2009). Marine two-headed sphingolipid-like compound rhizochalin inhibits EGF-induced transformation of JB6 P+ Cl41 cells. *Lipids*, *44*, 777–785.

Feng, X. J., Yang, X. Y., Luo, Y., Li, X., Tang, W., Zuo, J. P., et al. (2012). Synthesis and immunomodulating activity of new analogues of fingolimod. *Archiv der Pharmazie*, *345*, 93–100.

Fox, T. E., Finnegan, C. M., Blumenthal, R., & Kester, M. (2006). The clinical potential of sphingolipid-based therapeutics. *Cellular and Molecular Life Sciences*, *63*, 1017–1023.

French, K. J., Schrecengost, R. S., Lee, B. D., Zhuang, Y., Smith, S. N., Eberly, J. L., et al. (2003). Discovery and evaluation of inhibitors of human sphingosine kinase. *Cancer Research*, *63*, 5962–5969.

French, K. J., Upson, J. J., Keller, S. N., Zhuang, Y., Yun, J. K., & Smith, C. D. (2006). Antitumor activity of sphingosine kinase inhibitors. *The Journal of Pharmacology and Experimental Therapeutics*, *318*, 596–603.

Freyer, A. J., Patil, A. D., Killmer, L., Troupe, N., Mentzer, M., Carte, B., et al. (1997). Three new pseudodistomins, piperidine alkaloids from the ascidian Pseudodistoma megalarva. *Journal of Natural Products*, *60*, 986–990.

Fujita, T., Inoue, K., Yamamoto, S., Ikumoto, T., Sasaki, S., Toyama, R., et al. (1994). Fungal metabolites. Part 11. A potent immunosuppressive activity found in Isaria sinclairii metabolite. *The Journal of Antibiotics*, *47*, 208–215.

Furuya, H., Shimizu, Y., & Kawamori, T. (2011). Sphingolipids in cancer. *Cancer Metastasis Reviews*, *30*, 567–576.

Futerman, A. H., & Hannun, Y. A. (2004). The complex life of simple sphingolipids. *EMBO Reports*, *5*, 777–782.

Fyrst, H., & Saba, J. D. (2010). An update on sphingosine-1-phosphate and other sphingolipid mediators. *Nature Chemical Biology*, *6*, 489–497.

Gangoiti, P., Camacho, L., Arana, L., Ouro, A., Granado, M. H., Brizuela, L., et al. (2010). Control of metabolism and signaling of simple bioactive sphingolipids: Implications in disease. *Progress in Lipid Research*, *49*, 316–334.

Garnier-Amblard, E. C., Mays, S. G., Arrendale, R. F., Baillie, M. T., Bushnev, A. S., Culver, D. G., et al. (2011). Novel synthesis and biological evaluation of enigmols as therapeutic agents for treating prostate cancer. *ACS Medicinal Chemistry Letters*, *2*, 438–443.

Garrido, L., Zubia, E., Ortega, M. J., Naranjo, S., & Salva, J. (2001). Obscuraminols, new unsaturated amino alcohols from the tunicate Pseudodistoma obscurum: Structure and absolute configuration. *Tetrahedron*, *57*, 4579–4588.

Gnewuch, C., Jaques, G., Havemann, K., & Wiegandt, H. (1994). Re-assessment of acidic glycosphingolipids in small-cell-lung-cancer tissues and cell lines. *International Journal of Cancer. Supplement*, *8*, 125–126.

Gouaze, V., Liu, Y. Y., Prickett, C. S., Yu, J. Y., Giuliano, A. E., & Cabot, M. C. (2005). Glucosylceramide synthase blockade down-regulates P-glycoprotein and resensitizes multidrug-resistant breast cancer cells to anticancer drugs. *Cancer Research*, *65*, 3861–3867.

Gouaze, V. V., Yu, J. Y., Bleicher, R. J., Han, T. Y., Liu, Y. Y., Wang, H., et al. (2004). Overexpression of glucosylceramide synthase and P-glycoprotein in cancer cells selected for resistance to natural product chemotherapy. *Molecular Cancer Therapeutics*, *3*, 633–640.

Gouaze-Andersson, V., & Cabot, M. C. (2006). Glycosphingolipids and drug resistance. *Biochimica et Biophysica Acta*, *1758*, 2096–2103.

Grammatikos, G., Teichgraber, V., Carpinteiro, A., Trarbach, T., Weller, M., Hengge, U. R., et al. (2007). Overexpression of acid sphingomyelinase sensitizes glioma cells to chemotherapy. *Antioxidants & Redox Signaling*, *9*, 1449–1456.

Gustafsson, K., Sander, B., Bielawski, J., Hannun, Y. A., & Flygare, J. (2009). Potentiation of cannabinoid-induced cytotoxicity in mantle cell lymphoma through modulation of ceramide metabolism. *Molecular Cancer Research*, 7, 1086–1098.

Hakomori, S. (1985). Aberrant glycosylation in cancer cell membranes as focused on glycolipids: Overview and perspectives. *Cancer Research*, *45*, 2405–2414.

Hakomori, S., & Handa, K. (2002). Glycosphingolipid-dependent cross-talk between glycosynapses interfacing tumor cells with their host cells: Essential basis to define tumor malignancy. *FEBS Letters*, *531*, 88–92.

Hanada, K. (2003). Serine palmitoyltransferase, a key enzyme of sphingolipid metabolism. *Biochimica et Biophysica Acta*, *1632*, 16–30.

Hanai, N., Nakamura, K., & Shitara, K. (2000). Recombinant antibodies against ganglioside expressed on tumor cells. *Cancer Chemotherapy and Pharmacology*, *46*(Suppl.), S13–S17.

Hanessian, S., Charron, G., Billich, A., & Guerini, D. (2007). Constrained azacyclic analogues of the immunomodulatory agent FTY720 as molecular probes for sphingosine 1-phosphate receptors. *Bioorganic & Medicinal Chemistry Letters*, *17*, 491–494.

Hannun, Y. A., & Obeid, L. M. (2008). Principles of bioactive lipid signalling: Lessons from sphingolipids. *Nature Reviews. Molecular Cell Biology*, *9*, 139–150.

Harvey, A. L. (1999). Medicines from nature: Are natural products still relevant to drug discovery? *Trends in Pharmacological Sciences*, *20*, 196–198.

Henon, E., Dauchez, M., Haudrechy, A., & Banchet, A. (2008). Molecular dynamics simulation study on the interaction of KRN 7000 and three analogues with human CD1d. *Tetrahedron*, *64*, 9480–9489.

Henry, B., Moller, C., Dimanche-Boitrel, M. T., Gulbins, E., & Becker, K. A. (2011). Targeting the ceramide system in cancer. *Cancer Letters*, 2011 Jul 23. [Epub ahead of print].

Hiraiwa, N., Fukuda, Y., Imura, H., Tadano-Aritomi, K., Nagai, K., Ishizuka, I., et al. (1990). Accumulation of highly acidic sulfated glycosphingolipids in human hepatocellular carcinoma defined by a series of monoclonal antibodies. *Cancer Research*, *50*, 2917–2928.

Hogenauer, K., Hinterding, K., & Nussbaumer, P. (2010). S1P receptor mediated activity of FTY720 phosphate mimics. *Bioorganic & Medicinal Chemistry Letters*, *20*, 1485–1487.

Humpf, H.-U., Schmelz, E.-M., Meredith, F. I., Vesper, H., Vales, T. R., Wang, E., et al. (1998). Acylation of naturally occurring and synthetic 1-deoxysphinganines by ceramide synthase. *The Journal of Biological Chemistry*, *273*, 19060–19064.

Igarashi, N., Okada, T., Hayashi, S., Fujita, T., Jahangeer, S., & Nakamura, S. (2003). Sphingosine kinase 2 is a nuclear protein and inhibits DNA synthesis. *The Journal of Biological Chemistry*, *278*, 46832–46839.

Jareserijman, E. A., Bapat, C. P., Lithgowbertelloni, A., Rinehart, K. L., & Sakai, R. (1993). Crucigasterins, new polyunsaturated amino-alcohols from the Mediterranean tunicate Pseudodistoma-Crucigaster. *The Journal of Organic Chemistry*, *58*, 5732–5737.

Jarvis, W. D., Fornari, F. A., Traylor, R. S., Martin, H. A., Kramer, L. B., Erukulla, R. K., et al. (1996). Induction of apoptosis and potentiation of ceramide-mediated cytotoxicity by sphingoid bases in human myeloid leukemia cells. *The Journal of Biological Chemistry*, *271*, 8275–8284.

Jeon, H., Bae, H., Baek, D. J., Kwak, Y.-S., Kim, D., & Kim, S. (2011). Syntheses of sulfur and selenium analogues of pachastrissamine via double displacements of cyclic sulfate. *Organic and Biomolecular Chemistry*, *9*, 7237–7242.

Jimenez, C., & Crews, P. (1990). Novel marine sponge amino acids, 10. Xestoaminols from Xestospongia sp. *Journal of Natural Products*, *53*, 978–982.

Johnson, K. R., Johnson, K. Y., Crellin, H. G., Ogretmen, B., Boylan, A. M., Harley, R. A., et al. (2005). Immunohistochemical distribution of sphingosine kinase 1 in normal and tumor lung tissue. *The Journal of Histochemistry and Cytochemistry*, *53*, 1159–1166.

Kandyba, A. G., Kobliakov, V. A., Kozlov, A. M., Nagaev, I. Y., Shevchenko, V. P., & Dyatlovitskaya, E. V. (2002). Dihydroceramide desaturase activity in tumors. *Biochemistry (Mosc)*, *67*, 597–599.

Kandyba, A. G., Somova, O. G., Kozlov, A. M., Zubova, E. S., Dudnik, L. B., Alessenko, A. V., et al. (2000). Sphingolipids of transplantable rat nephroma-RA. *Biochemistry (Mosc)*, *65*, 703–706.

Karahatay, S., Thomas, K., Koybasi, S., Senkal, C. E., Elojeimy, S., Liu, X., et al. (2007). Clinical relevance of ceramide metabolism in the pathogenesis of human head and neck squamous cell carcinoma (HNSCC): Attenuation of C(18)-ceramide in HNSCC tumors correlates with lymphovascular invasion and nodal metastasis. *Cancer Letters*, *256*, 101–111.

Kaur, R., Chen, J., Dawoodji, A., Cerundolo, V., Garcia-Diaz, Y. R., Wojno, J., et al. (2011). Preparation, characterisation and entrapment of a non-glycosidic threitol ceramide into liposomes for presentation to invariant natural killer T cells. *Journal of Pharmaceutical Sciences*, *100*, 2724–2733.

Kawamori, T., Osta, W., Johnson, K. R., Pettus, B. J., Bielawski, J., Tanaka, T., et al. (2006). Sphingosine kinase 1 is up-regulated in colon carcinogenesis. *The FASEB Journal*, *20*, 386–388.

Khanal, P., Kang, B. S., Yun, H. J., Cho, H.-G., Makarieva, T. N., & Choi, H. S. (2011). Aglycon of rhizochalin from the Rhizochalina incrustata induces apoptosis via activation of AMP-activated protein kinase in HT-29 colon cancer cells. *Biological and Pharmaceutical Bulletin*, *34*, 1553–1558.

Kiuchi, M., Adachi, K., Kohara, T., Minoguchi, M., Hanano, T., Aoki, Y., et al. (2000). Synthesis and immunosuppressive activity of 2-substituted 2-aminopropane-1,3-diols and 2-aminoethanols. *Journal of Medicinal Chemistry*, *43*, 2946–2961.

Knapp, P., Baranowski, M., Knapp, M., Zabielski, P., Blachnio-Zabielska, A. U., & Gorski, J. (2010). Altered sphingolipid metabolism in human endometrial cancer. *Prostaglandins & Other Lipid Mediators*, *92*, 62–66.

Ko, R. Y. Y., Chu, J. C. K., & Chiu, P. (2011). Synthesis of fluorinated analogues of the immunosuppressive drug FTY720. *Tetrahedron*, *67*, 2542–2547.

Kobayashi, J., Cheng, J., Ishibashi, M., Walchli, M. R., Yamamura, S., & Ohizumi, Y. (1991). Penaresidin-a and Penaresidin-B, 2 novel azetidine alkaloids with potent actomyosin atpase-activating activity from the Okinawan marine sponge Penares sp. *Journal of Chemical Society, Perkin Transactions*, *1*, 1135–1137.

Kobayashi, J., Naitoh, K., Doi, Y., Deki, K., & Ishibashi, M. (1995). Pseudodistomin-C, a new piperidine alkaloid with unusual absolute-configuration from the Okinawan tunicate Pseudodistoma-Kanoko. *The Journal of Organic Chemistry*, *60*, 6941–6945.

Koch, M. A., Schuffenhauer, A., Scheck, M., Wetzel, S., Casaulta, M., Odermatt, A., et al. (2005). Charting biologically relevant chemical space: A structural classification of natural products (SCONP). *Proceedings of the National Academy of Sciences of the United States of America, 102*, 17272–17277.

Koch, M., Stronge, V. S., Shepherd, D., Gadola, S. D., Mathew, B., Ritter, G., et al. (2005). The crystal structure of human CD1d with and without alpha-galactosylceramide. *Nature Immunology, 6*, 819–826.

Kolesnick, R. C. (2002). The therapeutic potential of modulating the ceramide/sphingomyelin pathway. *The Journal of Clinical Investigation, 110*, 3–8.

Kolesnick, R., & Fuks, Z. (2003). Radiation and ceramide-induced apoptosis. *Oncogene, 22*, 5897–5906.

Kombarov, R., Altieri, A., Genis, D., Kirpichenok, M., Kochubey, V., Rakitina, N., et al. (2010). BioCores: Identification of a drug/natural product-based privileged structural motif for small-molecule lead discovery. *Molecular Diversity, 14*, 193–200.

Kong, F. H., & Faulkner, D. J. (1993). Leucettamol-A and leucettamol-B, 2 antimicrobial lipids from the calcareous sponge leucetta-microraphis. *The Journal of Organic Chemistry, 58*, 970–971.

Kossuga, M. H., MacMillan, J. B., Rogers, E. W., Molinski, T. F., Nascimento, G. G. F., Rocha, R. M., et al. (2004). (2S,3R)-2-aminododecan-3-ol, a new antifungal agent from the ascidian Clavelina oblonga. *Journal of Natural Products, 67*, 1879–1881.

Kraveka, J. M., Li, L., Szulc, Z. M., Bielawski, J., Ogretmen, B., Hannun, Y. A., et al. (2007). Involvement of dihydroceramide desaturase in cell cycle progression in human neuroblastoma cells. *The Journal of Biological Chemistry, 282*, 16718–16728.

Kuroda, I., Musman, M., Ohtani, I. I., Ichiba, T., Tanaka, J., Gravalos, D. G., et al. (2002). Pachastrissamine, a cytotoxic anhydrophytosphingosine from a marine sponge, Pachastrissa sp. *Journal of Natural Products, 65*, 1505–1506.

Kutuk, O., Pedrech, A., Harrison, P., & Basaga, H. (2005). Pramanicin induces apoptosis in Jurkat leukemia cells: A role for JNK, p38 and caspase activation. *Apoptosis, 10*, 597–609.

Ledroit, V., Debitus, C., Lavaud, C., & Massiot, G. (2003). Jaspines A and B: Two new cytotoxic sphingosine derivatives from the marine sponge Jaspis sp. *Tetrahedron Letters, 44*, 225–228.

Lemonnier, L. A., Dillehay, D. L., Vespremi, M. J., Abrams, J., Brody, E., & Schmelz, E. M. (2003). Sphingomyelin in the suppression of colon tumors: Prevention versus intervention. *Archives of Biochemistry and Biophysics, 419*, 129–138.

Levi, M., Meijler, M. M., Gomez-Munoz, A., & Zor, T. (2009). Distinct receptor-mediated activities in macrophages for natural ceramide-1-phosphate (C1P) and for phosphoceramide analogue-1 (PCERA-1). *Molecular and Cellular Endocrinology, 314*, 248–255.

Li, R., Villacreses, N., & Ladisch, S. (1995). Human tumor gangliosides inhibit murine immune responses in vivo. *Cancer Research, 55*, 211–214.

Lim, K. G., Sun, C., Bittman, R., Pyne, N. J., & Pyne, S. (2011). (R)-FTY720 methyl ether is a specific sphingosine kinase 2 inhibitor: Effect on sphingosine kinase 2 expression in HEK 293 cells and actin rearrangement and survival of MCF-7 breast cancer cells. *Cellular Signalling, 23*, 1590–1595.

Lim, K. G., Tonelli, F., Li, Z., Lu, X., Bittman, R., Pyne, S., et al. (2011). FTY720 analogues as sphingosine kinase 1 inhibitors: Enzyme inhibition kinetics, allosterism, proteasomal degradation and actin rearrangement in MCF-7 breast cancer cells. *The Journal of Biological Chemistry, 286*, 18640.

Liu, Y., Chen, Y., Momin, A., Shaner, R., Wang, E., Bowen, N. J., et al. (2010). Elevation of sulfatides in ovarian cancer: An integrated transcriptomic and lipidomic analysis including tissue-imaging mass spectrometry. *Molecular Cancer, 9*, 186.

Liu, Y. Y., Han, T. Y., Giuliano, A. E., & Cabot, M. C. (1999). Expression of glucosylceramide synthase, converting ceramide to glucosylceramide, confers

adriamycin resistance in human breast cancer cells. *The Journal of Biological Chemistry, 274,* 1140–1146.
Liu, Y. Y., Han, T. Y., Giuliano, A. E., & Cabot, M. C. (2001). Ceramide glycosylation potentiates cellular multidrug resistance. *The FASEB Journal, 15,* 719–730.
Liu, Y. Y., Han, T. Y., Yu, J. Y., Bitterman, A., Le, A., Giuliano, A. E., et al. (2004). Oligonucleotides blocking glucosylceramide synthase expression selectively reverse drug resistance in cancer cells. *Journal of Lipid Research, 45,* 933–940.
Lu, X., Sun, C., Valentine, W. J., Shuyu, E., Liu, J., Tigyi, G., et al. (2009). Chiral vinylphosphonate and phosphonate analogues of the immunosuppressive agent FTY720. *The Journal of Organic Chemistry, 74,* 3192–3195.
Maceyka, M., Harikumar, K. B., Milstien, S., & Spiegel, S. C. (2012). Sphingosine-1-phosphate signaling and its role in disease. *Trends in Cell Biology, 22,* 50–60.
Maceyka, M., Sankala, H., Hait, N. C., Le Stunff, H., Liu, H., Toman, R., et al. (2005). Sphk1 and Sphk2: Sphingosine kinase isoenzymes with opposing functions in sphingolipid metabolism. *The Journal of Biological Chemistry, 280,* 37118–37129.
Makarieva, T. N., Guzii, A. G., Denisenko, V. A., Dmitrenok, P. S., & Stonik, V. A. (2008). New two-headed sphingolipid-like compounds from the marine sponge Oceanapia sp. *Russian Chemical Bulletin, 57,* 669–673.
Mandala, S. M., Thornton, R. A., Frommer, B. R., Dreikorn, S., & Kurtz, M. B. (1997). Viridiofungins, novel inhibitors of sphingolipid synthesis. *The Journal of Antibiotics, 50,* 339–343.
Mao, Z., Sun, W., Xu, R., Novgorodov, S., Szulc, Z. M., Bielawski, J., et al. (2010). Alkaline ceramidase 2 (ACER2) and its product dihydrosphingosine mediate the cytotoxicity of N-(4-hydroxyphenyl)retinamide in tumor cells. *The Journal of Biological Chemistry, 285,* 29078–29090.
Maurer, B. J., Metelitsa, L. S., Seeger, R. C., Cabot, M. C., & Reynolds, C. P. (1999). Increase of ceramide and induction of mixed apoptosis/necrosis by N-(4-hydroxyphenyl)-retinamide in neuroblastoma cell lines. *Journal of the National Cancer Institute, 91,* 1138–1146.
Merchant, T. E., Kasimos, J. N., de Graaf, P. W., Minsky, B. D., Gierke, L. W., & Glonek, T. (1991). Phospholipid profiles of human colon cancer using 31P magnetic resonance spectroscopy. *International Journal of Colorectal Disease, 6,* 121–126.
Merrill, A. H., Jr. (2002). De novo sphingolipid biosynthesis: A necessary, but dangerous, pathway. *The Journal of Biological Chemistry, 277,* 25843–25846 Epub 22002 May 25813.
Merrill, A. H., Jr., Sullards, M. C., Wang, E., Voss, K. A., & Riley, R. T. (2001). Sphingolipid metabolism: Roles in signal transduction and disruption by fumonisins. *Environmental Health Perspectives, 109*(Suppl. 2), 283–289.
Merrill, A. H., Jr., van Echten, G., Wang, E., & Sandhoff, K. (1993). Fumonisin B1 inhibits sphingosine (sphinganine) N-acyltransferase and de novo sphingolipid biosynthesis in cultured neurons in situ. *The Journal of Biological Chemistry, 268,* 27299–27306.
Meyer, E. V. S., Holt, J. J., Girard, K. R., Ballie, M. T., Bushnev, A. S., Lapp, S., et al. (2012). Sphingolipid analogues inhibit development of malaria parasites. *ACS Medicinal Chemistry Letters, 3,* 43–47.
Mishra, B. B., & Tiwari, V. K. (2011). Natural products: An evolving role in future drug discovery. *European Journal of Medicinal Chemistry, 46,* 4769–4807.
Mitra, P., Maceyka, M., Payne, S. G., Lamour, N., Milstien, S., Chalfant, C. E., et al. (2007). Ceramide kinase regulates growth and survival of A549 human lung adenocarcinoma cells. *FEBS Letters, 581,* 735–740.
Modrak, D. E., Cardillo, T. M., Newsome, G. A., Goldenberg, D. M., & Gold, D. V. (2004). Synergistic interaction between sphingomyelin and gemcitabine potentiates ceramide-mediated apoptosis in pancreatic cancer. *Cancer Research, 64,* 8405–8410.

Modrak, D. E., Lew, W., Goldenberg, D. M., & Blumenthal, R. (2000). Sphingomyelin potentiates chemotherapy of human cancer xenografts. *Biochemical and Biophysical Research Communications, 268*, 603–606.

Morichika, H., Hamanaka, Y., Tai, T., & Ishizuka, I. (1996). Sulfatides as a predictive factor of lymph node metastasis in patients with colorectal adenocarcinoma. *Cancer, 78*, 43–47.

Morita, M., Motoki, K., Akimoto, K., Natori, T., Sakai, T., Sawa, E., et al. (1995). Structure-activity relationship of alpha-galactosylceramides against B16-bearing mice. *Journal of Medicinal Chemistry, 38*, 2176–2187.

Morjani, H., Aouali, N., Belhoussine, R., Veldman, R. J., Levade, T., & Manfait, M. (2001). Elevation of glucosylceramide in multidrug-resistant cancer cells and accumulation in cytoplasmic droplets. *International Journal of Cancer, 94*, 157–165.

Morokuma, K., Takahashi, K., Ishihara, J., & Hatakeyama, S. (2005). Total synthesis of viridiofungin A. *Chemical Communications (Cambridge, England)*, 2265–2267.

Mullen, T. D., Hannun, Y. A., & Obeid, L. M. (2012). Ceramide synthases at the centre of sphingolipid metabolism and biology. *The Biochemical Journal, 441*, 789–802.

Murakami, M., Ito, H., Hagiwara, K., Yoshida, K., Sobue, S., Ichihara, M., et al. (2010). ATRA inhibits ceramide kinase transcription in a human neuroblastoma cell line, SH-SY5Y cells: The role of COUP-TFI. *Journal of Neurochemistry, 112*, 511–520.

Nagaoka, Y., Otsuki, K., Fujita, T., & Uesato, S. (2008). Effects of phosphorylation of immunomodulatory agent FTY720 (fingolimod) on antiproliferative activity against breast and colon cancer cells. *Biological and Pharmaceutical Bulletin, 31*, 1177–1181.

Nakayama, S., Uto, Y., Tanimoto, K., Okuno, Y., Sasaki, Y., Nagasawa, H., et al. (2008). TX-2152: A conformationally rigid and electron-rich diyne analogue of FTY720 with in vivo antiangiogenic activity. *Bioorganic & Medicinal Chemistry, 16*, 7705–7714.

Natori, T., Koezuka, Y., & Higa, T. (1993). Agelasphins, novel alpha-galactosylceramides from the marine sponge *Agelas mauritianus*. *Tetrahedron Letters, 34*, 5591–5592.

Natori, T., Morita, M., Akimoto, K., & Koezuka, Y. (1994). Agelasphins, novel antitumor and immunostimulatory cerebrosides from the marine sponge *Agelas mauritianus*. *Tetrahedron, 50*, 2771–2784.

Nava, V. E., Hobson, J. P., Murthy, S., Milstien, S., & Spiegel, S. (2002). Sphingosine kinase type 1 promotes estrogen-dependent tumorigenesis of breast cancer mcf-7 cells. *Experimental Cell Research, 281*, 115–127.

Nicholas, G. M., Eckman, L. L., Newton, G. L., Fahey, R. C., Ray, S., & Bewley, C. A. (2003). Inhibition and kinetics of mycobacterium tuberculosis and mycobacterium smegmatis mycothiol-S-conjugate amidase by natural product inhibitors. *Bioorganic & Medicinal Chemistry, 11*, 601–608.

Nicholas, G. M., Hong, T. W., Molinski, T. F., Lerch, M. L., Cancilla, M. T., & Lebrilla, C. B. (1999). Oceanapiside, an antifungal bis-alpha, omega-amino alcohol glycoside from the marine sponge Oceanapia phillipensis. *Journal of Natural Products, 62*, 1678–1681.

Nicholas, G. M., Li, R., MacMillan, J. B., & Molinski, T. F. (2002). Antifungal activity of bifunctional sphingolipids. Intramolecular synergism within long-chain alpha, omega-bis-aminoalcohols. *Bioorganic & Medicinal Chemistry Letters, 12*, 2159–2162.

Ogretmen, B., & Hannun, Y. A. (2001). Updates on functions of ceramide in chemotherapy-induced cell death and in multidrug resistance. *Drug Resistance Updates, 4*, 368–377.

Ohfune, Y., & Shinada, T. (2005). Enantio- and diastereoselective construction of alpha, alpha-disubstituted alpha-amino acids for the synthesis of biologically active compounds. *European Journal of Organic Chemistry*, 2005, 5127–5143.

Ohshita, K., Ishiyama, H., Takahashi, Y., Ito, J., Mikami, Y., & Kobayashi, J. (2007). Synthesis of penaresidin derivatives and its biological activity. *Bioorganic & Medicinal Chemistry, 15*, 4910–4916.

Ohta, H., Sweeney, E. A., Masamune, A., Yatomi, Y., Hakomori, S., & Igarashi, Y. (1995). Induction of apoptosis by sphingosine in human leukemic HL-60 cells: A possible endogenous modulator of apoptotic DNA fragmentation occurring during phorbol ester-induced differentiation. *Cancer Research*, *55*, 691–697.

Oki, S., Chiba, A., Yamamura, T., & Miyake, S. (2004). The clinical implication and molecular mechanism of preferential IL-4 production by modified glycolipid-stimulated NKT cells. *The Journal of Clinical Investigation*, *113*, 1631–1640.

Othman, A., Ruetti, M. F., Ernst, D., Saely, C. H., Rein, P., Drexel, H., et al. (2012). Plasma deoxysphingolipids: A novel class of biomarkers for the metabolic syndrome? *Diabetologia*, *55*, 421–431.

Payne, S. G., Milstien, S., & Spiegel, S. (2002). Sphingosine-1-phosphate: Dual messenger functions. *FEBS Letters*, *531*, 54–57.

Perry, D. K. (2002). Serine palmitoyltransferase: Role in apoptotic de novo ceramide synthesis and other stress responses. *Biochimica et Biophysica Acta*, *1585*, 146–152.

Pitsinos, E. N., Wascholowski, V., Karaliota, S., Rigou, C., Couladouros, E. A., & Giannis, A. (2003). Synthesis and evaluation of three novel scyphostatin analogues as neutral sphingomyelinase inhibitors. *Chembiochem*, *4*, 1223–1225.

Ponnusamy, S., Meyers-Needham, M., Senkal, C. E., Saddoughi, S. A., Sentelle, D., Selvam, S. P., et al. (2010). Sphingolipids and cancer: Ceramide and sphingosine-1-phosphate in the regulation of cell death and drug resistance. *Future Oncology*, *6*, 1603–1624.

Pruett, S. T., Bushnev, A., Hagedorn, K., Adiga, M., Haynes, C. A., Sullards, M. C., et al. (2008). Thematic review series: Sphingolipids. Biodiversity of sphingoid bases ("sphingosines") and related amino alcohols. *Journal of Lipid Research*, *49*, 1621–1639.

Pyne, S., Bittman, R., & Pyne, N. J. (2011). Sphingosine kinase inhibitors and cancer: Seeking the golden sword of hercules. *Cancer Research*, *71*, 6576–6582.

Pyne, N. J., & Pyne, S. (2010). Sphingosine 1-phosphate and cancer. *Nature Reviews. Cancer*, *10*, 489–503.

Pyne, N. J., Tonelli, F., Lim, K. G., Long, J. S., Edwards, J., & Pyne, S. (2012). Sphingosine 1-phosphate signalling in cancer. *Biochemical Society Transactions*, *40*, 94–100.

Reina, E., Camacho, L., Casas, J., Van Veldhoven, P. P., & Fabrias, G. (2012). Determination of sphingosine-1-phosphate lyase activity by gas chromatography coupled to electron impact mass spectrometry. *Chemistry and Physics of Lipids*, *165*, 225–231.

Ribes, C., Falomir, E., Murga, J., Carda, M., & Marco, J. A. (2009a). Convergent, stereoselective syntheses of the glycosidase inhibitors broussonetines C, O and P. *Tetrahedron*, *65*, 10612–10616.

Ribes, C., Falomir, E., Murga, J., Carda, M., & Marco, J. A. (2009b). Convergent, stereoselective syntheses of the glycosidase inhibitors broussonetines D and M. *Organic and Biomolecular Chemistry*, 7, 1355–1360.

Riboni, L., Campanella, R., Bassi, R., Villani, R., Gaini, S. M., Martinelli-Boneschi, F., et al. (2002). Ceramide levels are inversely associated with malignant progression of human glial tumors. *Glia*, *39*, 105–113.

Rives, A., Ladeira, S., Levade, T., Andrieu-Abadie, N., & Génisson, Y. (2010). Synthesis of cytotoxic aza analogues of jaspine B. *The Journal of Organic Chemistry*, *75*, 7920–7923.

Ruckhaberle, E., Karn, T., Hanker, L., Gatje, R., Metzler, D., Holtrich, U., et al. (2008). Prognostic relevance of glucosylceramide synthase (GCS) expression in breast cancer. *Journal of Cancer Research and Clinical Oncology*, *135*, 81–90.

Ruckhaberle, E., Karn, T., Rody, A., Hanker, L., Gatje, R., Metzler, D., et al. (2009). Gene expression of ceramide kinase, galactosyl ceramide synthase and ganglioside GD3 synthase is associated with prognosis in breast cancer. *Journal of Cancer Research and Clinical Oncology*, *135*, 1005–1013.

Ruckhaberle, E., Rody, A., Engels, K., Gaetje, R., von Minckwitz, G., Schiffmann, S., et al. (2008). Microarray analysis of altered sphingolipid metabolism reveals prognostic significance of sphingosine kinase 1 in breast cancer. *Breast Cancer Research and Treatment, 112*, 41–52.

Ryland, L. K., Fox, T. E., Liu, X., Loughran, T. P., & Kester, M. (2011). Dysregulation of sphingolipid metabolism in cancer. *Cancer Biology & Therapy, 11*, 138–149.

Rylova, S. N., Somova, O. G., & Dyatlovitskaya, E. V. (1998). Comparative investigation of sphingoid bases and fatty acids in ceramides and sphingomyelins from human ovarian malignant tumors and normal ovary. *Biochemistry (Mosc), 63*, 1057–1060.

Sakakibara, N., Gasa, S., Kamio, K., Makita, A., Nonomura, K., Togashi, M., et al. (1991). Distinctive glycolipid patterns in Wilms' tumor and renal cell carcinoma. *Cancer Letters, 57*, 187–192.

Salcedo, M., Cuevas, C., Alonso, J. L., Otero, G., Faircloth, G., Fernandez-Sousa, J. M., et al. (2007). The marine sphingolipid-derived compound ES 285 triggers an atypical cell death pathway. *Apoptosis, 12*, 395–409.

Salma, Y., Ballereau, S., Ladeira, S., Lepetit, C., Chauvin, R., Andrieu-Abadie, N., et al. (2011). Single- and double-chained truncated jaspine B analogues: Asymmetric synthesis, biological evaluation and theoretical study of an unexpected 5-endo-dig process. *Tetrahedron, 67*, 4253–4262.

Salma, Y., Lafont, E., Therville, N., Carpentier, S., Bonnafe, M. J., Levade, T., et al. (2009). The natural marine anhydrophytosphingosine, Jaspine B, induces apoptosis in melanoma cells by interfering with ceramide metabolism. *Biochemical Pharmacology, 78*, 477–485.

Sanchez, A. M., Malagarie-Cazenave, S., Olea, N., Vara, D., Cuevas, C., & Diaz-Laviada, I. (2008). Spisulosine (ES-285) induces prostate tumor PC-3 and LNCaP cell death by de novo synthesis of ceramide and PKCzeta activation. *European Journal of Pharmacology, 584*, 237–245.

Santana, P., Pena, L. A., Haimovitz-Friedman, A., Martin, S., Green, D., McLoughlin, M., et al. (1996). Acid sphingomyelinase-deficient human lymphoblasts and mice are defective in radiation-induced apoptosis. *Cell, 86*, 189–199.

Sauane, M., Su, Z. Z., Dash, R., Liu, X., Norris, J. S., Sarkar, D., et al. (2010). Ceramide plays a prominent role in MDA-7/IL-24-induced cancer-specific apoptosis. *Journal of Cellular Physiology, 222*, 546–555.

Schiffmann, S., Sandner, J., Birod, K., Wobst, I., Angioni, C., Ruckhaberle, E., et al. (2009). Ceramide synthases and ceramide levels are increased in breast cancer tissue. *Carcinogenesis, 30*, 745–752.

Schiffmann, S., Sandner, J., Schmidt, R., Birod, K., Wobst, I., Schmidt, H., et al. (2009). The selective COX-2 inhibitor celecoxib modulates sphingolipid synthesis. *Journal of Lipid Research, 50*, 32–40.

Selzner, M., Bielawska, A., Morse, M. A., Rudiger, H. A., Sindram, D., Hannun, Y. A., et al. (2001). Induction of apoptotic cell death and prevention of tumor growth by ceramide analogues in metastatic human colon cancer. *Cancer Research, 61*, 1233–1240.

Sensken, S. C., Bode, C., & Graler, M. H. (2009). Accumulation of fingolimod (FTY720) in lymphoid tissues contributes to prolonged efficacy. *The Journal of Pharmacology and Experimental Therapeutics, 328*, 963–969.

Serra, M., & Saba, J. D. (2010). Sphingosine 1-phosphate lyase, a key regulator of sphingosine 1-phosphate signaling and function. *Advances in Enzyme Regulation, 50*, 349–362.

Shibano, M., Kitagawa, S., & Kusano, G. (1997). Studies on the constituents of Broussonetia species. I. Two new pyrrolidine alkaloids, broussonetines C and D, as beta-galactosidase and beta-mannosidase inhibitors from Broussonetia kazinoki Sieb. *Chemical and Pharmaceutical Bulletin, 45*, 505–508.

Shibano, M., Tsukamoto, D., & Kusano, G. (2002). Polyhydroxylated alkaloids with lipophilic moieties as glycosidase inhibitors from higher plants. *Heterocycles, 57*, 1539–1553.

Sibirtsev, J. T., Shastina, V. V., Menzorova, N. I., Makarieva, T. N., & Rasskazov, V. A. (2011). Ca(2+), Mg(2+)-dependent DNase involvement in apoptotic effects in spermatozoa of sea urchin strongylocentrotus intermedius induced by two-headed sphingolipid rhizochalin. *Marine Biotechnology, 13*, 536–543.

Signorelli, P., Munoz-Olaya, J. M., Gagliostro, V., Casas, J., Ghidoni, R., & Fabrias, G. (2009). Dihydroceramide intracellular increase in response to resveratrol treatment mediates autophagy in gastric cancer cells. *Cancer Letters, 282*, 238–243.

Smith, E. L., & Schuchman, E. H. (2008a). Acid sphingomyelinase overexpression enhances the antineoplastic effects of irradiation in vitro and in vivo. *Molecular Therapy, 16*, 1565–1571.

Smith, E. L., & Schuchman, E. H. (2008b). The unexpected role of acid sphingomyelinase in cell death and the pathophysiology of common diseases. *The FASEB Journal, 22*, 3419–3431.

Soriano, J. M., Gonzalez, L., & Catala, A. I. (2005). Mechanism of action of sphingolipids and their metabolites in the toxicity of fumonisin B1. *Progress in Lipid Research, 44*, 345–356.

Stevenson, C. E., Takabe, K., Nagahashi, M., Milstien, S., & Spiegel, S. (2011). Targeting sphingosine-1-phosphate in hematologic malignancies. *Anti-Cancer Agents in Medicinal Chemistry, 11*, 794–798.

Strader, C. R., Pearce, C. J., & Oberlies, N. H. (2011). Fingolimod (FTY720): A recently approved multiple sclerosis drug based on a fungal secondary metabolite. *Journal of Natural Products, 74*, 900–907.

Sugiura, M., Kono, K., Liu, H., Shimizugawa, T., Minekura, H., Spiegel, S., et al. (2002). Ceramide kinase, a novel lipid kinase. Molecular cloning and functional characterization. *The Journal of Biological Chemistry, 277*, 23294–23300.

Symolon, H., Bushnev, A., Peng, Q., Ramaraju, H., Mays, S. G., Allegood, J. C., et al. (2011). Enigmol: A novel sphingolipid analog with anti-cancer activity against cancer cell lines and in vivo models for intestinal and prostate cancer. *Molecular Cancer Therapeutics, 10*, 648–657.

Tanaka, M., Nara, F., Suzuki Konagai, K., Hosoya, T., & Ogita, T. (1997). Structural elucidation of scyphostatin, an inhibitor of membrane-bound neutral sphingomyelinase. *Journal of the American Chemical Society, 119*, 7871–7872.

Thamilselvan, V., Li, W., Sumpio, B. E., & Basson, M. D. (2002). Sphingosine-1-phosphate stimulates human Caco-2 intestinal epithelial proliferation via p38 activation and activates ERK by an independent mechanism. *In Vitro Cellular & Developmental Biology. Animal, 38*, 246–253.

Thiel, P. G., Marasas, W. F., Sydenham, E. W., Shephard, G. S., & Gelderblom, W. C. (1992). The implications of naturally occurring levels of fumonisins in corn for human and animal health. *Mycopathologia, 117*, 3–9.

Thiericke, R., Stellwaag, M., Zeeck, A., & Snatzke, G. (1987). The structure of manumycin. III. Absolute configuration and conformational studies. *The Journal of Antibiotics, 40*, 1549–1554.

Tonelli, F., Lim, K. G., Loveridge, C., Long, J., Pitson, S. M., Tigyi, G., et al. (2010). FTY720 and (S)-FTY720 vinylphosphonate inhibit sphingosine kinase 1 and promote its proteasomal degradation in human pulmonary artery smooth muscle, breast cancer and androgen-independent prostate cancer cells. *Cellular Signalling, 22*, 1536–1542.

Tsukamoto, S., Takeuchi, T., Rotinsulu, H., Mangindaan, R. E. P., van Soest, R. W. M., Ukai, K., et al. (2008). Leucettamol A: A new inhibitor of Ubc13-Uev1A interaction isolated from a marine sponge, Leucetta aff. microrhaphis. *Bioorganic & Medicinal Chemistry Letters, 18*, 6319–6320.

Uchida, Y., Murata, S., Schmuth, M., Behne, M. J., Lee, J. D., Ichikawa, S., et al. (2002). Glucosylceramide synthesis and synthase expression protect against ceramide-induced stress. *Journal of Lipid Research, 43*, 1293–1302.

Van Brocklyn, J. R., Jackson, C. A., Pearl, D. K., Kotur, M. S., Snyder, P. J., & Prior, T. W. (2005). Sphingosine kinase-1 expression correlates with poor survival of patients with glioblastoma multiforme: Roles of sphingosine kinase isoforms in growth of glioblastoma cell lines. *Journal of Neuropathology and Experimental Neurology*, *64*, 695–705.

Wang, F., Van Brocklyn, J. R., Edsall, L., Nava, V. E., & Spiegel, S. (1999). Sphingosine-1-phosphate inhibits motility of human breast cancer cells independently of cell surface receptors. *Cancer Research*, *59*, 6185–6191.

Wascholowski, V., & Giannis, A. (2001). Neutral sphingomyelinase as a target for drug design. *Drug News & Perspectives*, *14*, 581–590.

Wascholowski, V., & Giannis, A. (2006). Sphingolactones: Selective and irreversible inhibitors of neutral sphingomyelinase. *Angewandte Chemie International Edition*, *45*, 827–830.

Wascholowski, V., Giannis, A., & Pitsinos, Emmanuel N. (2006). Influence of the scyphostatin side chain on the mode of inhibition of neutral sphingomyelinase. *ChemMedChem*, *1*, 718–721.

Watanabe, K., Oguchi, T., Takizawa, T., Furuuchi, M., Abe, H., & Katoh, T. (2007). Synthesis of scyphostatin analogs possessing various saturated fatty acid side-chains. *Heterocycles*, *73*, 263–268.

Weiss, M., Hettmer, S., Smith, P., & Ladisch, S. (2003). Inhibition of melanoma tumor growth by a novel inhibitor of glucosylceramide synthase. *Cancer Research*, *63*, 3654–3658.

Williams, R. (2009). Discontinued drugs in 2008: Oncology drugs. *Expert Opinion on Investigational Drugs*, *18*, 1581–1594.

Wu, B. X., Clarke, C. J., & Hannun, Y. A. (2010). Mammalian neutral sphingomyelinases: Regulation and roles in cell signaling responses. *Neuromolecular Medicine*, *12*, 320–330.

Yester, J. W., Tizazu, E., Harikumar, K. B., & Kordula, T. (2011). Extracellular and intracellular sphingosine-1-phosphate in cancer. *Cancer Metastasis Reviews*, *30*, 577–597.

Yeung, B. K. (2011). Natural product drug discovery: The successful optimization of ISP-1 and halichondrin B. *Current Opinion in Chemical Biology*, *15*, 523–528.

Yin, J., Miyazaki, K., Shaner, R. L., Merrill, A. H., Jr., & Kannagi, R. (2010). Altered sphingolipid metabolism induced by tumor hypoxia - new vistas in glycolipid tumor markers. *FEBS Letters*, *584*, 1872–1878.

Yoshikawa, M., Yokokawa, Y., Okuno, Y., Yagi, N., & Murakami, N. (1995). Syntheses, immunosuppressive activity, and structure-activity relationships of myriocin analogs, 2-epi-myriocin, 14-deoxomyriocin, Z-14-deoxomyriocin, and nor-deoxomyriocins. *Chemical and Pharmaceutical Bulletin*, *43*, 1647–1653.

Yoshimitsu, Y., Oishi, S., Miyagaki, J., Inuki, S., Ohno, H., & Fujii, N. (2011). Pachastrissamine (jaspine B) and its stereoisomers inhibit sphingosine kinases and atypical protein kinase C. *Bioorganic & Medicinal Chemistry*, *19*, 5402–5408.

Yu, K. O., & Porcelli, S. A. (2005). The diverse functions of CD1d-restricted NKT cells and their potential for immunotherapy. *Immunology Letters*, *100*, 42–55.

Zajonc, D. M., Cantu, C., 3rd, Mattner, J., Zhou, D., Savage, P. B., Bendelac, A., et al. (2005). Structure and function of a potent agonist for the semi-invariant natural killer T cell receptor. *Nature Immunology*, *6*, 810–818.

Zeeck, A., Frobel, K., Heusel, C., Schroder, K., & Thiericke, R. (1987). The structure of manumycin. II. Derivatives. *The Journal of Antibiotics*, *40*, 1541–1548.

Zeeck, A., Schroder, K., Frobel, K., Grote, R., & Thiericke, R. (1987). The structure of manumycin. I. Characterization, structure elucidation and biological activity. *The Journal of Antibiotics*, *40*, 1530–1540.

Zheng, T., Meng, X., Wang, J., Chen, X., Yin, D., Liang, Y., et al. (2010). PTEN- and p53-mediated apoptosis and cell cycle arrest by FTY720 in gastric cancer cells and nude mice. *Journal of Cellular Biochemistry*, *111*, 218–228.

Zhong Wu, X., Honke, K., Long Zhang, Y., Liang Zha, X., & Taniguchi, N. (2004). Lactosylsulfatide expression in hepatocellular carcinoma cells enhances cell adhesion to vitronectin and intrahepatic metastasis in nude mice. *International Journal of Cancer, 110*, 504–510.

Zhou, B.-N., Mattern, M. P., Johnson, R. K., & Kingston, D. G. I. (2001). Structure and stereochemistry of a novel bioactive sphingolipid from a Calyx sp. *Tetrahedron, 57*, 9549–9554.

Zhu, R., Snyder, A. H., Kharel, Y., Schaffter, L., Sun, Q., Kennedy, P. C., et al. (2007). Asymmetric synthesis of conformationally constrained fingolimod analogues—Discovery of an orally active sphingosine 1-phosphate receptor type-1 agonist and receptor type-3 antagonist. *Journal of Medicinal Chemistry, 50*, 6428–6435.

Zitomer, N. C., Mitchell, T., Voss, K. A., Bondy, G. S., Pruett, S. T., Garnier-Amblard, E. C., et al. (2009). Ceramide synthase inhibition by fumonisin B1 causes accumulation of 1-deoxysphinganine: A novel category of bioactive 1-deoxysphingoid bases and 1-deoxydihydroceramides biosynthesized by mammalian cell lines and animals. *The Journal of Biological Chemistry, 284*, 4786–4795.

INDEX

Note: Page numbers followed by "*f*" indicate figures, and "*t*" indicate tables.

L

M

N

O

P

R

T

U

V

X

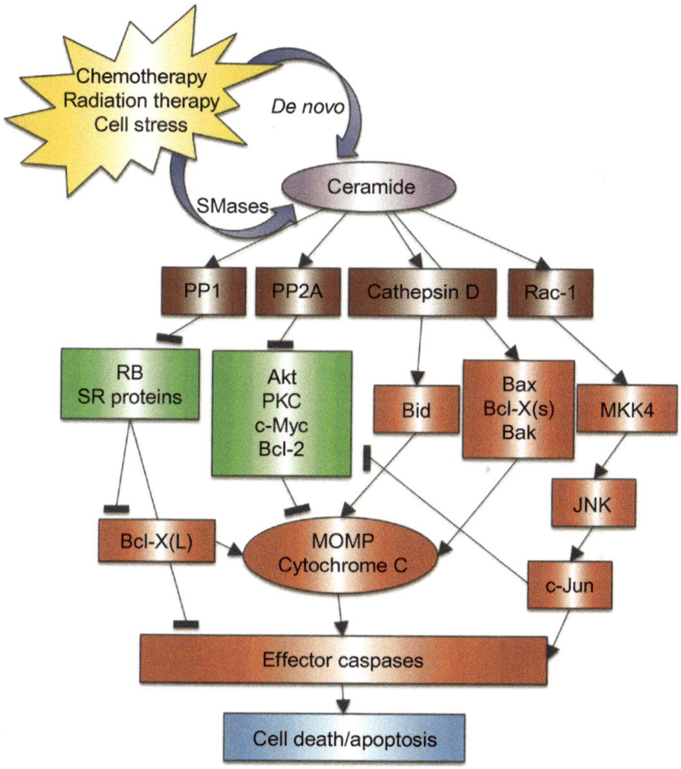

Figure 1.1, Thomas H. Beckham *et al.* (See Page 3 of this volume.)

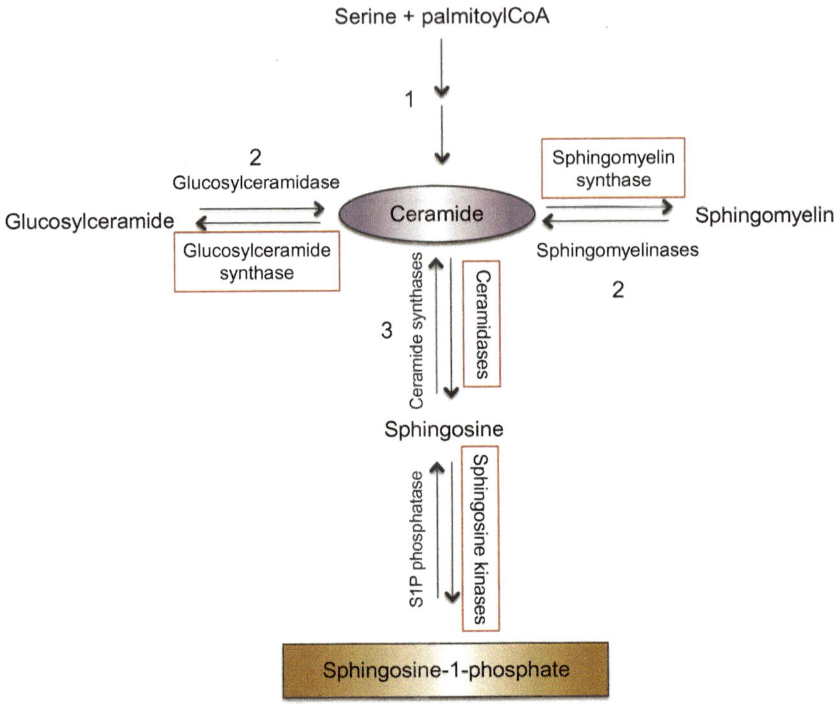

Figure 1.2, Thomas H. Beckham *et al.* (See Page 5 of this volume.)

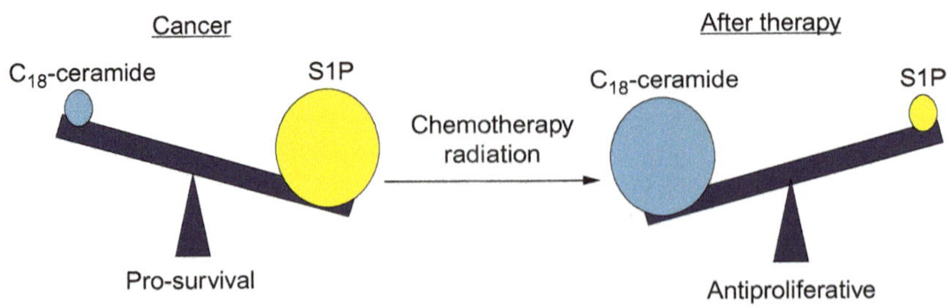

Figure 2.3, Sahar A. Saddoughi and Besim Ogretmen (See Page 45 of this volume.)

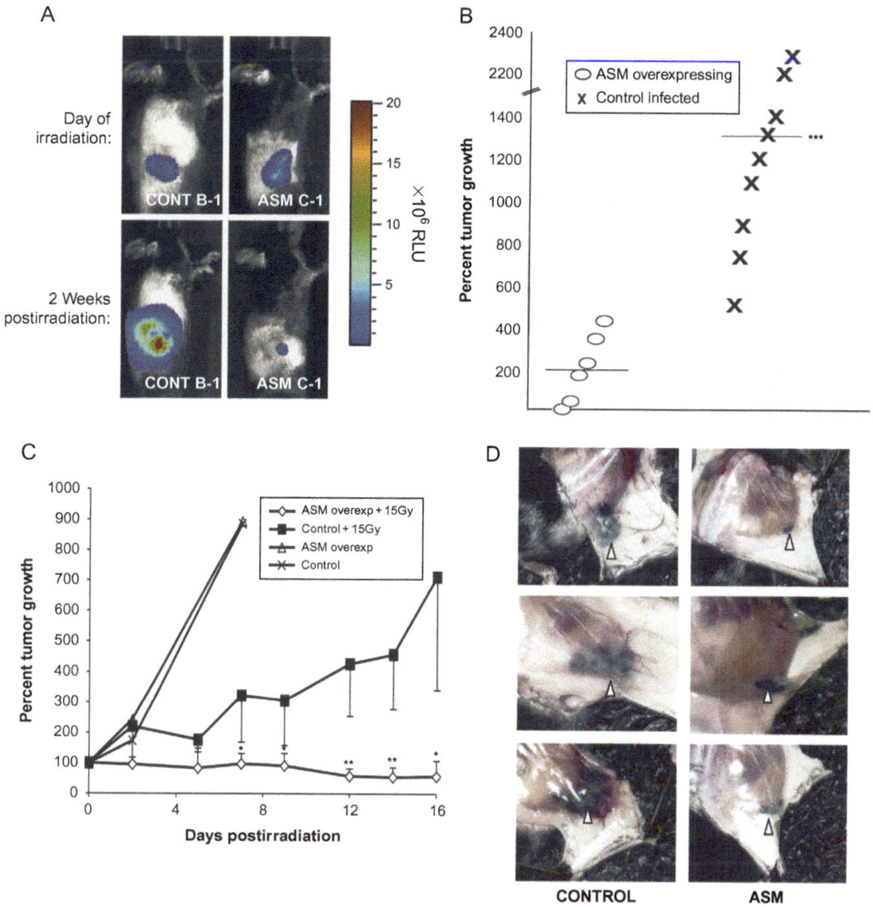

Figure 4.2, Radoslav Savic and Edward H. Schuchman (See Page 104 of this volume.)

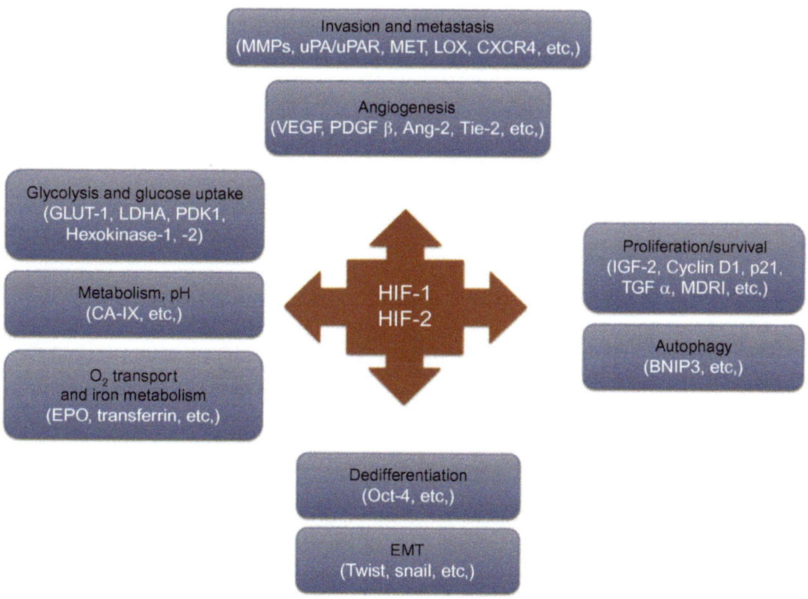

Figure 5.1, Olivier Cuvillier *et al.* (See Page 121 of this volume.)

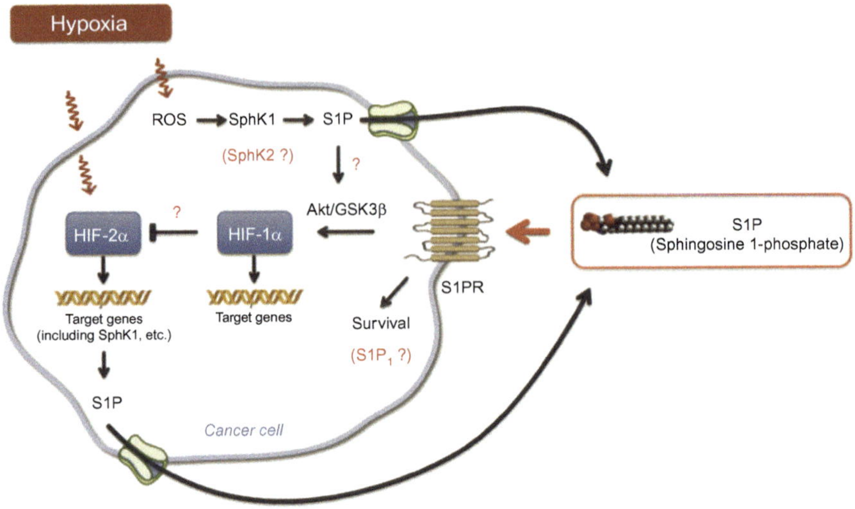

Figure 5.3, Olivier Cuvillier *et al.* (See Page 125 of this volume.)

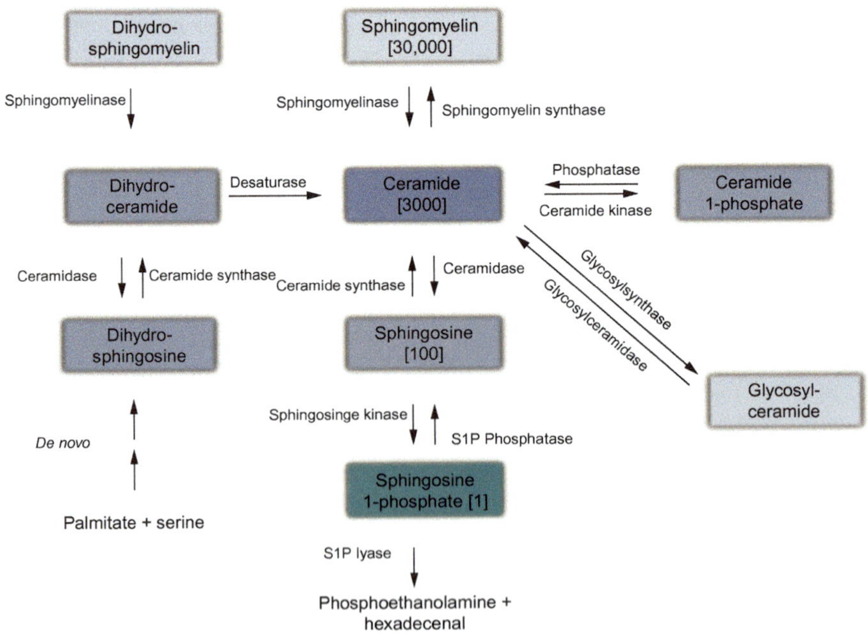

Figure 6.1, Heba Alshaker *et al.* (See Page 145 of this volume.)

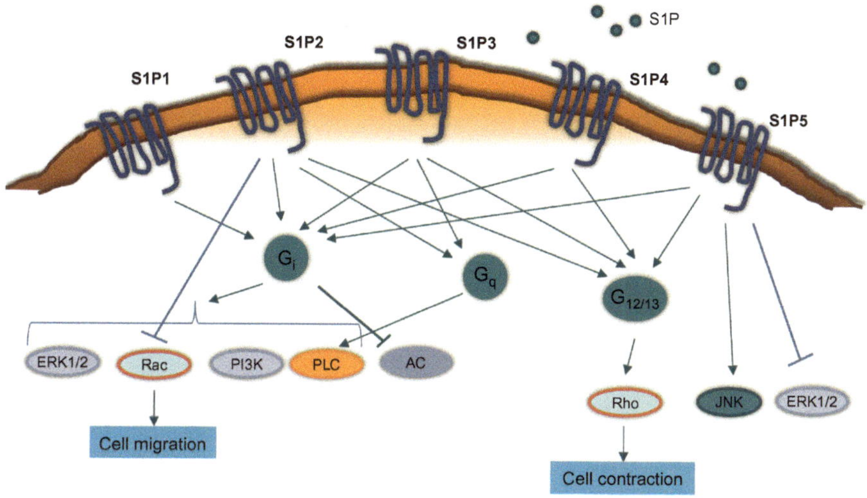

Figure 6.2, Heba Alshaker *et al.* (See Page 147 of this volume.)

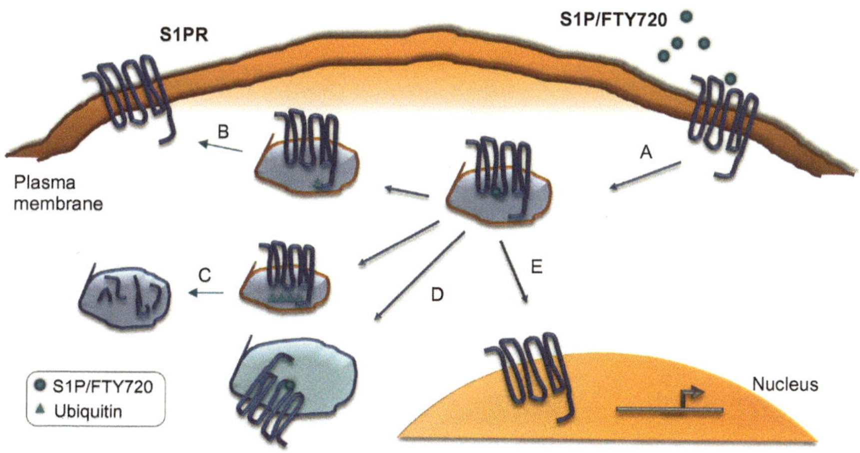

Figure 6.3, Heba Alshaker *et al.* (See Page 149 of this volume.)

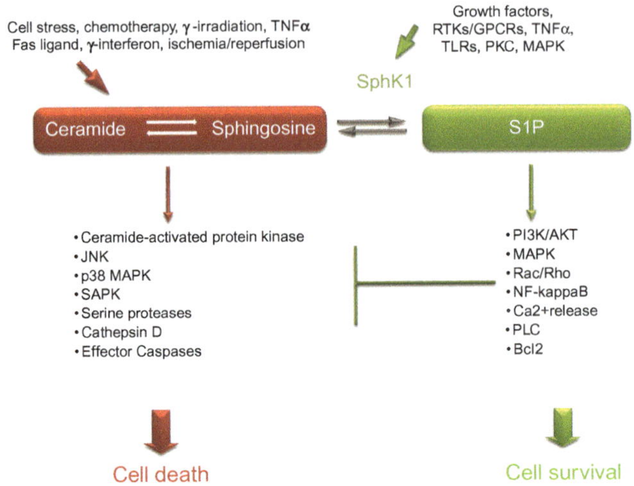

Figure 6.4, Heba Alshaker *et al.* (See Page 153 of this volume.)

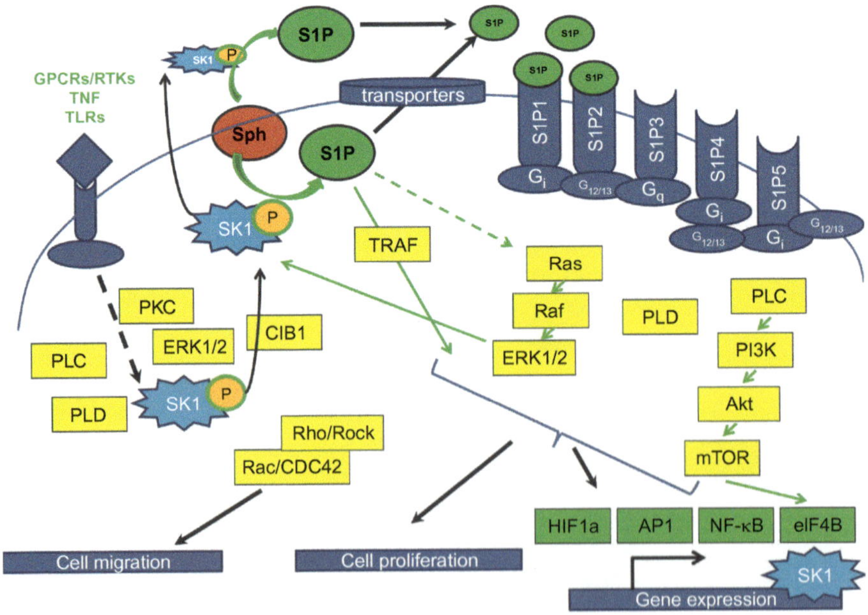

Figure 6.5, Heba Alshaker *et al.* (See Page 154 of this volume.)

FSC
www.fsc.org
MIX
Papier | Fördert
gute Waldnutzung
FSC® C083411

Kolibri 360 GmbH
Ferdinand-Jühlke-Straße 7
99095 Erfurt, Deutschland
produktsicherheit@kolibri360.de